Evil in Healthcare

From the Opioid Crisis to Human Experimentation

Cindy Stanfield

www.IntellectPublishing.com

FV-9Q

ISBN: 978-1-969589-10-2 Paperback

ISBN: 978-1-969589-11-9 Hardback

www.EvilinHealthcare.com

Dedication

To my family, James Stanfield, Steve Thurston, Joe Thurston, Pat Duryee, Mike Thurston, Janice Dilbeck, and John Thurston, and all my nieces and nephews.

Acknowledgments

I would like to thank my students for making me a better person. We are all learners and continue to learn from each other.

I thank Andrew Thurston for his creation of the cover artwork. Thank you to James Stanfield, Michelle Dilbeck, Pat Duryee, Janice Dilbeck, and John Thurston for reading drafts of the book and providing critical feedback.

I thank John O'Melveny Woods for his publishing support. I would have never found my way without your guidance.

Lastly, I thank Dr. Pat Capps-Covey for entrusting me to serve the University of South Alabama not only as a professor, but as a prehealth advisor. You gave me the opportunity to promote student growth in ways I never dreamed possible.

Evil in Healthcare: From the Opioid Crisis to Human Experimentation

Evil in Healthcare

From the Opioid Crisis to Human Experimentation

Chapter 1: Introduction

Why Evil?

In the summer of 1972, a ten year old girl sat on the carpet a few feet from the television watching the Munich Olympics. These were the Olympics where American swimmer Mark Spitz won seven gold medals while breaking seven world records. These were the Olympics where Russian gymnast Olga Korbut earned admiration from fans around the world as she demonstrated poise after performing well in team events, then struggled in individual competition, only to come back and win gold in the individual apparatus competition. But this little ten-year-old girl was not watching sports. She was watching hooded terrorists on the balcony of a Munich Olympic Village room where Israeli athletes were staying.

On September 5, 1972, eight members of the Palestinian terrorist group Black September killed two Israeli athletes and took nine others hostage in the hotel. The news showed the terrorists walking on the balconies with large guns and bringing out Israeli athletes for show. The leader of the terrorist group, Luttif Afif, demanded Israel release 234 Palestinian prisoners and West Germany release two. Israel had established a policy of not negotiating with terrorists, and they did not give in to that policy. Israel Prime Minister, Golda Meir, appealed to the international community to, "save our citizens and condemn the unspeakable criminal acts committed…if we should give in, then no Israeli anywhere in the world shall feel that his life is safe ... it's blackmail of the worst kind."

Negotiators offered the terrorists money and deceived the terrorists into believing that Israel was considering releasing the prisoners. The Munich police attempted a rescue, but the attempt was being filmed live on television and the terrorists could see it coming. The terrorists forced the police to retreat. The terrorists next demanded they be flown, with the hostages, to Cairo. Germany pretended to agree to fly the terrorists and hostages by helicopter to a NATO airbase, where they would board a plane to Cairo. German police officers planned to ambush the terrorists at the airport.

Five West German police officers positioned themselves as snipers at the airport, although none had special sniper training. The ambush failed, and the nine hostages, five terrorists, and a West German policeman were killed. Germany arrested the other three terrorists but then released them in October following the hijacking of Lufthansa Flight 615. The Israeli Central Institute for Intelligence and Special Operations, Mossad, eventually tracked down and killed two of the terrorists along with others involved in the planning of taking hostages during the Olympics.

When I was ten years old, I watched these events unfold. Although I am the youngest of eight children, I watched alone, unable to take my eyes off the evil I was seeing live on television. My first exposure to evil left a lasting impression as I still vision the terrorists on the balcony and the police trying to carry out a rescue. But this was not the last exposure I had to evil.

Where Were You on 9/11?

For my parents' generation, everybody could tell you where they were when Japan bombed Pearl Harbor or when President John F. Kennedy was assassinated. For my generation, we know where we were on September 11, 2001. I was on faculty at the University of South Alabama, teaching in the Department of Biomedical Sciences. Most of our students wanted to become physicians. On campus, we had a chapter of AED the Prehealth Honor Society, run by a Chemistry Professor, Dr. T.G. Jackson. Dr. Jackson also served as the National President of AED. I assisted Dr. Jackson with advising the local chapter.

The student officers of the organization showed their dedication by meeting at 7:00 a.m. every Tuesday morning on the floor of the Chemistry Department office. During one of these mornings, Dr. Jackson received a phone call from his wife. She told him a plane had just crashed into the World Trade Center, and he passed the message on to me and the officers. At the time, we assumed it was an accident. We continued with our meeting until he received a second call from his wife that the crash may have been an act of terrorism. Our meeting ended shortly thereafter as some students had to get to 8:00 classes. I went back to my office and heard that the Communications Department had the news on television in their lobby. I immediately went there, and as on September 5, 1972, I spent the morning of September 11, 2001 watching evil show itself once again on live television.

The events of 9/11 tore at my heart as I tried to understand how terrorists could kill innocent people. Then the second plane crashed. The towers later collapsed. Then the third plane crashed into the pentagon. And then we heard of a lost flight, later found in a field in Pennsylvania. All this evil in one day, carried out in the name of Allah.

The backlash against Muslim Americans showed evil on the other side. I had many Muslim students, who were good people, many going on to become physicians or other healthcare providers. Most of the students felt safe, but we offered refuge at our house if they needed it – none did. We frequently ate at a Mediterranean restaurant in Mobile, Alabama, called Jerusalem Café. It was vandalized in retaliation for 9/11. It was not run by Muslims. The confusion about evil grew deeper in me, as evil begot evil in the form of revenge. Could the cycle end?

I started reading about the three monotheistic religions, Christianity, Islam, and Judaism, to try to understand why we have not lived together in peace. I believe that if people tried to understand each others, we could get along better. The United States believes in religious tolerance, although Christianity clearly influenced the writers of the Constitution and our laws, but the Founders also recognized the importance of separating religion and state. Some countries, however, do not have religious tolerance, and one religion dominates the country's rule.

Though the terrorists who carried out the attacks on 9/11 were from different countries, mostly Saudi Arabia, they believed they were fighting tyranny over their religion and over the evil influences the United States had on the world, especially in the Middle East. These Al Qaeda terrorists believed they massacred thousands of infidels and thus died martyrs. They were doing good by killing the evil Americans. This brings up a fundamental problem as we discuss cases that could be considered evil by some, but others may believe that the acts were justified, and maybe even good.

Before answering questions on good and evil, we need to understand what they mean not only in our culture but in others.

Early in this book, I define/describe evil, discuss the concept of cultural relativism, and raise the question as to whether there is an absolute truth of what is good and what is evil.

Why Healthcare?

Evil people come in many forms. There are evil men and evil women. There are evil construction workers and evil preachers. There are evil teachers and evil engineers. So why write a book on evil healthcare providers.

I became the prehealth advisor at the University of South Alabama, advising students who wanted to go on to careers in healthcare that required a degree beyond a bachelor level. As I advised students, I always considered internally whether or not I would want that student as my healthcare provider. As of this moment, my primary care physician, dentist, and pharmacist are all former students.

While advising, I continued my teaching responsibilities in the Department of Biomedical Sciences. As the number of students majoring in Biomedical Sciences increased, we had to cover more classes. In addition to teaching my specialties of physiology and neuroscience, I taught a course called Issues in Biomedical Sciences. The course focused on ethical discussions regarding controversial cases in biomedical research and healthcare. The course was an eye opener for students, and for myself, as we discussed physician assisted suicide, use of animals in research, use of humans in research, genetic engineering, and many more topics. What amazed me most was how much discrepancy there was on which acts were right or wrong. I thought a group of students studying science to hopefully one day themselves become healthcare providers would lend itself to narrow points of view. In contrast, a class with science, math, philosophy, English, history, etc. students would bring in many more viewpoints. But we had enough controversy to generate demonstrative discussions, and I had a busy job as the moderator.

Teaching this class opened my eyes to cultural differences in how we view the morality of acts. In one of the most touching cases (and this same event occurred in a couple of my classes), we discussed Down Syndrome and how the fetus can be tested for the genetic defect that causes Down Syndrome. This gave the parents an option to abort the baby, resulting in 90% fetuses with Down Syndrome aborted. While several students discussed the decreased stress on the family in that they do not have to raise a Down Syndrome child, which can be quite expensive, one student in each of the two classes where this occurred talked about a sibling born with Down Syndrome. They talked about their siblings as being loving and how they are so happy to have him/her.

I learned a lot from my students: About their reasons for wanting to become doctors, what they would give up to become a doctor, and how they visioned a doctor's responsibility. Most students were altruistic in their views, wanting to provide a service to people who are hurting. Some students wanted to build their ego. A few students wanted to please their parents. No students indicated a desire to hurt others. But we will see that a few doctors and other healthcare providers have that goal.

Chapter 2

Health Care as Careers of Compassion

In the United States, we expect certain professions to uphold higher standards than others. For example, we expect teachers to care about our children and protect them from harm. Several teachers demonstrated their concern for students when they protected their students during mass school shootings. Victoria Soto, teacher at Sandy Hook Elementary School in Connecticut, died December 14, 2012, protecting her students from an active shooter by hiding the students in the closet and using her body to shield them from the gun man [1]. May 25, 2018, Jason Seaman, science teacher at Noblesville High School in Indiana, was shot while tackling a school shooter preventing more deaths on the campus [2]. These teachers, and others like them, performed heroic acts by going beyond normal expectations.

While many teachers are heroes, others are villains. For example, some teachers, such as Brittany Zamora, sexually assault underage students (even of age students would be wrong). In March 2018, 27-year-old Brittany Zamora, an Arizona teacher, was arrested for sexually assaulting her 6th grade male student. The student claimed he had sex with the married teacher several times in the classroom and in Mrs. Zamora's car [3]. She was convicted and sentenced to 20 years in prison [4]. In another sexual assault case, Assistant Principal Dyana Thomas pled guilty to sexual battery of a 17 year old student in Mississippi. He was sentenced to 30 years [5]. Teachers like Mrs. Zamora betray the trust of society. How did she, and others like her, become a teacher?

Police officers must also follow higher standards as their motto says, "to protect and to serve." We expect the police to use the least amount of force necessary to stop and arrest a criminal while protecting themselves and innocent civilians. Police officers literally put their lives on the line to protect the public. Without them, chaos would consume our streets and ordinary citizens would fear for their lives. But there are *bad cops* who abuse their authority in various ways. They may partake in criminal activities of their own, or they may use excessive force on the criminals they arrest.

Historically, conviction of police officers for actions committed while on duty are rare. However, after several incidents in officer-involved shootings of black male suspects, the police have been scrutinized more closely. In October 2015, Corey Jones' car stalled while driving home in Palm Beach Gardens, Florida at 3:15 a.m. A plain clothes police officer, Nouman Raja, approached Jones at his car and questioned Jones about his presence without identifying himself as a police officer. Jones pulled out his gun (he had a concealed weapon carry permit), and may have been trying to escape Officer Raja. After Raja fired three shots at Jones, Jones discarded his gun but kept running. Raja fired three more times, killing Jones with a bullet through the heart. Raja was convicted of manslaughter and attempted murder; the first police officer in Florida to be convicted of an on-duty shooting for 30 years [6].

But there is another side to officer-involved shootings. According to the FBI Uniform Crime Reports, from 1980-2018, an average of 85 law enforcement officers were killed annually by criminal actions (as opposed to accidents) while on duty [7]. Thus, police officers must be vigilant as they make arrests of suspects, and when pulling over a person who just ran a stop sign.

In recent years, over 900 nonfederal police officers per year shot suspects while on-duty. From 2005 to 2019, 104 charges of manslaughter or murder against police officers for on-duty shootings resulted in 36 convictions [7]. Whether the convicted officers were evil and shot the suspect knowing it was not necessary or the officers made an error in judgment is truly only known by the officer involved. I would like to believe the latter is true in the majority of cases. But at least one officer, in my opinion, demonstrated intent.

On May 25, 2020, Officer Derek Chauvin killed George Floyd in Minneapolis, Minnesota, with three other officers present. Chauvin did not shoot Mr. Floyd, a black man suspected of using a counterfeit $20 bill. After hand-cuffing Floyd, the officers struggled putting him in the police vehicle and took him down to the ground to subdue him. Officer Chauvin put his knee on Floyd's neck and kept it there for 8 minutes and 46 seconds, while Floyd pleaded for his life, claimed he could not breathe, and called out for his mother. Standers-by also pleaded with the police to quit choking him. In fact, one of the other officers on the scene asked Chauvin if they should roll Floyd over, but Officer Chauvin denied the request and kept his knee in place even when Floyd lost consciousness. Several videos of the incident showed the death of Mr. Floyd.

In response to a by-stander phone video of the incident, the Minneapolis mayor immediately fired all four officers involved in the death of George Floyd. Protests erupted on the streets throughout the United States, some turning into riots and looting. In the midst of the Covid-19 pandemic, the Black Lives Matter Movement (which began in 2013 after the acquittal of George Zimmerman for the death of Trayvon Martin) took to the streets of America to protest the unjust treatment of black people by police departments. On April 21, the Minneapolis Star

Tribune front page read in large bold print on the top of the front page, "CONVICTION" [8]. On April 20, after about ten hours of deliberation, the jury found Chauvin guilty of second-degree unintentional murder, third-degree murder, and second-degree manslaughter [8]. Chauvin was sentenced to serve 22.5 years in prison. The other three officers involved in the death of Floyd (Tou Thao, Thomas Lane, and J. Alexander Kueng) were all found guilty of violating Floyd's civil rights. Kueng also pled guilty to aiding and abetting second-degree manslaughter and was sentenced to 3.5 years in prison. Lane was sentenced to 3 years in prison and Thao to 4.75.

Excessive violence is not the only way some police officers have abused their authority. While off duty, Solomon Nhiwatiwa, a 34-year-old Cleveland police officer, urinated in random cars in Euclid, Ohio. An hour later, he saw a 12-year-old girl waiting for the school bus and asked if she wanted a ride. When she declined, he parked his car, exposed himself to the girl and filmed himself urinating on her [9,10]. The prosecutor assigned to this case, Michael O'Malley, posted the following statement on social media, a statement I believe explains a lot of evil behavior in respectable occupations [10]:

> This is just an individual who is just a bad man,
> who just happens to work as a police officer.

Somehow, Nhiwatiwa got through the screening process for police officers, just as Zamora did for teachers. I wonder how many of the "rogue" teachers and police officers took their respective jobs knowing that it would put them in a position to find victims, and how many more gained the desire to assault victims based on the opportunity afforded them after taking their jobs?

Even the most devout profession, the clergy, betray their trust (and their faith). Regardless the religion, if a person believes in God, he often turns to the priest, minister, rabbi, or imam for personal, as well as spiritual, guidance. A person shares their innermost thoughts with clergy, such as the sharing of sins in the confessional of the Catholic Church. I grew up Catholic and went to St. Bede's Church, headed by Monsignor George Francis. Monsignor Francis joined St. Bede's in 1957 and served there until 1988. He baptized me, gave me my first communion, and took my childhood confessions on a regular basis. I later found out Monsignor Francis sexually assaulted at least 6 girls [11]. Two of the victims provided details of their assaults. The first victim was first assaulted in 1959, when she was 7 years old. The second victim, first assaulted in 1979 when she was 5 years old, claimed she was victimized by "abuse of every sexual nature" at the hands of the respected clergyman [11].

Monsignor Francis took the second victim on several unsupervised day and weekend trips, during which she was assaulted. In addition, he abused her every other Friday at the rectory over a three-year period. Her abuse was described as follows [12]:

> [Victim] said she was raped by Francis, her parish priest, over a four-year period, beginning when she was about 6. The abuse included ritualistic assaults in hotel rooms involving ropes and religious objects, she said.
>
> After the rape, she said he would sprinkle holy water on her.
>
> She said the priest warned her never to tell or God would kill someone in her family.

The victim also stated, "He [Francis] was the evil, God is not" [11].

The victim was compensated $3 million from the Oakland Diocese, one of the higher single person case pay outs by the Catholic Church at the time.

I denounced my baptism by Monsignor Francis and was rebaptized in the Episcopal Church. I did not have to do this, as the Episcopal Church recognizes all baptisms, but I agree with his victim and consider Monsignor Francis evil for his actions, and I considered the water I was baptized with as *unholy* water. My Episcopal priest explained to me that the blessed water is recognized by the church as holy water, even if not blessed by the holiest of priests. I could not accept that.

We now know that many priests have sexually assaulted children (usually boys). BishopAccountability.org contains an extensive database of sexual assault accusations against Catholic clergy [12]. The data show that between 1950 and 2018, 7,002 clerics have been accused of sexually abusing more than 20,000 minors in the United States. The clerics include 35 bishops, 5,808 priests, 119 nuns, 370 brothers, 108 deacons, and 47 seminarians. Other sources show considerably higher numbers of priests. These data suggest that about 6% of Catholic priests in the United States are pedophiles, similar to data reported for the general population. But should we expect those who enter the priesthood to have higher moral standards than the general population? Are not those who dedicate their lives to God and God's mission the most trustable people? If you cannot trust a priest or other minister of faith, who can you trust?

Let us now turn to the main subject of the book, healthcare providers. When we think of health care providers, we think of compassionate people desiring to help others. We put a lot of faith in health care providers; in fact, our lives and the lives of our loved ones often depend on the provider. But just like teachers, police officers, and clergy, we cannot trust all health care providers to have the desire to care for us in the manner we expect. Before going into specific cases, I first provide a very basic overview of different health professions, the training and screening process, and the oversight procedures.

Training of the Healthcare Professional

Healthcare providers vary in level of training from non-degree certificates to doctorates, and they have different responsibilities with the patient. This book is not meant to provide a thorough description of all health careers. For detailed information on health careers, I recommend https://explorehealthcareers.org. Instead, the focus will be on the more traditional health professions, knowing that much could be extrapolated to other careers. In addition, generalizations often are used because different states have different health care regulations, different schools have different training programs, and healthcare systems vary considerably from one to another. This chapter provides a basic understanding of who gets into healthcare programs and the education process. These two factors determine who and how our bodies will be cared for when needed.

Some healthcare careers require only on-the-job training or technical school such as nurse's aides and nurse's assistants, home care assistant, dental laboratory technician, pharmacy assistant and limited phlebotomist. Most of their work involves preparing materials used during treatment and taking vital signs. Most aides and assistants do not provide treatment other than maybe providing ice or heat for a patient receiving occupational or physical therapy.

Certified technicians have more formal training and must pass a licensing test. Technicians tend to have more patient interaction. For example, an ophthalmic medical technician, who works directly with an ophthalmologist (medical doctor specializing in the eye and vision) takes medical histories, performs eye exams, and assists with patient procedures.

Other certified technicians include pharmacy technicians, X-ray technicians, emergency medical technicians, and paramedics. Occasionally, the pharmacist (who now has a doctorate degree) greets you at the counter; but more commonly, the pharmacy technician greets you. The pharmacy technician completes some approved prescription orders. But if a patient has questions, the pharmacist answers them in a private consultation. X-ray technicians do exactly what the name suggests – take patient x-rays. Emergency Medical Technicians and Paramedics ride on ambulances or fire trucks. The paramedics receive certification through two-year training programs, and can perform certain life-saving procedures and establish intravenous (IV) lines for saline or drug administration. Emergency Medical Technicians receive several weeks of training and are authorized to perform minimal procedures such as taking vital signs.

Many health careers require bachelor's degrees, including ultrasound technicians, MRI technicians, and respiratory therapists. The critical need for respiratory therapists, and many health care providers, became evident during the Covid-19 pandemic. Many former bachelor degree programs now require masters or doctoral degrees, usually requiring 2 to 4 more years of training.

Nurses form the most heterogeneous group of health care providers under one title. Licensed Practical Nurses (LPN) and Licensed Vocational Nurses (LVN) require approximately 1 year of vocational training and passing a national licensing exam. LPNs and LVNs typically work under the guidance of a Registered Nurse (RN) at a variety of medical settings. An RN either completes an Associates Degree in Nursing (2 years) or completes two years of training at a technical school and receives an RN Diploma. RNs typically work in hospitals providing bedside patient care. A Bachelor of Science in Nursing (4 years; BSN) often specializes to work in cardiac, pediatrics, emergency, etc. Nurse Practitioners either complete a Master's Degree or more recently a Doctorate Degree in Nursing, but to enter a Nurse Practitioner program, the student must have a BSN and a few years of work experience. Unlike other levels of nursing, Nurse Practitioners can be independent of doctors, and some have their own practices.

And what about doctors. They receive the highest level of training; or do they? Many health care training programs now offer doctorate degrees. For example, physical therapy moved from a bachelor level program to a master's program in the 1990's, and today, physical therapy is a doctoral program. While Occupational Therapy is still a master's program at a few schools, it too will become a doctoral program. Pharmacy also went from a bachelor level program to a doctoral program in less than 25 years. As mentioned above, Nurse Practitioners have doctorate degrees.

Few master's level health professions still exist. One that is likely to stay a master's degree is Physicians Assistant. A physician assistant receives three years of training consisting of didactic training similar to medical school and medical rotations. Duke University started the first physician assistant program in

1965 to provide career paths for army medics coming home from Vietnam and to meet an increasing need for general practitioners in medicine. Unlike nurse practitioners, physician assistants only work under the management of a physician, and although they initially filled the gap in general practice, many work in specialty fields, such as surgery.

Medical doctors receive four years of medical school training followed by a minimum of three years residency in their chosen health profession. What most people normally think of as medical school are allopathic medical schools, where a graduate receives a Medical Doctorate degree (MD). However, schools of osteopathic medicine have been around since 1892, and due to a shortage of physicians as the *baby boomers* age, osteopathic medical schools have grown considerably in the number of schools and the number of doctors they graduate with a Doctor of Osteopathic Medicine degree (DO). With the physician shortage hitting primary care the greatest, the majority of DOs pursue careers in primary care, although that is changing. Dental schools similarly have two types of degree programs, Doctor of Dental Surgery (DDS) and Doctor of Medicine in Dentistry (DMD). Dentists with a DDS or a DMD provide the same dental services to patients.

Healthcare programs are competitive, but those at the doctoral level much more so than others. College students compete for acceptance into medical and dental schools, with approximately 1 seat for every 2 students applying per year. Applicants usually have a bachelor's degree to be competitive, though not always required, and they must complete certain coursework in math and science. Applicants must also take an admissions test (MCAT for medical school, DAT for dental school). Admissions Committees at the medical and dental schools consider the applicant's overall grade point average (GPA), math and science GPA, and admissions test score as significant indicators of the likelihood of the applicant's success in the program.

However, Admission Committees seek out students with more than the ability to complete their training program. There is a buzz word in health education today: holistic. Health professional schools are seeking well-rounded students whom they believe will be good healthcare providers, not just technically, but also in how they interact with patients.

Logically, a patient would desire a physician who shows compassion. However, in the Issues in Biomedical Sciences course I taught, I asked my students the following question: If you had a choice between a caring surgeon with adequate skills or an insensitive surgeon with excellent surgical skills, who would you choose for your surgery? The choice of the majority was always for the surgeon with excellent skills. Ideally, a surgeon would be both.

Compassion in Health Care

Having served on a medical school admissions committee, physician assistant admissions committee and as an advisor for students desiring to become a healthcare professional, I have seen the qualities desired of health professional school students, and these qualities should come as no surprise. First, the student must have strong academic abilities to successfully complete the rigorous curricula. Second, the student must be skilled at taking standardized tests to successfully become licensed after successfully completing the program. Third is the hardest – the student must demonstrate he/she has the qualities desired in a physician or other healthcare provider.

The American Association of Medical Colleges Group on Student Affairs Committee defined 17 core competencies expected of medical school students.[13] While schools do not expect students to have all the following qualities, they do expect the students to have the majority.

The competencies include:

Professional Competencies:
1) Commitment to Learning and Growth
2) Cultural Awareness
3) Cultural Humility
4) Empathy and Compassion
5) Ethical Responsibility to Self and Others
6) Interpersonal Skills
7) Oral Communication
8) Reliability and Dependability
9) Resilience and Adaptability
10) Service Orientation
11) Teamwork and Collaboration

Thinking and Reasoning Competencies:
12) Critical Thinking
13) Quantitative Reasoning
14) Scientific Inquiry
15) Written Communication

Science Competencies:
16) Human Behavior
17) Living Systems

Most of us would agree that we want health care providers with these qualities. But how do professional schools analyze applicants for these primarily subjective qualities? Most medical schools require applicants to submit a personal statement and letters of recommendation. Students often write personal statements based on what they think the medical school wants to see as opposed to what their reasonings are for wanting to be a doctor. As their advisor, I often steered students away from the "I want to help people" personal statement to what impacted their decisions to become a doctor. Letters of recommendation can be helpful to the committee if they truly represent the student. Many professors hesitate to write negative comments about a student for fear of lawsuits or that their letters will not be taken seriously. To help overcome this bias toward students, many medical schools prefer committee letters or packages prepared by the prehealth advisor. For example, I would have students select 5 professors to complete a rating of the 17 competencies above and provide a letter of evaluation. I would send these letters and ratings to the professional schools along with a cover letter from me that would summarize what the professors said and add my own knowledge from interactions with the students.

Having read the faculty letters, only one (out of a few thousand) recommended the student not be accepted to medical school. Fear of lawsuits hinders our ability to write honest letters of evaluation. Although students often waive their right to see the letters so the letters can be confidential, more than once I heard back from students that one of the interviewers at the medical school told them what was in their letters.

Admission committees make tough decisions, and some students who should not be doctors slip through the cracks. Students who desire to be a healthcare professional have a check list of things they should do to be competitive for the profession they desire. Other than the grades and standardized test scores (intellectual abilities), students can easily complete the items on the check list, whether they desire to or not. Often, students provide community service to *pad* their applications as opposed to wanting to provide service. Professional schools generally rely on interviews to distinguish between the truly compassionate individual who wants to help patients from those who are more interested in the prestige, power, or money associated with the profession. Unfortunately, as you read this book, you will find that some of the more devious students become doctors, and some of them do evil things.

It is very important to state that the majority of medical students (and other healthcare professional students) are compassionate people and want to serve their patients the best they can, just like most teachers, police officers, and ministers of faith want to serve people. In most professional schools, education includes training in ethics and how to interact with patients with empathy. Nowhere has this been more evident than during the Covid-19 pandemic, where healthcare providers became national heroes for their personal sacrifices to help patients. Unfortunately, many of these heroes died from their actions, making the ultimate sacrifice. We must remember the good of most healthcare providers, but be aware that evil lurks within any career.

When Health Care Providers go Rogue

While the book is more focused on physicians, the bad apple is found in many health professions. For example, some nurses have been able to cause as much harm as a physician, primarily because they have heavy patient contact. We also want to distinguish between *evil* healthcare providers from those who make mistakes, and those who commit fraudulent activities. Distinguishing between evil and just bad is subjective. I do not expect you to always agree with me, and sometimes I have difficulty deciding if a certain bad act by a healthcare provider is incompetence or evil. You may face the same dilemma.

The next few chapters discuss ethical principles of good and evil, giving the foundation for evaluating cases in subsequent chapters. The bulk of the book describes various types of atrocities carried out by healthcare providers. For example, two chapters focus on the opioid crisis: one on the contribution of pharmaceutical companies to the crisis and the other on the contribution of physicians and researchers. Other chapters focus on healthcare professionals who kill their patients, either out of sympathy (mercy killings/evil?) or for the pleasure it brings (evil). Several chapters focus on human experimentation, including the notorious experiments of the Japanese and Nazi doctors during World War II. One chapter describes the impact scientific misconduct in biomedical research has on health care, with an emphasis on the improper study that suggested a link between autism and the MMR vaccine.

Chapter 3

Evil Personified

"When I look inside myself?
I see everything. I see all.
I've seen the good, bad, the evil.
I see the whole thing."
— Charles Manson

The Merriam-Webster dictionary defines evil as "morally reprehensible" or "arising from actual or imputed bad character or conduct [1]." One problem with definitions is when they are subjective. What makes something reprehensible? What makes character or conduct bad?

We all have notions of who or what we consider evil. But we do not always agree. Take, for example, the Al Qaeda attack on the World Trade Center on September 11, 2001. I consider the attacks evil. But someone living in the Middle East who considers the United States as the enemy of Islam, might consider the terrorists who carried out the attack heroes. How can the same act be considered evil by some and heroic by others?

Evil is highly subjective and varies widely between different philosophies, cultures, and religions. In the next few chapters, several views of evil will be presented including the possibility that evil is in our genes. But no one view has been proven, and the world is far from all people agreeing on right versus wrong.

In this chapter, we look at a non-medical example of an act and person (and his followers) most agree are evil.

The Manson Murders

On the morning of August 9th, 1969, Winifred Chapman finds the word "pig" written in blood on the outside door of 10050 Cielo Drive, where Hollywood producer Roman Polanski lived [2,3]. When Polanski's maid Chapman entered the house, she saw a grisly scene including the bloody body of Polanski's wife, actress Sharon Tate, sprawled on the floor with a white rope around her neck (likely hanged). Tate was 8 months pregnant at the time. The killers stabbed her sixteen times and slashed her two times. One of the killers used Tate's blood to write "pig" on the door. The police found four other bodies on the property, but not that of Polanski, who was in Europe producing a movie. On the floor, tied to Tate and with a white hood over his head, lay the body of Jay Sebring, a hairdresser once engaged to Tate. The killers beat Sebring, stabbed him seven times and shot him once. Outside the house, Abigail Folger (heiress of Folger's coffee) and Wojciech Frokowski (writer) lay dead on the front lawn. The killers stabbed Abigail twenty-one times (some news reports indicate twenty-eight) with four of the stabs to her face. The killers stabbed Wojciech fifty-one times, beat him with the butt of the gun, and shot him once or twice according to the coroner. In a car, the body of Steven Parent, friend of the gardener, lay dead from three gunshot wounds.

The mastermind behind the attack, Charles Manson, considered the killings sloppy and ordered his "family" members to go out the next day and kill again. This time, he went along to make sure they did it right. When the group drove up to the residence of Pasqualino Antonio LaBianca and his wife Rosemary, Charles entered the house and bound the two together. He then sent three to the house to kill the LaBiancas. They stabbed Pasqualino ten times with a bayonet and fourteen times with a bifurcated fork. The fork was left in his abdomen and a knife was left in his throat. The killers stabbed Rosemary forty-one times with the same bayonet used on her husband. Half of the stabbings were post-mortem. The killers soaked a rag in Pasqualino's blood and wrote "HEALTER SKELTER" on the refrigerator, "DEATH TO PIGS" and "RISE" on the wall, and they drew a paw print on the wall to make people believe that the Black Panther's (a black revolutionary organization headed by Huey Newton) carried out the murders. (HEALTER SKELTER was in reference to the Beatles song Helter Skelter, which Charles Manson believed referred to a race war.) As the three were killing their victims, Manson and two others drove off in the car looking for another victim – they failed [4,5].

Born to 16 year-old Kathleen Maddox-Manson on November 12, 1934, Charles Milles Maddox lived with his mother in Cincinnati, Ohio. Kathleen neglected Charles, often leaving him home alone while she went out drinking with her brother. But she and her brother did more than drink; they were arrested for armed robbery (pretending a ketchup bottle was a gun). Kathleen was convicted and sent to prison when Charles was five [4].

During Kathleen's imprisonment, Charles moved in with his Aunt Glenna, Uncle Bill, and cousin JoAnn in West Virginia, just a few miles from the prison where his mother served her time. Even as a child, Charles showed signs of sociopathy – an antisocial personality disorder in which the person shows a lack of respect for right versus wrong and disinterest in the feelings of others. He wanted to be the center of attention, often lying to gain sympathy and he manipulated other children, especially girls, to follow his guidance to hurt others [5]. His cousin JoAnn described an incident where seven-year-old Charles attacked her with a sickle. Being three years older, she was able to fend him off and lock him outside, only to have him cut through the screen door to attack her again. Fortunately, JoAnn's parents returned home and stopped Charles before he could stab her [6].

When Charles was eight, his mother was released from prison and Charles went back to live with her for a year before she was arrested again. At the age of nine, Charles became a ward of the state and sent to Gibault School for Boys, a Catholic-run school for delinquents in Terre Haute, Indiana. He quickly developed a long history with the juvenile department. Charles ran away from Gibault and lived on the streets, turning to crime for survival, and ultimately landing himself in a correctional facility in Nebraska, beginning a cycle from crime to correctional facilities to running away, then back to crime only to be sent to another facility. At some facilities, Charles raped other boys, at least once at knife point. His records showed warning signs of antisocial behavior and psychic trauma. Charles never showed remorse for his actions. Yet, he was released for good behavior in 1954 [4].

After his release, he married and had a child. However, before the birth, Charles was arrested again and his wife divorced him. He was released from prison in 1958 and turned to solicitation of women to make a living. He got on the FBI watch list. Arrested yet again, Manson served a ten-year prison sentence, during which he took up the guitar, wrote folk music, and practiced Scientology [4].

After his release from prison, Charles went to the Haight Ashbury district of San Francisco, where Charles met some of his soon to be followers. Charles was in his early thirties and picked up teenagers and young adults, mostly female, to accompany him. They travelled south, where Charles pursued a career in music [4].

Charles Manson headed a commune at Spahn Ranch near Chatsworth, California, a site where episodes of westerns, such as Bonanza and the Lone Ranger, were filmed. He also had a second site in Death Valley, the Barker Ranch, where the commune could seek refuge if needed. Manson's control over his followers, referred to as the "Manson Family," grew over time as he became a father figure who gave them freedom from rules, lots of sex, and lots of LSD. Some members left the group, but Charles' hold on those who stayed became powerful enough he could manipulate them into carrying out his evil deeds [4].

What led Manson to murder Sharon Tate and her friends and the LaBianca's? Several books have been written about the Manson murders, the most popular, **Helter Skelter**, written by the prosecutor of the case, Bugliosi [7]. According to Bugliosi, Manson tried to catalyze a race war by committing murders and setting up the crime scene as if the Black Panthers carried out the murders, specifically the writings and paw prints in blood at the scenes.

Tom O'Neill investigated the Manson murders for over twenty years and presented several possible theories in his book, **Chaos: Charles Manson, the CIA, and the Secret History** [4]. In his search for a career in music, Manson befriended Dennis Wilson of the Beach Boys, giving Manson access to music producer Terry Melcher. Manson met with Terry Melcher and performed for him, with his *girls* providing backup. Terry complimented Manson on his music, which meant to Manson that he was going to get a deal. When that deal never happened, Charles Manson sought revenge. Charles had been to Terry Melcher's house before, and remembered the address, 10050

Cielo Drive. Melcher sold the house to Roman Polanski, and whether Manson knew this is controversial. O'Neill theorized Manson sent his family after Melcher, whereas Bugliosi theorized that Polanski and Tate were the targets as they reminded Manson of the *establishment* [4,7].

O'Neill provided evidence that since Manson's release from prison in 1958 to the Tate and LaBianca murders in 1969, local law enforcement and the federal parole system provided leniency to Manson and his family [4]. Charles was released on parole, and he violated the conditions of his parole several times. In addition, the Los Angeles Police Department and Southern California Sheriff's Office arrested Charles and some of his followers on several occasions, including a major raid that took place 7 days after the Tate murders. However, Manson's parole was never revoked and he was never charged with the crimes he committed. During his investigation, O'Neill found circumstantial evidence that the CIA may have used Manson in their mind control experiments, especially those that studied the effects of LSD as part of the MKUltra Project, a twenty-year program that violated federal law [4].

Whatever the motive for the murders, Charles Manson proved himself to be the epitome of evil.

It is unknown whether Charles Manson killed anybody with his own hands, only that he directed others to do so. In July, 1969, Manson cut off the ear of Gary Hinman, a drug dealer who refused to give Manson money, and had his *family* kill the dealer and they wrote "POLITICAL PIGGY" in Hinman's blood on the wall, as well as a paw print. Family members Robert "Bobby" Beausoleil and Bruce Davis were arrested for that murder and are serving life sentences. In addition, Manson and family members Bruce Davis and Steve Grogan were convicted of the murder of Donald "Shorty" Shea, ranch hand at the Spahn Ranch. The police raided the Spahn Ranch on August 16, 1969, following a tip that stolen cars were on the property. Manson believed that Shea informed the police. The body was not found until 1977, delaying the prosecution [4].

The Manson family claims to have killed 35 people. For their acts the nights of August 9th and 10th, Manson and his followers were tried in the summer of 1970. When Manson first appeared in court, he had carved an X into his forehead; his followers did the same. (Manson later changed his X to a swastika.) Manson and four followers were found guilty and sentenced to death. However, California rescinded the death penalty in 1972, and all death sentences were converted to life sentences.

Almost everyone would consider Charles Manson an evil person. But why was he evil? Was he born with an "evil gene"? Did his lack of a loving childhood make him evil? Did the devil make him do it? Maybe it was the drugs or the CIA. Or could it have been a combination of factors?

The killings by Manson's followers were also evil. Could one person have turned six others into monsters? Or were his followers predisposed to become evil if given the opportunity? Remarkably, those who followed Manson's orders to kill had healthy upbringings, showing no tendencies to violence. The one exception may be Susan Atkins.

Susan Atkins murdered Sharon Tate. Her parents were alcoholics, and after her mother died of cancer, her father abandoned her. She met Manson in 1967 and joined his "family." She believed Manson was Jesus, although Manson claimed he was "Jesus and the devil in one. [4]"

We do not know why Charles Manson was evil. The next few chapters describe some theories on what is evil, why is there evil in the world, and what makes one individual evil compared to others.

Chapter 4

Philosophical Doctrines of Evil

I learned Greek mythology in elementary school, where Zeus was king of the Gods and lived on Mount Olympus with his seven wives, Poseidon ruled the oceans, Ares was God of wars, and Pegasus flew the skies. At the time, I considered Greek "mythology" to mean Greek fairy tales. I never considered this a history lesson in that people actually believed these myths. People generated myths for a purpose: to explain the mysteries of life.

In the 6th century BCE, Greek philosophy became the new source of answers to life's mysteries. Philosopher Thales of Miletus explained the state of our existence based on observations as opposed to Gods, Goddesses and mythical creatures [1]. Socrates (469 – 399 BCE) was one of the greatest Greek philosophers, and he was the teacher of Plato, 42 years younger, who became a great philosopher in his own light [1]. One of Plato's techniques in presenting an argument involved studying the disagreement from the perspective of two intelligent men having a fictitious dialogue. In terms of evil, Plato wrote *Protagoras*, a fictitious conversation between his teacher, Socrates, and Protagoras, the supposed wisest man of the times [2]. Plato has Socrates describe evil in that no intelligent man believes that anybody ever willingly errs or willingly does base and evil deeds. In the dialogue, Protagoras agrees with Socrates, and Plato also believes Socrates' account of evil [2].

Socrates believed that a person who carries out an evil act is ignorant of the evil nature of his act. Socrates would say the act is not evil if it provides the person happiness, money, food, or other necessities [3]. I disagree and believe that most evil people know what they are doing is wrong and causes harm, but they are only interested in the riches or pleasures it brings to them. I would also say that those who steal food out of hunger are not evil, but forced to steal to survive.

Today, evil is considered morally reprehensible. However, people of different cultures share different moral values, a concept called *cultural relativism*.

Cultural Relativism

Culture simply refers to a group of people who share common beliefs, practices, behaviors, and morals. A single person could be part of several cultures. For example, a black African American with children could be part of the black culture, mother culture, a religious culture, and any of a number of qualities/values she shares with others. For that matter, a white woman with children could belong to the black culture if she shares qualities and values with them. Currently, the United States faces a crisis as the two main political parties, republicans and democrats, have gone so far to the right and left, respectively, of common opinion, that political agreements on right and wrong are nearly impossible and compromise unlikely. Even though cultural relativism suggests that we should accept each other's cultures, that is not always possible without major conflict.

On another scale, *western culture* varies considerably from *eastern culture*. Consider the animal rights activist groups of the west. They greatly impacted the use of animals in medical research and development of cosmetics, the treatment of enclosed animals at zoos, and even how people treat their own pets. While this movement slowly heads east, Yulin, China still celebrates the Dog Meat Festival, and western research organizations hire eastern companies to perform animal research not legal in their own countries.

We tend to take on the morals of those we know best and are most like us. Cultural relativism can result in small differences in morality or in large differences, such as you will read in the chapter on euthanasia. As a current example of conflict on cultural ideas, female circumcision is a practice denounced by western culture and the World Health Organization (WHO), but it is still considered a right of passage to womanhood in many countries including some in northern and central Africa, and it has been performed on over 500,000 women in the United States. One might be thinking that males are often circumcised, so what is the issue? WHO provides four classifications of female circumcision [4]:

> Type I: Partial or total removal of the clitoris and/or the prepuce (clitoridectomy).
>
> Type II: Partial or total removal of the clitoris and the labia minora, with or without excision of the labia majora (excision).
>
> Type III: Narrowing of the vaginal orifice with creation of a covering seal by cutting and appositioning the labia minora and/or the labia majora, with or without excision of the clitoris (infibulation).
>
> Type IV: All other harmful procedures to the female genitalia for non-medical purposes, for

example: pricking, piercing, incising, scraping and cauterization.

The purpose of female circumcision varies in different cultures, but it primarily decreases the sexual enjoyment for a woman to decrease her likelihood of promiscuity. Circumcision of the male, on the other hand, is for cleanliness.

Time also influences morality, as can be seen in the different generations. People born between 1925 and 1945 form the *Silent Generation*. Having parents who lived through the Great Depression and growing up during World War II affected their views on life. Most in this generation accepted the government as it was, even allowing McCarthyism to run its course in fighting communism. They also lived a quiet life simply hoping to survive, and they encouraged their children to work hard.

People born between 1946 and 1964 are the *Baby Boomers*. The Vietnam War, Civil Rights Movement, Sexual Revolution, and Cold War were major events shaping the moralities of the Baby Boomers. Baby Boomers criticized the government and fought for changes through protests and riots. Teenagers rebelled against their parents. The hippie movement brought about changes in sexual promiscuity and drug use.

People born between 1965 and 1980 are *Generation X*, who faced increased divorced rates and working mothers. Major events included the end of the cold war and President Nixon's impeachment and resignation over the Watergate scandal. With parents divorced or both parents working, children learned to be more independent. Generation X is thought to be the first generation that will not earn more than their parents' generation.

We also have *Generation Y* (born 1981 to 1996; also called *Millenials*) and *Generation Z* (born 1987 to 2012), who are more independent than their predecessors, even becoming self-reliant. Computer technology advanced rapidly and the introduction of the internet and cell phone changed lives dramatically, with less direct social interaction, and less outside play time for children. Childhood obesity, thus, became a problem. Fewer people attended organized religious ceremonies. Trust of the government declined.

As we go from the Silent Generation to Generation Z, the trend is for people to want more freedom and independence, ultimately changing society's moral standards. This is nowhere more evident than in the entertainment industry. Curse words were largely banned from television and radio, unless you had Cable TV (which used to be commercial free). TV couples went from sleeping in separate beds in the 1960s to having sexual intercourse (mostly under a sheet) in the 2000s during prime time when children would be watching. Criminal activity is often glorified in shows, such as in the **Sons of Anarchy** or **Breaking Bad**. Many computer games have extreme violence and bloodshed. But what drives changes in morality?

In the 20th century, most gay people in the United States stayed in the closet, as they knew the majority of people would not accept them. Today, gay people are more comfortable being themselves in the open as the majority of people recognize their rights. Does the acceptance of gay marriages make us more or less moral? Your answer possibly depends on your sexual orientation, in which generation you were born, your religion, and if you have personal experience with a gay person.

Accepting cultural relativism decreases the ability to define an act as evil, but defining evil people is even more difficult. I believe that some good people carry out evil acts when put in dire circumstances; and maybe sometimes the circumstances under which an act is carried out may determine whether or not the act is evil. Consider the many victims of Hurricane Katrina left without homes, food, water, and quick government assistance. Many people looted stores for food and water – a necessity if they were to survive. Normally, looting during a catastrophe is evil, but one could argue that taking food and water for survival is justified. However, looting of private property or big screen TVs cannot be justified.

As another example, consider a group of soldiers sitting around a fire to keep warm. A person throws a grenade at them killing the entire group. This was an evil act. Or was it? What if the person who threw the grenade was an enemy soldier following orders? Killing is allowed during wartime, if the rules of the Geneva Convention are followed (an example of cultural relativism; not all countries follow the Geneva Convention). But what if this enemy soldier was ordered to go with his platoon and capture the soldiers around a fire, but the enemy went rogue and decided he wanted them dead? Clearly this is an evil act, but is the enemy soldier evil? Maybe this group of soldiers killed the enemy soldier's brother in a most brutal way and the horrors of war mentally broke down this enemy's sense of morality – he was only interested in revenge. After the war, he became an outstanding law-abiding citizen of his country. And what if the soldier who threw the grenade was an ally – he threw the grenade at his comrades? Circumstances clearly change the context of an act.

As cultural relativism muddies the comparison of right and wrong actions, philosophy tries to clarify the actions. Following are some of the major doctrines of morality which, unfortunately, do not agree with each other.

Philosophy comes from the Greek word *philosophia*, which means "love of wisdom." Philosophers use reason to seek answers to questions of why we exist, and what is our relation to the world and others. These are complicated questions, and philosophers do not always agree on the answers, nor do they have all the answers. We will briefly describe a few of the most prominent philosophical principles about morality.

Act and Rule Utilitarian

The basic concept of Act and Rule Utilitarianism is that an action or a rule should bring about the greatest amount of good [5]. Basically, you weigh the good versus the bad of an action and if the good outweighs the bad, the act is moral. In terms of measuring good, all people are treated as equally important. Simply stated, utilitarianism indicates that the end justifies the means. If the end is good, the entire act was good. Utilitarianism is a branch of the broader concept of *consequentialism*.

Take for example, the atomic bombing of Hiroshima and Nagasaki at the end of World War II. The end of the war saved millions of lives, far more than those lost in the two destroyed cities. (However, it is likely that Russia would have conquered Japan within a few months had the United States not dropped the atomic bombs.) Followers of utilitarianism view this action as moral because it brought more good than the harm caused in the process. But the process was evil, and should an evil act be used to bring about good? The immediate vaporization of men, women and children, followed by the long slow painful death of many more in these cities was horrific. It reminds me of Vlad the Impaler, who in his battles placed enemy prisoners on stakes while alive and raised them up perpendicular to the ground for the enemy to see. The tactic created fear in the enemy giving

Vlad an advantage in war, and making him an early example of a terrorist (or was he a genius at war strategies).

Deontology

Deontology derives from the Greek words *deon* (duty) and *logos* (study). To be moral, one has a duty to follow certain rules considered morally appropriate. These rules must be obeyed even if the consequences are negative. Some rules, or obligations, apply only to certain people. For example, deontology obligates parents to care for their children. Deontology categorically forbids intended evils, even if the outcome might bring about more positive results than negative. One obligation, for example, is not to kill innocent people, even if it would save the lives of many more people [5]. Thus, in the example of the bombing of Hiroshima and Nagasaki, deontologists would be opposed to the act because innocent people were killed, and we do not kill innocents.

An example of deontology is the *social contract*, which is an agreed upon set of rules a society follows to exist in peace, such as the United States Constitution. U.S. citizens are obligated to follow the laws set forth in the Constitution; if a person breaks a law, she can be fined or imprisoned. But the jury system provides latitude in the laws – sometimes a jury finds a person innocent of a crime because of the circumstances under which that crime was committed. Laws can also be changed in the United States, as views of what is right or wrong change over time.

Virtue Ethics

Virtue ethics considers the moral character of the individual as important as opposed to the act itself [5]. A virtue, by definition, is a high moral standard. A moral person not only expresses the moral standard in his actions, but also believes in the virtue as a character people should have. Some virtues include honesty, humility, beneficence, justice, altruism, and compassion. Thus, according to deontology, lying breaks a rule; but one could also state that honesty is a trait of moral people.

Natural Law Theory

When people hear "natural law", they often think of either science and nature or God's law, neither of which is correct. To make it even more confusing, there are two tenets of natural law: natural law - moral theory, and natural law – legal theory [6]. Here, I present the natural law - moral theory, which will simply be called natural law.

Natural Law explains morality based on what people using reason would agree upon as being right or wrong. The Dominican Monk Thomas Aquinas (1225 – 1274) studied the teachings of Aristotle and applied those philosophies to explain Christianity in his **Summa Theologiae** [7]. Aquinas wrote the Summa Theologiae as a series of questions (Q) one would raise to a particular subject Aquinas is trying to prove. Aquinas follows the questions with answers (A), utilizing reason as his tool.

Aquinas defined a law as a certain rule and measure of acts whereby man is induced to act or is restrained from acting [7]. He defined "four kinds of law: (1) eternal law; (2) natural law; (3) human law; and (4) divine law." [7] Eternal Law refers to the ordered universe created by God [7]. Natural Law is how humans behave within the eternal law, and in the process, humans should do good and avoid evil [7]. Human Law refers to the rules formulated by humans using reason, and adapted to particular geographic, historical, and social circumstances (i.e. subject to cultural relativism). Divine Law "comes only from revelation or scripture, hence biblical law, and is necessary for human salvation." [7] Divine Law follows the Ten Commandments.

Aquinas considers Natural Law independent of cultural relativism as there are certain laws that all people using reason agree upon [7]. One example is the purposeful killing of an innocent person. Although this might seem reasonable to all cultures today, how do cultures define an innocent person? If the culture follows Sharia Law, then a women who has intercourse outside of marriage (even if she is raped) should be killed by a family member, reasoning that such a women is not innocent. If we look at historical societies, the Mayans practiced human sacrifice to praise their Gods. Thus, I question whether Natural Laws exist under Aquinas' definition. There is one law that might work, the Golden Rule in either the Christian version: " So whatever you wish that others would do to you, do also to them, for this is the Law and the Prophets." (King James Bible, Matthew 7:12), or as presented by Confuscius: "Don't do unto others what you don't want done unto you." [8]

The Trolley Dilemma

Ethics teachers often use the trolley dilemma to demonstrate conflicts in moral decisions:

> There is a runaway trolley barreling down the railway tracks. Ahead, on the tracks, there are five people tied up and unable to move. The trolley is headed straight for them. You are a bystander some distance off in the train yard, next to a lever that controls which way the trolley goes. Currently, the trolley will stay on the track heading for the five people. However, if you pull this lever, the

trolley will switch to a different track. Unfortunately, you notice that there is one person on this other track.

You are left with only two options: (1) You may do nothing, in which case the trolley will kill the five people on the main track. Or (2) you may pull the lever, diverting the trolley to the side track where it will kill only one person. Which is the more ethical option?

According to Utilitarianism, you should pull the lever, because killing one person is better than letting five die. According to Deontology, however, you should not pull the lever because if you pull the lever, you are causing the death of one innocent, breaking the rules. However, if you let the train move on and five people die, you have done nothing wrong. Nature took its course.

Could a person who pulled the lever be virtuous while causing the death of one person? What about a person who did not pull the lever and let 5 people die?

Natural Law would require a person not to purposely kill an innocent person. Pulling the lever would result in the purposeful killing of one person. However, not pulling the lever would result in the allowing of five people to die. Natural Law refers to both actions and inactions. Thus, Natural Law does not provide an answer to the trolley dilemma.

Many contemporary philosophers provided explanations for evil. A couple of the more famous are presented next.

Hobbes and the Social Contract

Thomas Hobbes (1588-1679) was an English philosopher best known for presenting modern ideas of politics. Thomas Hobbes developed the Social Contract Theory, whereby for a society to exist peacefully, it needed a social contract to provide the rules by which the society must live. The writers of the Constitution of the United States followed much of Hobbes' ideas.

In terms of evil, Hobbes wrote [9]:

> The Desires…are in themselves no Sin. No more are the Actions…till they know a law that forbids them; which till laws be made they cannot know; nor can any law be made, till they have agreed upon the person that can make it.

Thus Hobbes believes that humans created evil as there is no such thing as good or evil outside of a "society". Moral actions become the construct of the social contract, and evil is going against that contract [9].

Michel de Montaigne

Michel de Montaigne (1533-1592) was a French writer and philosopher. Montaigne wrote essays providing his philosophies, including what contributes to evil and how to avoid being evil. Montaigne suggested that people need to be moderate and should strive to live simple orderly lives, as going to extremes and living complex lives put people at risk of carrying out evil acts [5]. He also suggested that people's creativity and imagination can lead to evil [5].

Montaigne also expressed concerns about religions contributing to evil [5]. Some religious beliefs, according to Montaigne, cause us to be cruel to others. In addition, religious extremism leads to much of the evil in the world, which can be seen in acts of terrorism today, or in the crusades of long ago. Thus, even religions that preach peace can promote evil.

Conclusion

As you read this book, you may feel you do not have all the answers about evil—none of us do, not even the great minds of philosophy. Even right versus wrong seems to be subjective, not absolute. In the next chapter, we take a religious look at evil, where more absolute statements could be made, but not agreed upon between religions.

Chapter 5

Religious Doctrines of Evil

Religions are based on beliefs that something supernatural in the world not only exists, but helps shape life by providing a foundation for our existence and often explaining right from wrong. Most religions promote peace, although religious-based wars have occurred as long as religions have existed. Many religions believe in a deity (god), and they could believe in one god (monotheistic), or many gods (polytheistic). Those who do not believe in a god are atheists.

There are over 2000 religions practiced in the world today, and over 85% of the people in the world believe in one. Only five religions are widely distributed. Religions often define morality for its people, explaining right from wrong and what punishment should arise if you choose to do wrong. While this sounds much like government, religions go deeper and relies on its followers to have faith in the leader(s) and to hold the leader(s) in esteem in the hope the leader will take care of him. The leader could be a God, prophet, a wise person, or an idea.

In 2015, Christians made up the largest population of the world's religions at 31.2%, followed by Muslims at 24.1%, unaffiliated at 16%, Hindus at 15.1%, Buddhists at 6.9%, and Jews at 0.2% [1]. By 2060, the PEW Research Center predicts the world population will increase by 32% to 9.6 billion people [1]. The PEW study suggests the number of Christians will decline due to Christians leaving their faith while the number of Muslims will increase, these two religions will both make up approximately 31% of the world population [1].

Monotheistic Religions

Christianity, Judaism, and Islam follow the doctrine of one God (monotheistic), and they all believe in the God of Abraham. They share many beliefs as they share parts of the Old Testament, which all three religions believe contains the words of God, albeit not verbatim. Judaism follows the Torah, which contains the first five books of the Old Testament (Books of Moses). Christianity and Islam consider the Old Testament to be the words of God, but Islam considers it to have errors. Because they share the Old Testament, the three religions also share the story of Abraham, who had two sons, Ishmael (born to a slave woman) and Isaac (born to Abraham's wife). Mohammad is a descendent from Ishmael, whereas Jesus is a descendent of Isaac. After the Old Testament, the three religions depart.

Within the three monotheistic religions, each has different sects with differing beliefs. Thus, the following are generalizations of the religions and do not include the beliefs of all sects.

The New Testament tells the story of Christianity, or Jesus as the son of God. Judaism disregards the New Testament as scripture, as in Judaism, the savior has yet to come. Judaism believes in the existence of Jesus, but as a person whose teachings often went against Judaic principles. In addition, some Christians do not believe the New Testament is scripture. Islam believes in the New Testament and the teachings of Jesus as a prophet, but not the son of God. As Christians are the only religions that believe Jesus is the son of God, they solely believe in the Holy Trinity (God the father, God the son, and God the Holy Spirit).

The Quran, a religious text specific to Islam, contains the writings of Mohammad as dictated to him by the angel Gabriel.

Muslims believe the Quran contains God's exact words. The Hadith, another important text of Islam, records the sayings and actions of Mohammad. The Hadith guides Muslims on the Quran and the understanding of Islamic laws.

Rabbis wrote the Tahmud in the third century CE to interpret the Torah and provide Jewish rules for living in the world. The writing came two centuries after Jerusalem fell leading to the Diaspora (dispersion of Jews across the Mediterranean and near east). Many Jews consider the Diaspora more significant than the Holocaust, as they lost their homeland and temple [2].

Islam, Christianity, and Judaism share the Old Testament and thus the story of Adam and Eve in the Garden of Eden. The religions vary in their interpretation of the famous Genesis story, but they agree this is the beginning of evil. Islam and Christianity have similar interpretations in that the devil (Iblis in Islam and Satan in Christianity) tempts Eve to eat from the tree of wisdom against God's commands. Eve eats the fruit and shares it with Adam. For disobeying God's order, Adam and Eve are banned from the Garden of Eden. Most forms of Judaism do not believe in an external devil. Instead, the devil exists as bad thoughts in the mind of people.

The Christian and Islam devil is a fallen angel whom God bans from heaven. According to the 7th Sutra of the Quran, Allah (God) tells the angels to "'Prostrate to Adam'", and they prostrated, except Iblis, he refused to be of those who prostrate." Allah then says to Iblis, "What prevented you (O Iblis) that you did not prostrate, when I commanded you?" Iblis said: "I am better than him (Adam), You created me from fire, and him You created him from clay." Qamar states the 7th Sutra of the Quran means Iblis was jealous of Adam and humans, and jealousy is a main cause of evil [3]. Christianity has a slightly different interpretation, but both Islam and Christianity teach of the devil promising to drive people to be evil, and God responding that those who become evil will all go to hell with the devil. All three monotheistic religions believe the devil continues to tempt humans.

But who was Satan, and why did he bring evil to the world God created?

Thomas Aquinas wrote about the battle between God and Satan. Specifically, Satan wanted to be as powerful as God [4]:

> One may desire to be "like God" in a way that is not natural to one, as if one were to desire to create heaven and earth, which is proper to God: there is sin in this sort of desire. It was in this way that the devil desired to be like God; not that the devil desired to resemble God by being subject to no one else but himself absolutely—for in that case, he would be desiring his own not-being, since no creature can exist except by holding its existence under God; but rather, it was in another way that he desired to be like God: by desiring as his ultimate end of beatitude, something that he could attain by the power of his own nature, turning his appetite away from the supernatural beatitude that is attained by God's grace. Or, if he desired as his ultimate end that likeness of God that is bestowed by grace, he sought to acquire that likeness by the power of his own nature, and not from Divine aid as God ordained.

Is Satan tempting people to do evil? Charles Manson claimed he was both Jesus and the devil in one. Could he have been possessed to do such evil and to lead others to commit evil acts?

Islam teaches that humans are born with fitrah, the inner sense of knowing right from wrong. Christians also believe children know right from wrong, but do not have a full understanding or appreciation for its significance. Judaism, however, believes that children do not understand right from wrong until taught, and only at the time of the Bar Mitzvah (generally at age 12) are children considered knowledgeable to follow God's laws and do good [2].

Islam, Christianity, and Judaism all share a common conflict because of the very notion of one God. If God brings good and nurturing to the earth, including man, why does God allow evil to exist and not protect us from it? Hume raises this question in his dialogues [5]:

> Is [God] willing to prevent evil, but not able? Then he is impotent. Is he able, but not willing? Then he is malevolent. Is he both able and willing? Whence then is evil?

In 1710, German philosopher and mathematician Gottfried Leibniz produced theodicy, a philosophy addressing the conflict of a loving God and the existence of evil [6]. Since then, many theologians have developed their own theodicies, ranging from evil giving people a chance to grow spiritually as they face conflicts (i.e. to develop a soul) to God granting people free will such that they may decide their own paths [6].

Hindu

Hinduism typically recognizes more than 330 million Gods, but there seem to be different views on these Gods [7]. One view is that God takes on many forms as he is in everyone and everything. Thus, we are all part of God. Other Hindus believe in the trinity of Gods [7]: "Brahma the creator of the universe, Vishnu the preserver of the universe, and Shiva the destroyer of the universe."

There are varying thoughts about evil among the different branches of Hinduism. Some Hindus believe that evil comes about because of one's own actions. It is a natural part of life, coming from Karma, and thus not being imposed on humans by God(s) or demons.

Most Hindus consider their own people incapable of evil, a belief called *ahimsa* [7]. Hindus are expected to show kindness and peace to all people and animals. Thus, to be evil would contradict their religion. I believe most religions expect their followers to be good and not evil, but some go astray while still claiming they follow their religion.

Buddhism

Buddhism is based on the teachings of Siddhartha Gautama (a.k.a. Buddha), who lived from approximately 566 to 480 BCE. After growing up a child of privilege, in his early adulthood he ventured out to find a better understanding of the world. He became a monk and deprived himself of worldly possessions. While meditating under a tree, he had an epiphany on how to be free from suffering and achieve salvation. At that point, he became known as the Buddha (enlightened one) and spent the rest of his life traveling India to teach others.

Buddha's teachings center around the four noble truths: (1) truth of suffering, (2) truth of the cause of suffering (3) truth of the end of suffering, and (4) truth of the path to the end of suffering [8]. Thus, people will suffer, but they may find a path to end their suffering. Buddha teaches that suffering is brought about by greed, seeking pleasure, and seeking immortality. When a person loses their vices that bring suffering, the person either leaves this life and enters the spiritual world, or reaches Nirvana. When a person enters the spiritual world, they will be reincarnated and start the process of moving through the four noble truths once again. Based on their Karma ((the amount of good (positive karma) or bad (negative karma) during their earthly life), the person can be reincarnated at a higher level than their previous earthly life or a lower level, respectively. Eventually, a person may reach spiritual enlightenment (Nirvana) consisting of a transcendent state free from suffering and the cycle of birth and rebirth [8].

Buddhists believe that evil is neither an inherent part of people, nor a product of external factors (such as demons). Buddhism opposes the categorizing of people as good or evil. Life is simply actions and reactions, and one should strive for positive karma. Evil exists because of the "negative actions and beliefs of humans, including greed, anger, and ignorance." [9,10]

Demons

The concept of demons goes back to the Sumerians who lived in southern Mesopotamia from approximately 5000 to 1900 BCE. Sumerians used demonology to explain illnesses and severe storms. A demon possessing a person could cause seizures. The demons lived in the underworld and opposed the actions of the gods.

Although science can explain many illnesses and weather occurrences without using demons, when science fails, demonology raises its head. All five of the religions described above believe in some form of demon. Demons perform evil acts and encourage humans to do likewise.

In some forms of Christianity, Islam, and Judaism, demons, like the devil, are fallen angels. Demons tempt humans to disobey God and harm himself or others. Some religions believe that demons can possess a person, bringing evil into that individual.

According to Dr. Joe Pierre, 40-50% of the population believes in demonic possessions, and the number of cases is on the rise [11]. Many religious and cultural traditions believe people can be possessed by demons or cursed by witches based on physical signs not associated with a known "medical" condition [10]. Exorcisms are the spiritual practice of expelling demons from a possessed person, and they are currently on the rise. The ceremonies by the three monotheistic religions are similar in that a religious leader uses holy words to drive demons out of the possessed body.

In 1971, William Peter Blatty wrote a book, **The Exorcist**, which was based on a true story in which Jesuit priests performed an exorcism on a 14 year old boy called Roland Doe [12]. The exorcisms started in January, 1949 in Maryland, and then continued when the family moved to St. Louis until March of that same year. In 1973, a movie was produced based on the book. The Catholic Church (and other religions) still practices exorcisms today.

In Islam, demons are called Shayatin, and they are fallen angels who chose to follow Iblis [3]. People often think of Jinn as evil, but Jinn are born of free will and some choose to be evil, but others choose good. Those who choose evil are the Shayatin. Islam discourages interactions between humans and Jinn, and if a Jinn is evil, it can possess the human. Like Christian exorcisms, Muslims remove demons from possessed persons through the act of Ruqya. However, Ruqya can also be used to remove demons that are present, but not inside a person.

In Judaism, exorcisms are rare, but do occur. A Rabbi performs the ceremony with the assistance of minyans (10 trained adult males). The purpose is to expel evil spirits called Dybbuks.

Conclusion

Unfortunately, the amount of evil in the world seems to be increasing in number and severity of evil acts. I believe the higher population in the world contributes to criminal activity, as it increases density of people, especially in large cities, and the resources are not equally available to all creating more class distinction within a population. I believe that fewer people following a religion, which all promote some sort of morality, also contributes to more evil acts. Additionally, the evil taking place in the world today is much worse than when I was a child. We have teenagers (and adults) doing mass shootings, including at kindergartens and elementary schools, and at supermarkets and concerts. What could get more evil than killing innocent children? According to Christianity, evil will increase until the end of times as described in the Book of Revelations.

Chapter 6

Biological Basis of Evil

So far, neither philosophy nor religion provide a clear explanation of evil. Maybe science can provide some answers. Scientists focus on two primary causes of evil, nature and nurture. Are people evil because they have the "evil" gene, or does the environment people grow up in make them evil?

United States prisons are full of people who committed heinous acts, and scientists study these people and compare them to non-criminals to determine if physiological differences exist in the two groups. Most of these studies focused on the brain with the basic hypothesis that there are differences in the brains of criminals who commit violent crimes versus the brains of good people. These studies were often performed by psychologists (PhD) or psychiatrists (MD), or neuroscientists (PhD) or neurologists (MD).

The Role of Psychiatric Disorders in Evil

Experimental psychiatrists study whether mental illness causes violent behavior, and if it does, which disorder(s) as defined in The Diagnostic and Statistical Manual of Mental Disorders, fifth edition (a.k.a. DSM-5), lead to violence.

DSM-5 [1] presents numerous mental disorders which might be associated with higher incidence of violent behavior including schizophrenia, disruptive disorder, impulse-control disorder, or conduct disorder. People who have compulsive behaviors could do evil things if evil is part of their compulsion, but most compulsions are internal problems. Personality disorders might separate the good friendly people from the evil abusive people. However, personality disorders come in many flavors, and are not always obvious, especially to the untrained person.

Psychopathy, a word commonly used in criminal television shows, is noticeably missing from the DSM-5. According to the Merriam-Webster dictionary [2], psychopathy is defined as a "mental disorder especially when marked by egocentric and antisocial activity, a lack of remorse for one's actions, an absence of empathy for others, and often criminal tendencies." Psychiatrists utilizing the DSM-5 diagnose a "psychopath" as a person with *Antisocial Personality Disorder*, defined as "a pattern of disregard for, and violation of, the rights of others," and they might include *Narcissistic Personality Disorder*, a pattern of grandiosity, need for admiration, and lack of empathy 1. People studying behavior in serial criminals generally use the terms psychopath (sometimes sociopath) and provide a description of psychopaths that fits several of the people described in later chapters.

Currently, scientists recognize two forms of psychopathy: primary psychopathy and secondary psychopathy. Primary psychopathy seems more associated with emotional disturbances originating from neural dysfunctions, resulting in interpersonal and affective deficiencies [3]. In contrast, secondary psychopathy resembles the criminal psychopathy with their impulsivity, narcissism, and aggression. Some studies suggest that secondary psychopathy is more affected by lifestyle and antisocial deficiencies and thus a more environmentally- versus biologically-induced condition [3].

Whereas psychopaths make up only 1% of the general population, they make up 15-30% of the prison population. Kiehl and Hoffman [4] studied males over the age of 18 and showed that psychopaths comprised less than 1% non-institutionalized males versus 16% of males incarcerated, on parole, or on probation. These data indicate that 93% of adult male psychopaths are either incarcerated, on parole, or on probation [4].

Numerous scientists study the neurobiological basis of psychopathology. Although gaps in the knowledge exist, much evidence points to differences in the brains of psychopaths compared to normal people in several critical areas. Robert Hare, PhD, was one of the earlier scientists to study the structure and function of the psychopathic brain [5]. Structure can be studied utilizing common imaging techniques, such as magnetic resonance imagery (MRI), or even post-mortem dissection. Function requires measuring brain activity in response to stimuli via processes such as functional MRI, or positron emission topography (PET).

To compare psychopathic brains to other brains, one has to be able to define a person as a psychopath. Hare developed the Psychopathy Checklist (PCL) in the 1970s [5], and it is still used in a 1990's revised form, PCL-R (R for revised). Today, scientists, physicians, and lawyers utilize the PCL-R for research, diagnostic, and legal purposes. In terms of the latter, court systems utilize the PCL-R on some repeat offenders to help make sentencing decisions.

The PCL-R checklist contains twenty items that measure twenty elements of psychopathic behavior. The subject rates the items 0, 1, or 2 based on the degree to which the subject believes he demonstrates that behavior, 0 = none, 1 = some, 2 = a lot. With

20 questions, the maximum score is 40. A score of 0 to 5 indicates no psychopathic behavior. A score of 30 or higher indicates psychopathy and a need for professional help. But this test is not taken alone as a sole indicator of psychopathy. The test is given in an interview format, with two mental health professionals asking the questions on the checklist and observing behavioral responses (facial features, body posture, etc.) in addition to the verbal response to the question. As psychopaths tend to be excellent liars, these other signals provide important further insight into what the subject truly thinks.

In addition to the interview/checklist, another expert must review the patient's medical and criminal files (if a criminal is being tested). Thus, lay persons can determine nothing about an individual by giving him the Hare PCL-R test; it is meant to be given by experts. Because a self-analysis would be superfluous, the twenty items on the questionnaire are listed below. We all probably have relatives that would score above 30, but remember, the true test involves observing behavior during the test taking and reviewing other relevant files. So do not call grandma a psychopath if she scores a 31.

The 20 questions given during the test are: [5]

1) Do you have "excess glibness" or superficial charm?
2) Do you have a grandiose sense of self-worth?
3) Do you have an excessive need for stimulation or proneness to boredom?
4) Are you a pathological liar?
5) Are you conning or manipulative?
6) Do you display a lack of remorse or guilt?
7) Do you have a "shallow effect"?

8) Are you callous, or do you lack empathy?
9) Do you have a "parasitic lifestyle?"
10) Do you have poor behavioral controls?
11) Do you have a history of promiscuous sexual behavior?
12) Do you have a history of early behavioral problems?
13) Do you lack realistic long-term goals?
14) Are you overly impulsive?
15) Do you have a high level of irresponsibility?
16) Do you fail to accept responsibility for your own actions?
17) Have you had many short-term "marital" relationships?
18) Do you have a history of juvenile delinquency?
19) Have you ever experienced a "revocation of conditional release"?
20) Do you display "criminal versatility"?

Professionals estimate that 1% of the population are psychopaths [6], but prisoners do not make up 1% of the population. Are the psychopaths not in prison simply smart criminals who avoid getting caught? Or do some psychopaths live life on the right side of the law?

Not all psychopaths are criminals. Lack of empathy and a grandiose view of oneself, does not indicate criminality. For those psychopaths living amongst the general population, Kevin Dutton ranks the top 10 careers, as well as the lowest 10 for psychopaths: [7].

Table 1: Top and Bottom Careers for Psychopaths

Top 10 Careers	**Bottom 10 Careers**
CEO	Care Aide
Lawyer	Chef
Media (TV/Radio)	Doctor
Salesperson	Civil Servant
Surgeon	Accountant
Beautician/Stylist	Nurse
Journalist	Therapist
Charity Worker	Craftsperson
Police Officer	Teacher
Clergy Person	Creative Artist

Notice that psychopaths tend to favor professions associated with power, especially over other people. In contrast, psychopaths tend to shy away from careers that are more people-oriented or provide service to others. Remember that only 1% of the population are psychopaths, thus even though CEO is the top career choice of non-criminal psychopaths, the number of CEOs who are actually psychopaths is very low.

Also note that health care providers occur in both the top and bottom ten careers of psychopaths. Surgeons rate in the top 10 whereas those health professions that involve frequent patient interactions tend to be in the bottom 10.

When Hare developed the PCL-R, he was studying prison populations, which affected the questions he chose to include. Several other psychopathy tests have since been developed to apply more appropriately to the general population, including the Levenson Self-Report Psychopathy Scale, Psychopathy Spectrum Test, and the Dark Triad (which tests for three negative traits – psychopathy, narcissism, and Machiavellianism [8].

Because a relatively high population of psychopaths can be found in the prison population, scientists often identify their subjects based on criminal activity. Many scientific studies showed certain brain areas differ anatomically and/or functionally in the psychopathic or violent criminal brain, including the prefrontal cortex, hippocampus, amygdala, corpus callosum, and angular gyrus [9].

To understand some of the technical neurological terms, a brief introduction to the brain is provided. The main portion of the brain, the forebrain, consists of four lobes: frontal lobe in the front, two temporal lobes on the sides, and the occipital lobe in the back. (Figure 1) Between the forebrain and the spinal cord lies the brainstem, best known for the unconscious control of life sustaining functions (heart rate, blood pressure, respiration, sleep). Below the forebrain and behind the brainstem is the cerebellum, which is best known for control of body movements.

Figure 1

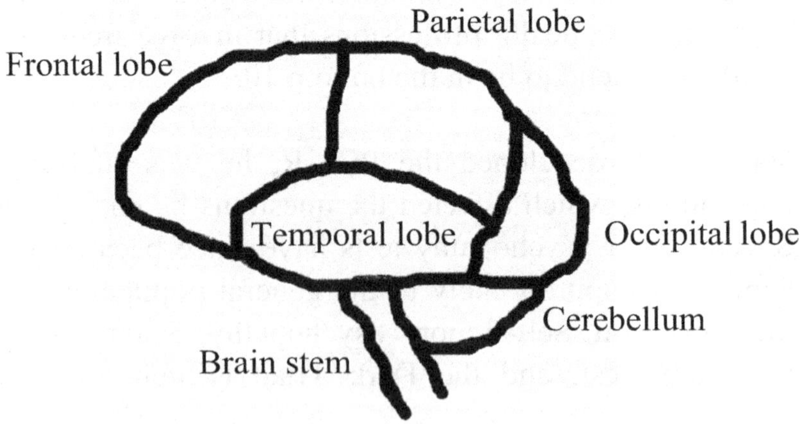

Another neural concept of importance concerns gray matter versus white matter. Communication and integration of information occurs between cells in the gray matter. White matter, in contrast, transmits information from one region of the brain to another.

The newest area of the brain on the evolution scale is the cortex, a thin (3-5 millimeters) outer layer of gray matter that performs the more complex brain functions. The prefrontal cortex refers to the anterior-most cortex of the frontal lobe (see Figure 1). Several functions associated with the prefrontal cortex include decision making, planning, impulse control, emotional responses, personality, and attention.

A person's brain does not completely develop until after birth, with the prefrontal cortex taking the longest. The neural connections that allow one to control their impulses finishes development about the age of 20 - 22. (Changes still occur in the brain, but all functionality is normally present at this age.) Thus, children and teenagers respond more by impulse than through controlled behavior compared to adults.

Psychopaths show a lack of impulse control and a lack of empathy. Two studies showed a negative correlation between psychopathy score on the PCL-R and volume of prefrontal cortex, that is, psychopaths had significantly less prefrontal cortex compared to controls [8,10]. In addition, brain imaging studies show that persons with antisocial personality disorders (not psychopathy specifically) and murderers show decreased activity in the prefrontal cortex compared to controls [9].

The orbital prefrontal cortex, located in the prefrontal cortex area behind the eyes, helps individuals control their impulses and make decisions more attuned to society norms. A well-known case study follows the changes in behavior of Phineas Gage, an 1840's railroad worker who had a tamping rod shoot through the frontal lobe of his brain following an explosion. Phineas went from a kind, even tempered person to a reckless, sexually promiscuous, and obnoxious person with a lack of concern for how his behavior affected others (i.e. a lack of empathy). His change in personality lasted about three years as he relearned how to control his impulses.

The prefrontal cortex is also involved in emotion processing, as is evident from frontal lobotomies, a process whereby a surgeon severs the pathways (white matter) from the prefrontal cortex to other areas of the brain. This technique was performed on "mental" patients to a great extent in the 1940's, with one physician, Dr. Freeman, performing more than 3500 lobotomies. One of the most famous lobotomy patients was Rosemary Kennedy, former President John F. Kennedy's sister. Following lobotomies, patients lost their sense of who they are (personality), empathy, lack of initiative, and a decrease in emotional responses. While never made illegal in the United States, the practice decreased to near zero in the 1980s.

The hippocampus is named for its seahorse shape. In Greek mythology, a *hippocamp* was a horse with a fish tail. The hippocampus is located in the medial aspects of both temporal lobes and regulates emotional responses, memory and learning [3]. Unlike the cortex, the hippocampus, and the next structure to be described, the amygdala, are older on the evolutionary scale and thus control more primitive functions (functions common to many animal species).

The amygdalae are also located in both temporal lobes just anterior to the hippocampus. The amygdala is named for its almond shape; *amygdale* is Greek for almond. Like the hippocampus, the amygdala regulates emotions, learning and memory. The amygdala is best known for its control of fear, a necessary emotion for survival. However, too much activity in the amygdala can lead to excessive fear, such as occurs in post-traumatic stress disorder (PTSD) [11]. The amygdala is also associated with anxiety, recognizing facial expressions, and regulating anger. The amygdala receives communication from the prefrontal cortex and hippocampus. Dysfunction of the amygdala or it's communication pathways with the prefrontal cortex and hippocampus are associated with psychopathic tendencies.

The corpus callosum consists of white matter providing lines of communication between the left and right cerebral hemispheres. While the two hemispheres may appear to be mirror images, there are different, but related, functions of structures located on the right and left sides. For example, the left brain has greater control (dominant) than the right brain in language in about 85-90% of people. Music, on the other hand, is more a right brain function.

Raine and his colleagues showed that psychopaths have 22% greater volume of corpus callosum compared to non-psychopaths [12]. While one might think that having greater volume of a brain structure would be positive, it indicates a problem occurred during development of the nervous system. Excessive pathways of communication can mean that incorrect information is transmitted to certain parts of the brain. For example, synesthesia occurs in individuals who have inappropriate communication between sensory systems. A person with synesthesia might see the number 7 as green and the number 8 as blue (even when typed with black ink), or taste steak when they see red. Thus, psychopaths may have faulty connections in their brains affecting their thoughts and/or behaviors.

The angular gyrus is located in the bottom part of each parietal lobe, just above the temporal lobe. The angular gyrus regulates language, attention, spatial recognition, memory retrieval, reasoning, and social cognition [13]. Psychopaths showed decreased angular gyrus activity while providing false information compared to non-psychopaths [14]. Thus, excessive activity of the angular gyrus may contribute to the ability of psychopaths to frequently lie and make it believable.

Clearly, the brains of psychopaths differ from those of non-psychopaths in many ways, with some of these differences contributing to their negative, and sometimes evil, behaviors. Does this indicate that genetics determine whether a person becomes a psychopath, and if so, should psychopaths be held responsible for their actions? Not completely.

Expression of genetic information during development (and to a lesser extent during adult life) depends on the environment. Because the brain does not completely develop until age 20 and it remains plastic (able to change) for life, there is plenty of time for the environment to negatively affect the brain. But one could say there is a genetic *tendency* toward psychopathy.

No two brains are exactly alike, not even those of identical twins. We all likely have over-developed and under-developed regions compared to "normal" which makes us unique. We still must be held accountable for our actions, especially when our actions affect others.

What about other psychiatric disorders? Schizophrenia causes people to perceive an alternate reality through visual and aural hallucinations and delusional thoughts. Schizophrenia can also disrupt attention, memory, and other cognitive processes [15]. Schizophrenia has a delayed onset, with symptoms starting between the ages of 16 and 30. Schizophrenia does run in some families showing a genetic component, but as in psychopathy, environment is also a factor. Further, variations in brain structure and function compared to non-schizophrenics exist [15]. But like psychopathy, if a person with schizophrenia is a danger to society, they must be kept away from society.

Unfortunately, our society does not have adequate means for housing and caring for those with mental illness, whether they are a threat to society or not. Described in detail in Chapter 12, patients institutionalized for mental illness or mental retardation lived in appalling conditions and were often abused in scientific experiments in the early twentieth century. In 1946, Congress passed the National Mental Health Act, which directed the formation of the National Institute of Mental Health (NIMH). The NIMH was established in 1949 and is part of the United States National Institutes of Health [16]. "The mission of the National Institute of Mental Health (NIMH) is to transform the understanding and treatment of mental illnesses through basic and clinical research, paving the way for prevention, recovery, and cure. [17]"

In the 1960's, the Joint Commission on Mental Illness and Health "received a mandate from Congress to survey the resources and to make recommendations for combating mental illness in the United States"[18]. The report detailed many inadequacies in the care of the mentally ill [16]. As a consequence, Congress passed the Community Mental Health Act of 1963, whereby "Kennedy and Congress sought to decrease the number of institutionalized individuals by incubating self-sufficient and local mental health care centers."[16] Erickson writes of President Kennedy's goals [16]:

> With his New Frontier platform, Kennedy sought improvements in the nation's mental and physical health. He also aimed to unburden society of chronically dependent persons. In particular, he hoped to liberate the population of confined mentally ill patients through advancements in psychopharmacology and supportive housing. He was emotionally drawn to issues of mental illness and intellectual disability because of his sister Rosemary, who underwent a lobotomy that significantly worsened her quality of life.

When nationally institutionalized mental patients were released to be cared for by local communities, inadequate funding meant too few mental health centers to handle the patients. Thus, many patients relied on family and friends to help them. Many ended up homeless. While intentions may have been good, the mentally ill have never been treated in the United States with the dignity and care warranted to all people. Because some of these people are a danger to society but do not belong in jail, more needs to be done.

The Role of Environmental Factors in Evil

Although the United States Declaration of Independence states that "all men are created equal" and have "certain unalienable Rights", all citizens of the United States do not live in equal conditions. As described above, the brain continues developing at least 20 years after birth, and the environment under which that development occurs affects the outcome.

The fetus is particularly sensitive to the environment provided by the mother. Consuming alcohol while pregnant woman can cause fetal alcohol syndrome, which affects the development of many organ systems including the brain. A child born with fetal alcohol syndrome may show signs of learning disabilities, attention disorders, memory difficulties, decreased language abilities, a low IQ, or many other dysfunctions [19]. Various illicit drugs, and some legal, also negatively affect fetal development. A pregnant woman's diet affects fetal brain development, especially the lack of various vitamins, including folic acid (vitamin B9) and vitamin B12.

The first few years of life are very critical for normal brain development. Toxins in the environment including mercury, lead, and certain pesticides affect childhood brain development, as do the foods children eat [20]. People of poverty are more likely to live in areas exposed to toxic chemicals, and they are more likely to not afford nutritious foods.

Another major environmental influence on brain development is stress. Merriam-Webster dictionary defines stress as "something that causes physical or emotional tension" [21]. Physiologically, stress takes the body out of a state of balance, shifting energy resources to fight the stress as opposed to growth and development. Lopatina, *et. al.*, consider early life stress "one of the most critical factors that could modify brain plasticity, memory and learning abilities, behavioral reactions, and emotional response in adulthood leading to development of different mental disorders" [22]. Stress of the pregnant mother can also affect fetal development.

Childhood stress increases the likelihood of criminal behavior and antisocial personality disorder as an adult [23]. Stress in the home (especially abuse or neglect by a parent) is particularly problematic as children see their parents as role models. Interactions with peers, teachers, others of authority, and those met on the internet also affect a child's developed behaviors, as explained by Akers Social Learning Theory [24]:

> The probability that persons will engage in criminal and deviant behavior is increased and the probability of their conforming to the norm is decreased when they differentially associate with others who commit criminal behavior and espouse definitions favorable to it, are relatively more exposed in-person or symbolically to salient criminal/deviant models, define it as desirable or justified in a situation discriminative for the behavior, and have received in the past and anticipate in the current or future situation relatively greater reward than punishment for the behavior.

According to the Social Learning Theory, the likelihood of developing deviant or criminal behavior is increased when exposed to others with a positive attitude toward such behavior and decreased when exposed to others who respect law and responsible behavior. Thus, growing up in a family or neighborhood where criminal behavior is common could lead to a cycle of criminal behavior. According to Akers and Jennings, the Social Learning Theory exists world-wide, and is one of the more consistent predictors of criminal and deviant behavior [24].

Nature or Nurture?

Brain development is controlled by our DNA, but we also know the environment affects how DNA is expressed. Thus, it is not "nature *or* nurture," but "nature *and* nurture." Krohn and Ward describe the sociogenetic approach to criminal behavior [25]:

> The research on discovering the genetic basis of criminal behavior is still in its relative infancy. However, most scholars suggest that if crime is partially explained by a person's genetic basis, that genetic basis represents either a predisposition for criminal behavior or a susceptibility to specific environments or environmental conditions generally. In either case, research is clear in suggesting that genes interact with social context to influence antisocial behavior (as opposed to having direct effects).

If criminal behavior is based on genetics and exposure to an unhealthy environment, then when that child gets older and becomes an adult criminal, should that criminal be held accountable for his actions? It is easy to say that a genetic predisposition for a trait, such as evil, can be controlled – it is a *pre*-disposition after all. But children do not have control of their environment, and thus certain children are more likely to grow up to be criminals because of their DNA. Can we hold a person responsible for having the wrong DNA expression?

Society must. The United States Constitution was based on a social contract. People living together in communities can only succeed if they all follow a basic set of standards (or laws). Criminals defy the contract and threaten the community in which they live.

Chapter 7

The Role of Pharmaceutical Companies and Professional Organizations in the Opioid Crisis

In the late 1990's, opioid prescriptions increased rapidly in the United States as pharmaceutical companies and others promoted the efficacy and safety of opioid use in the treatment of pain. Not surprisingly, the number of deaths by opioid overdose also increased.

As a graduate student studying pain, I joined the International Association for the Study of Pain (IASP) and the American Pain Society (APS). In the 1990's, OxyContin was promoted as the drug to cure all pain. I learned all about the benefits of OxyContin, an opioid analgesic, at APS and IASP conventions: *if a patient took OxyContin to treat pain, that patient would not become addicted.* One of the promoters of opioid use at these conventions was Dr. Russell Portenoy.

Increase in Prescription Drug Use in the 1990's

Dr. Portenoy most recently served as the Executive Director for the MJHS Institute for Innovation in Palliative Care; Chief Medical Officer, MJHS Hospice and Palliative Care; and Professor of Neurology at Albert Einstein College of Medicine. Earlier, Dr. Portenoy served as President of the American Academy of Hospice and Palliative Medicine and President of the American Pain Society. He received numerous Awards and served as Editor in Chief of the Journal of Pain and Symptom Management. He participated in the writing of 21 books, and published over 500 scientific papers on pain management and palliative care. Clearly, Dr. Portenoy is an expert on pain management.

Before the 1990's, physicians treated cancer pain with opioids, including high doses. But physicians hesitated to use opioids in non-malignant pain. In 1986, Dr. Portenoy and his colleague, Dr. Kathleen Foley, tested the effects of long-term (4 years) opioid treatment of non-malignant pain on 38 patients [1]. Twenty-four patients reported adequate pain relief, whereas only two reported management problems. These latter two patients had a history of prior drug abuse.

Dr. Portenoy promoted the use of opioids to treat chronic pain. Elisabeth Rosenthal interviewed him for a 1993 NY Times article about a patient successfully taking high doses of opioids to treat her pain [2]. Dr. Portenoy states in the article:

> There is a growing literature showing that these drugs can be used for a long time, with few side effects and that addiction and abuse are not a problem.

Dr. Portenoy also commented about how pain specialists view treating pain with opioids:

> Among pain specialists, the controversy about whether opiates cause addiction has waned, and now we debate the benefit in terms of comfort and function.

In a 1996 article, Dr. Portenoy proposed the need to further study the use of opioids in chronic, non-malignant, pain [3]. He cites results from treatment of cancer pain as evidence that opioids are safe to use in non-cancer patients as well. Portenoy also indicates pain is undertreated because of regulatory policies in the 1990's [3]:

> Regulatory policies contribute to the undertreatment of pain both directly, by impeding access to controlled prescription drugs, and indirectly, by negatively influencing prescribing behavior.

I attended several presentations at the American Pain Society and IASP meetings, including some by Drs. Portenoy and Foley, emphasizing the undertreatment of pain and how opioids could be the solution.

OxyContin Enters the Opioid Drug Market

In 1996, Purdue Pharma released OxyContin, an opioid with twice the potency of morphine, to treat chronic pain. A progressive marketing plan caused sales to skyrocket, increasing from $48 million in 1996 to $1.1 billion in 2000 [4]. Interestingly, OxyContin (a slow release version of oxycodone) "offered no advantage over appropriate doses of other potent opioids [5]." The Purdue Pharma marketing strategy included all expenses paid symposia for doctors, pharmacists and nurses; especially targeting those physicians who wrote the most opioid prescriptions [4]. Purdue Pharma also offered incentives to patients – a coupon to receive their first prescription to OxyContin free. Most important, Purdue Pharma understated the risk of addiction, citing a study of acute (short term) pain during which treatment did not produce addiction [6]. The marketing strategy lead to large increases in opioid sales, starting in small southern states then spreading nationwide. According to Zee [4]:

> With the growing availability of OxyContin prescriptions, the once regional problem began to spread nationally. By 2002, OxyContin accounted for 68% of oxycodone sales. Lifetime nonmedical use of OxyContin increased from 1.9 million to 3.1 million people between 2002 and 2004, and in 2004 there were 615,000 new nonmedical users of OxyContin. By 2004, OxyContin had become the most prevalent prescription opioid abused in the United States.

Zee continues to explain how OxyContin abuse lead to increases in abuse of other opioids as well [4]. The marketing strategy for OxyContin took away a stigma of opioids being potentially harmful. Thus, prescriptions for all therapeutic opioids increased from 1997 to 2002: Fentanyl prescriptions increased 226%, morphine prescriptions increased 73%, and oxycodone prescriptions increased 402%. In fact, abuse of prescription opioids occurred more frequently than abuse with cocaine and heroin.

Congress Reviews Purdue Pharma and OxyContin

A 2019 congressional report by Representatives Katherine Clark and Hal Rogers claims that Purdue Pharma deceived physicians about the efficacy of OxyContin and the so-called unlikeliness of addiction [7]. This report also named organizations that promoted opioid use, including the APS and the IASP. The report states that the APS received money from Purdue Pharma and from the Mayday Fund, a philanthropic organization whose board members overlap that of the APS and the American Pain Foundation (a Purdue Pharma funded organization). Specifically, former APS presidents (Russell Portenoy, Kathleen Foley, and James Campbell) had ties to the Mayday

Fund. In addition, former APS president Richard Payne has associations with Purdue Pharma, and former APS president Dennis Turk received funding from opioid manufacturers. Scott Fishman, former president of the American Pain Foundation, is the former chairman of the Mayday Advisory Board.

The Congressional Report provides further evidence of former APS presidents participating in questionable interactions with Purdue Pharma. Dr. Portenoy chaired the Advisory Committee for the Mayday Pain and Society Fund. In addition, Purdue Pharma paid Dr. Portenoy to speak at meetings. As of this writing, Dr. Portenoy has been sued in 18 federal cases for his role as an "opinion leader" on increasing the use of opioids. Dr. Foley currently served as Chair for Pain and Palliative Care, at the Center for Practical Bioethics. Purdue Pharma funded this Chair position with a $1.5 million endowment. Lastly, Dr. Payne received money from Purdue Pharma to speak at meetings.

Even more concerning, Purdue Pharma may have had direct influence on the WHO. In 2011, the WHO published a guideline for using opioids called *Ensuring Balance*, which promotes the safety of opioid analgesics [8]:

> It should be recognized that controlled medicines, when used rationally for medical purposes, are safe medicines. Opioid analgesics, if prescribed in accordance with established dosage regimens, are known to be safe and there is no need to fear accidental death or dependence. A systematic review of research papers concludes that only 0.43% of patients with no previous history of substance abuse treated with opioid analgesics to

> relieve pain abused their medication and only 0.05% developed dependence syndrome.

Contrary to the WHO report, a 2011 study from the National Institute on Drug Abuse reported "that substance use disorders among chronic pain patients treated with opioids ranged between 3 and 40 percent [9]."

Sells of OxyContin and other opioids soared as doctors prescribed it without fear of patient addiction. But I, like the doctors who started prescribing OxyContin more frequently, was fooled by Purdue Pharma, Drs. Portenoy and Foley, and WHO.

On May 22, 2019, a Congressional report "Corrupting Influence: Purdue and the WHO" claimed that the WHO, APS and IASP promoted the use of opioids, thereby contributing to the opioid crisis. The Congressional report describes the aspects of Purdue's marketing strategy that lead to the opioid crisis [10]:

> The quick and exponential rise in sales of OxyContin can be attributed to Purdue's deliberate three-pronged strategy that successfully encouraged doctors to increase the use and prescription of opioids:
>
> 1) Expand the use of OxyContin to reach patients suffering from non-cancer pain (such as chronic back pain) and replace the prescription of combination pain medication – i.e. ibuprofen and a weak opioid – with OxyContin;
> 2) Aggressively market OxyContin to clinicians by falsely minimizing the drug's addictive

potential and maximizing its effectiveness; and

3) Deliberately spread misinformation by building financial relationships with trusted physician groups, patient groups, and advocacy organizations.

As for WHO, the Congressional report claims that in the 1990's, Purdue Pharma developed a marketing strategy to increase sales of OxyContin by manipulating WHO. The report states that Purdue Pharma set out to "influence the WHO's recommendations on how health care providers and policy makers should administer prescription opioids [10]." A decade later, guidelines for opioid prescriptions developed by WHO included items of the Purdue Pharma marketing strategy. The Congressional report states [10]:

> In 2011, the WHO published a document called *Ensuring Balance in National Policies on Controlled Substances, Guidance for Availability and Accessibility of Controlled Medicines (Ensuring Balance).* ... *Ensuring Balance* corroborates the oft-repeated Purdue claim that dependence occurs in less than one percent of patients, despite no scientific evidence supporting this claim and a multitude of studies contradicting it. It states: "Opioid analgesics, if prescribed in accordance with established dosage regimens, **are known to be safe and there is no need to fear accidental death or dependence**."
>
> Following the publication of *Ensuring Balance*, the WHO published a second document in 2012 called *Pharmacological treatment of Persisting*

Pain in Children with Mental Illnesses (Persisting Pain in Children.). This guideline uses the marketing term coined by the opioid industry and utilized by Purdue: "opiophobia." The opioid industry defines opiophobia as a physician's "unreasonable fear" of prescribing opioids. *Persisting Pain in Children* tries to overcome "opiophobia" by emphasizing the safety of opioids. The WHO claims that **there is no maximum dosage of strong opioids, like OxyContin, *for children*.** The WHO published this claim despite the fact that U.S. public health agencies have found that fatal overdoses skyrocket in adult patients who are prescribed above 90 morphine milligram equivalents (MME) per day.

Pharmaceutical Companies Implicated in Promoting Opioid Use

Purdue Pharma had company in promoting opioid use. Several opioid-producing companies promoted their sales through donations to national organizations and leaders in the field of pain management. Table 1 shows the amount of money five pharmaceutical companies (Purdue Pharma L.P.; Insys. Therapeutics, Inc.; Depomed, Inc.; Janssen Pharmaceuticals, Inc.; and Mylan, N.V.) provided to non-profit organizations associated with pain management.

Company	2012	2013	2014	2015	2016	2017	Total
Purdue	824,228	$973,328	$812,452	$935,344	$558,068	$50,135	$4,163,554
Insys	14,040	$68,000	$35,200	$530,025	0	$2,500,000	$3,146,265
Depomed	$73,080	135,300	$113,600	$350,000	$318,257	$80,879	$1,071,117
Janssen	$239,903	99,250	$126,000	0	0	0	$465,153
Mylan	0	0	0	$15,000	$2,500	$2,750	$20,250
Total	$1,151,250	$1,275,878	$1,086,252	$1,830,369	$878,825	$2,633,764	$8,856,339

Data from Senate Report: Fueling an Epidemic [10]

Not all companies provided money to pain advocacy groups for the entire 2012-2017 period. Mylan came in late with relatively small donations in 2015, the same year it released its intermediate-dose fentanyl transdermal medication. Janssen showed the opposite trend in donations, stopping them in 2015, the same year it sold it's U.S. marketing rights to the Nucynta opioid line to Depomed. Ironically, Depomed increased it's donations that same year [10].

The top 5 recipients of this money are: U.S. Pain Foundation ($2,922,800; includes the $2.5 million from Insys in 2017), Academy of Integrative Pain Management ($1,265,567), American Academy of Pain Medicine ($1,199,410), APS ($962,725), and The National Pain Foundation ($562,500) [10].

In addition to donations to organizations, these pharmaceutical companies donated another $1.6 million to individuals associated with these organizations. Individuals associated with the National Pain Foundation received the most money ($839,849), followed by the American Academy of Pain Medicine Foundation ($314,176), and then the American Society of Pain Educators ($280,766) [10].

Clearly, pharmaceutical companies have strong financial ties to associations that contribute to policy-making regarding the prescriptions for pain medication. While most people see a conflict of interest, the Senate report goes further [10]. The Senate report accuses many of these organizations, and others not directly stated in the report, of promoting opioid sells in the following ways: (1) issuing guidelines that minimize the risks of opioid use, (2) lobbying efforts to defeat guidelines that restrict opioid usage, (3) criticizing the CDC guidelines for limiting opioid use put out in 2016, and (4) organizing legal efforts to protect physicians from being accountable for a patient abusing opioid prescriptions. This last item comes up again in Chapter 8, when we discuss the numerous physicians that either sold or traded opioids to patients who did not need the medication for pain.

Are Pharmaceutical Companies Evil?

Pharmaceutical companies provide medicines that allow many people to live longer, happier, and more productive lives. They do good. Can a company that does good be evil?

If pharmaceutical companies created drugs purely to provide a service to those in need of the medications, then we would owe them our gratitude. But pharmaceutical companies are a business, and like most businesses, are in it for the money. (There may be individual workers in it for humanitarian or scientific reasons, but the administrators and owners tend to want a profit.) And it has been said that money is the root of all evil; recall that the primary religions of the world all found greed a source of evil. Drug prices have risen to levels unapproachable for some patients. Is that a service to society? While pharmaceutical companies put a lot of money into research and development of medications, what good is a medication that patients cannot afford? I would argue that the pharmaceutical companies have some responsibility for drug costs, but the health care system has more. But the politics of healthcare are beyond the scope of this book.

Pharmaceutical companies also marketed medications inappropriately? In terms of opioids, Congress makes a strong case that pharmaceutical companies duped the American people. I suggest we are being duped by pharmaceutical company commercials as well. Your physician should choose medication based on what is best for you. Have you ever received a free sample from your physician, only to later get a prescription for the same medication? That sample came from the company marketing the drug.

As described in Chapter 3, many successful CEOs are psychopaths, and whereas most evil persons are psychopaths, the majority of psychopaths are not evil. Consider a car company that knows of a defect in a certain model that could result in a deadly accident. Correcting the defect could cost the company millions of dollars in profit. The CEO decides not to correct the defect and sell the cars as is. When people die in accidents because of the defect, what happens? Lawsuits.

The same happened to the pharmaceutical companies. On June 10, 2019, Insys Therapeutics became the first pharmaceutical company to file for bankruptcy following numerous lawsuits over the marketing of opioids. In August, 2019, Oklahoma won a $572 million dollar lawsuit against Janssen Pharmaceuticals (a subsidiary of Johnson & Johnson). Over 2,000 lawsuits were filed against Purdue Pharma and its owners, the Sackler family. On August 20, 2019, Purdue Pharma offered $10 to 12 billion to settle the lawsuits. About a month later, Purdue Pharma declared bankruptcy. Last, while Mylan is facing lawsuits for opioid marketing, greater lawsuits have been filed for their recent over-pricing of epi-pens.

Some pharmaceutical companies face federal criminal charges for the opioid crisis. As of the writing of this book, many of those cases are still pending. However, a federal court did find Insys Therapeutics founder John Kapoor and four other executives guilty of racketeering in the bribery of doctors to prescribe opioids [11].

Are Physicians and Scientists Who Promoted Opioid Use Evil?

Physicians assisted pharmaceutical companies in promoting opioids as "safe" and "non-addicting". But did the physician or scientist know they were part of a lie? If physicians knew, then they broke their oath to their profession and their patients, all in the name of money. If scientists knew, they participated in falsely utilizing data to deceive the public and should lose their title as scientist. Willfully corroborating in deception of people that results in addiction to opioids and deaths is an evil act.

Dr. Portenoy, originally a promoter of opioid use, changed sides once the lawsuits started. Dr. Portenoy made a deal with Federal prosecutors to testify against pharmaceutical companies in exchange for dropping criminal charges against him. Dr. Portenoy may have changed his opinion as he admitted during the past few years that opioids cause addiction in patients taking them for pain relief. However, because of earlier deceptions, one could question Dr. Portenoy's sincerity.

Most scientists develop theories based on the data, especially the data *they* gathered. Dr. Portenoy had early data supporting his claim that opioids could be used to treat pain without concern of addiction. However, those studies used small sample sizes, and pushing opioids as safe was premature. As the database on opioid use grew, any good scientist would analyze the additional data and make a new informed theory – opioids cause addiction.

When Trust is Lost

Deception of one pharmaceutical company, one physician, or one scientist can have long-lasting effects on the trust the public has in health care. For example, you will learn about the critical study contributing to the false claim autism is linked to vaccines in another chapter. This claim by one research group escalated the anti-vax movement, and many diseases nearly eradicated are making a comeback because parents fear vaccinating their children.

In 2020, pharmaceutical companies around the world worked vigorously to create a vaccine for the COVID-19 virus, which killed over a million people in the United States alone. Then President of the United States, Donald Trump, asked companies to work at "warp speed" to produce a vaccine. The last poll as to whether or not Americans would take the vaccine when it first came out showed that only 25% would take the vaccine, mostly out of concern for safety.

Without getting into the politics, the public lost confidence in the Federal Drug Administration when it suggested providing vaccine by an Emergency Use Authorization, a mandate generally applied to using a therapeutic agent to help treat a disease in very sick patients. However, a vaccine would be given to healthy people, and if rushed into use prior to proper testing, there could be detrimental effects. Three companies developed COVID-19 vaccines in record time. One had to be removed later due to too many detrimental effects (all had some negative effects in some people). The other two vaccines succeeded in decreasing the spread of COVID-19; however, many refused to take the vaccine allowing the disease to remain a pandemic for longer than necessary.

The loss of trust in the pharmaceutical companies (and the government) negatively impacted our healthcare system and lead to many more COVID-19 related deaths than necessary. But trust in the pharmaceutical companies on opioids also lead to many deaths. The evil in health care must be eliminated to build a positive healthcare system that puts patients above profit.

Fentanyl, the New Drug of Addiction and Death

Today, drug addicts die moreso from overdoses of fentanyl and fentanyl analogs than OxyContin. While physicians used morphine to treat pain for centuries, scientists continued looking for something better – more potent and fewer side effects. Dr. Paul Janssen and the Janssen Company of Beerse, Belgium, first synthesized fentanyl in 1960 [12]. In 1963, countries in Europe approved the use of fentanyl as an intravenous analgesic. The US approved fentanyl for the same use in 1968. Fentanyl is 100 – 200 times more potent than morphine and acts much more quickly. However, it is also much more addictive and deadly [13].

Figure 1 shows the rates of opioid addiction by type of opioid from 2002-2022 [14]. Prescription opioids include morphine, oxycodone, hydrocodone, OxyContin, etc., Synthetic opioids include fentanyl and fentanyl analogs. Heroin is a street drug. Methadone is a slow acting opioid with prolonged effects used in treating addiction – patients get less of a high with methadone, but the craving for opioids and withdrawal symptoms go away.

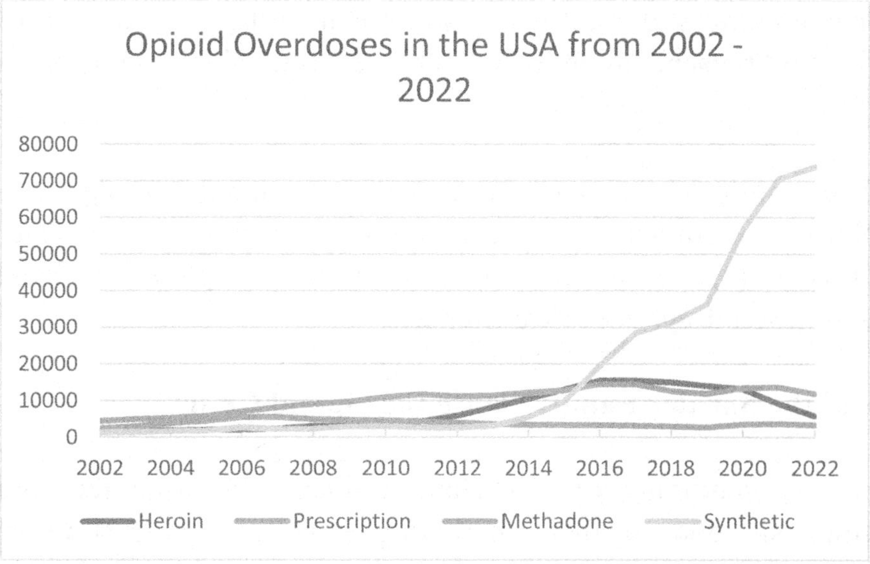

Opioid overdoses result from respiratory depression, which leads to hypoxia (low blood oxygen) and possibly death. Other symptoms include pupillary constriction and stupor [13]. If a person overdoses, naloxone (an opioid antagonist) must be delivered quickly to reverse the actions of the opioid. Today, overdoses are so common that Narcan (a nasal form of naloxone) is available for purchase over-the-counter, and many first responders carry Narcan in their vehicles.

We saw above how pharmaceutical companies pushed the distribution of OxyContin and other opioids, but what about fentanyl? Although a pharmaceutical company first produced fentanyl, it is relatively easy and cheap to make and drug cartels and other illicit drug distributors are producing not only fentanyl, but also derivatives of fentanyl called fentanyl analogues. These drugs are often mixed into other more commonly used illicit drugs such as heroin, cocaine, and other opioids; or the fentanyl may be made into a pill that looks like a prescription drug like OxyContin or Xanax. Often, the buyer is unaware of the fentanyl content. In addition, because illicit drug manufacturers do not have strict standards, the amount of fentanyl present varies considerably, even within the same batch [15].

Thus, while drug addiction started over a hundred years ago with abuse of clinical medicines such as morphine and later OxyContin, the current drug crisis is more highly associated with illicit drugs sold by criminals.

Did the Sackler Family Cause the Opioid Crisis?

At the beginning of this chapter, Purdue Pharma received emphasis due to their marketing style and promotion of OxyContin as a safe medicine. The details behind Purdue Pharma's business practices warrant a closer look.

Isaac Sackler emigrated with his family from the Austrian Empire, arriving in New York in 1906, He married Sophie Greenberg in 1912 and they had three sons, Arthur (born 1913), Mortimer (born 1916), and Raymond (born 1919). Isaac and his brother opened a grocery store in Williamsburg called the Sackler Bros. Their store did well, allowing the brothers to move to a busier neighborhood in Flatbush and Isaac invested in real estate. Unfortunately, the Great Depression struck and Isaac lost much of his property [16].

Isaac and Sophie wanted more for their children and thus sent the three sons to an excellent school, Erasmus Hall High School. Arthur Sackler was very ambitious and worked hard in school and out, as he held several jobs. Arthur was able to help his family with some of his earnings. After graduating from Erasmus, Arthur went to college to become a doctor. As an undergraduate and then a medical student at New York University, Arthur continued to work hard in class and at various jobs and extracurricular activities. He thrived at keeping busy. Arthur encouraged his younger brothers to follow in his footsteps, and they also became doctors [16].

When Arthur was a fourth-year medical student, he worked with the chief of surgical services, an elderly man whose surgical skills were declining to the point where patients were dying and staff referred to him as the "Angel of Death." Arthur took his oath to do no harm seriously and tried to dissuade patients from having surgery under this physician. Arthur Sackler graduated from medical school in 1937 [16].

Arthur completed his residency at Creedmoor State Hospital in Queens, New York, where he founded a research institute, The Institute of Psychobiologic Studies in 1949. He also worked for William Douglas McAdams, Inc., a medical advertising firm, which he ultimately headed. His brothers followed his steps, completing residencies at Creedmoor State Hospital and then working at the Institute. The Sackler Brothers all became psychiatrists and researched pharmaceutical products to replace harsh therapies they saw in the state hospital, such as electroconvulsive shock therapy. Together, they published over 150 research articles [16].

In 1952, the Sackler brothers bought Purdue-Frederick Company, a pharmaceutical business specializing in making laxatives and ear wax remover. It would later become Purdue Pharma. In 1966, the brothers bought Napp Pharmaceuticals Limited. They no longer practiced medicine, opting instead for selling drugs the legal way.

Arthur Sackler changed how pharmaceutical companies advertised, targeting physicians while utilizing deception and overzealous promotion of drug effectiveness. The Medical Advertising Hall of Fame selected Arthur in 1998, stating [16],

> No single individual did more to shape the character of medical advertising than the multi-talented Dr. Arthur Sackler. His seminal contribution was bringing the full power of advertising and promotion to pharmaceutical marketing.

Today, we all see the results of what Arthur started in the numerous pharmaceutical commercials on television.

In the 1960's, Arthur Sackler used his marketing skills to promote valium sales by Hoffman-LaRoche. The ads promoted valium for common stressors, not just severe anxiety, and negated the addictive nature of valium. For example, valium was targeted to the average college student, the over-worked under-appreciated housewife, and to those with neurotic tension. The successful promotions resulted in valium being the first $100 million drug, Arthur Sackler becoming rich, and 15% of the United States became addicted to valium.

Under ownership of the Sackler Brothers, Purdue Pharma developed a controlled-release form of morphine called MS Contin (MS = morphine sulfate). Purdue Pharma successfully marketed MS Contin, making substantial gains during the twenty-year patent period. As their exclusive rights to sell MS Contin was nearing an end, Purdue Pharma decided to improve on its own product to start another twenty-year gain with the release of OxyContin, a controlled-release form of oxycodone. The move made billionaires of the Sackler brothers [16].

Purdue Pharma marketed MS Contin primarily as a treatment for cancer pain. Because of its controlled-release, MS Contin would not produce the same "high" as morphine. In contrast, Purdue Pharma marketed OxyContin for many types of chronic pain, and targeted physicians who wrote the prescriptions [16].

Purdue Pharma hired pharmaceutical sales representatives and trained them in how to promote OxyContin to physicians. In addition, IMS Health, a data acquisition company partially owned by Arthur Sackler, analyzed data to determine which doctors should be targeted to increase sells of OxyContin. The results indicated Purdue Pharma should target general practitioners as opposed to cancer or pain specialists to better sell Oxycontin as a non-addictive pain reliever, and that is what the sales teams did. They also targeted physicians who wrote more opioid prescriptions than average [16].

Purdue Pharma also provided physicians with free samples of OxyContin which they could provide their patients, and Purdue Pharma provided coupons to patients who could then ask physicians for prescriptions.

Meanwhile, Purdue Pharma utilized physicians in their marketing schemes, including providing lists of physicians (who did not exist) who endorsed its medications [16].

Nobody argues that the aggressive marketing strategy used by Purdue Pharma increased the number of opioid prescriptions. Does that make the company, and its owners, responsible for the opioid crisis? I would argue yes. Thus, the Sackler brothers must be evil people, correct?

The Sackler brothers are better known for philanthropy than the opioid crisis. Arthur Sackler had a strong interest in the arts and donated very expensive items to numerous museums and universities. All three brothers donated to many science and medical programs at universities and had rooms, buildings and endowments named after them. For example, Yale University had the Sackler Institute for Biological, Physical and Engineering Sciences and two professorships endowed by the family: the David A. Sackler Professor of Pharmacology and the Richard Sackler and Jonathan Sackler Professorship in Internal Medicine. Harvard University had the Sackler Building which is used by the Faculty of Arts and Sciences, and The Arthur M. Sackler Museum. These are just a few places that adopted the Sackler name in exchange for donations. Most institutions have now removed the Sackler name because of the association of the family with the opioid crisis.

I believe the Sackler brothers are an example of how greed can make good people go bad.

Chapter 8

Physicians or Drug Dealers

I grew up a tomboy, playing sports with my brothers and their friends. I still love to play, and my favorite sport is softball. When I moved to Alabama in 1992, I was immediately recruited to play on a women's city league softball team. We went to the state tournament within three months of my moving to Alabama and won first place. Although I still play softball, I never played on a championship team again. In the spring of 1996, I tore the meniscus in my right knee in a coed softball game.

My orthopedic doctor ordered an MRI, which showed the meniscus tear. I had surgery later that year. A few years after my surgery, my surgeon spent two years in jail for the illegal distribution of hydrocodone. Specifically, he traded drugs for sex. On August 25, 1999, the surgeon plead guilty to illegal distribution of opioids, one of 57 counts to which he was charged. On October 27, he voluntarily surrendered his license to practice medicine, and on November 15, he was sentenced to two years in prison, three years supervised release, and he received a fine for $50,000 [1].

The surgeon regained his license to practice medicine January 24, 2003, and his DEA license to dispense controlled substances on September 24, 2003 [1]. He now practices medicine about 30 miles from where he performed my surgery.

According to the Center for Disease Control, from 1999 to 2017 nearly 218,000 people died in the United States from overdoses related to opioid prescriptions [2]. While these deaths are not necessarily linked to *inappropriately* prescribed opioids, there are health care providers clearly violating their oaths and knowingly providing opioids to drug abusers.

Prescription Drug Classification and the Controlled Substance Act

Congress passed the Comprehensive Drug Abuse Prevention and Control Act in 1970 in part to comply with the international Single Convention on Narcotic Drugs of 1961 to control drug trafficking. Title II of the Act is the Controlled Substance Act, which regulates the manufacture, importation, possession, use and distribution of certain narcotics, stimulants, hallucinogens, depressants, anabolic steroids and other potential substances of abuse [3]. Enforcement of the Controlled Substance Act was assigned to the Drug Enforcement Agency (DEA), established in 1973.

The Controlled Substance Act designates five categories or Schedules of drugs, chemicals, or other substances (such as plants or fungi) based on eight criteria [4]:

(1) Its actual or relative potential for abuse
(2) Scientific evidence of its pharmacological effect, if known
(3) The state of current scientific knowledge regarding the drug or other substance
(4) Its history and current pattern of abuse
(5) The scope, duration, and significance of abuse
(6) What, if any, risk there is to the public health
(7) Its psychic or physiological dependence liability

(8) Whether the substance is an immediate precursor of a substance already controlled under this subchapter

Schedule I drugs and chemicals have no accepted medical use and have a high potential for abuse. This category includes drugs such as heroin, marijuana, and LSD (lysergic acid diethylamide). Physicians generally do not have a license that includes Schedule 1 drugs. Biomedical researchers can be authorized by the DEA to use Schedule 1 drugs for research purposes only. Schedule II drugs and chemicals have a high potential for abuse, and use can potentially lead to addiction. Examples of Schedule 2 drugs include oxycodone, hydrocodone, and Ritalin. Most illegal prescriptions by physicians are for Schedule 2 drugs. Schedule III drugs and chemicals have low to moderate potential for abuse and likelihood of addiction. Examples of Schedule 3 drugs include ketamine, anabolic steroids, and low doses of codeine. Schedule IV drugs and chemicals have a low potential for abuse or addiction and include Tramadol, Ambien, and Xanax. Schedule V drugs and chemicals have the lowest potential for abuse or addiction, and include cough syrup with codeine.

In 2006, the DEA issued a policy statement for dispensing prescriptions to patients for the control of pain: "a registered physician acting within the usual course of professional practice must issue a prescription for a legitimate medical purpose. [5,6]" The DEA applied this policy in the cases described later in this chapter.

To regulate physician prescriptions for pain medications, over 30 years ago the American Medical Association endorsed the 4D Model to evaluating prescription writing. The four Ds are dated, duped, disabled, and dishonest [7]. Under the 4D model, a dated physician is out of date regarding knowledge of pharmacology and the different diagnosis and management of chronic pain. A duped physician is deceived by the patient and writes a prescription inappropriate for that patient, often a drug abuser. Disabled physicians are impaired by a substance abuse disorder of their own or some other medical disability. A dishonest physician is one who writes prescriptions for personal gain. Distinguishing between the 4 Ds is not always an easy task.

Dr. James Graves: First of Many Doctors Arrested for Prescription-related Crimes

In February 2002, Dr. James Graves became the first physician found guilty of manslaughter in the overdose deaths of four of his patients. The jury also found him guilty of racketeering and unlawful delivery of a controlled substance [8,9,10]. Graves had been the number one OxyContin supplier in the state of Florida.

Dr. Graves is a former Navy doctor who later worked as a general practitioner. In 1998, he specialized in pain management and worked in a Pensacola pain center. His employer at the clinic, Dr. Hutchinson, fired Dr. Graves after he sold pills to patients in parking lots and failed to record prescriptions in patient records. Dr. Graves later opened his own office. According to an AP report [11]:

> Witnesses testified that Graves wrote prescriptions from his front porch before he opened his practice. He allegedly told authorities he saw 50 to 60 patients a day at his Pensacola-area practice and wrote 200 prescriptions every day. His wife estimated he had 1,000 patients at the time authorities closed down his practice.

Dr. Graves claimed not knowing of the drug abuse patterns of his patients [8]. He also claimed no one would have died if the medications were taken as directed. OxyContin is often used by drug addicts to get a heroin-like high by crushing the pills and injecting the drug into a vein. His attorney, H.E. Ellis Jr., stated, “OxyContin is a good drug if taken properly. Pharmacy companies don’t spend billions of dollars developing drugs if they are going to kill people.” Ellis showed naivety – think of all the commercials for prescription drug lawsuits. Yes, pharmaceutical companies do not plan on developing drugs that kill people, but some drugs still do.

Dr. Graves often prescribed a group of drugs to a single patient (later referred to as “Graves cocktails”). Some Florida pharmacists refused to fill his prescriptions because they believed the combination of drugs was dangerous. One “cocktail” he prescribed included OxyContin, the painkillers Lortab and Vioxx, the tranquilizer Xanax and the muscle relaxer Soma, totaling 12 pills per day.

During his trial, Dr. Graves admitted to treating drug addicts because they are human and deserve treatment. However, he denied knowing that those patients who died were addicts, even though at least three had track marks from injections. One of his patients, Angela Cabaniss, reported Dr. Graves to the DEA when two of her friends, also patients of Dr. Graves, died of overdoses. During her testimony, Ms. Cabaniss stated, “I'm an addict and whenever we hear there's a doctor who would write something, we basically know what to say…People were shooting dope; they were selling it…It was out of hand” [10]. According to her testimony, Ms. Cabaniss sold the prescription drugs she received. Several other patients sold drugs obtained through prescriptions written by Dr. Graves, including Joanna Barnes who testified that she made between $1500 and $2000 a month selling those drugs. She also used some of the drugs herself [11].

After his conviction, Dr. Graves was sentenced to almost 63 years in prison (he was 55 at the time). In response to his conviction, DEA head of Diversion Control Laura Nagel stated, "It's going to send a very strong message to the medical community that they treat these very potent drugs, to include OxyContin, ... with respect [12]" As we see below, several physicians did not get that message.

Pain Clinic in Mobile Alabama Closed Due to Illegal Prescriptions

On May 20, 2015, the FBI and DEA raided Physician's Pain Specialists of Alabama clinic and arrested two physicians, Dr. John Couch and Dr. Xiulu Ruan, for illegal distribution of pain medicine. They were indicted for "(1) conspiracy to distribute controlled substances outside the usual course of professional practice and not for a legitimate medical purpose, and (2) conspiracy to commit healthcare fraud" [13]. The physicians also owned C&R Pharmacy, located next to one of two branches of the pain clinic. After the raid, the pain clinic, the largest in the state of Alabama, closed down.

WKRG, a local news station, obtained an affidavit for the closing of the clinic and reported the excessive and irresponsible amount of opioid prescriptions written [14]. Dr. Ruan wrote a prescription once every 3.5 minutes. Dr. Couch averaged writing a prescription once every 4 minutes. Previous employees at the clinic became informants, with one stating the two doctors would compete as to who could write the most prescriptions—Dr. Ruan won. Other informants claimed the authorization to see patients in lieu of Dr. Couch or Dr. Ruan and write prescriptions on pre-signed prescription pads. In addition, they were authorized to forge Dr. Couch's signature on prescriptions. Informants also stated that two-thirds of the patients were not legitimate patients, but simply seeking prescription drugs, sometimes for themselves and sometimes to sell to others.

The affidavit also stated some employees of the clinic were abusing drugs themselves, including physician assistants under Dr. Couch. Some evidence suggests Dr. Couch participated in the use of illicit drugs [14].

The affidavit also contained e-mail communications between the two doctors. Dr. Ruan wrote an e-mail to Dr. Couch addressing the need to keep their clinic off the DEA radar [14]:

> Pat, have you been watching the news? AL is the leading state with the most opioid Rx written! … Now, everyone in the nation knows that AL state prescribes the most painkillers in the nation, we will need to adjust our routine regimen a bit. One of the things I have done is to wean off on Benzo, or ask their PCP [Primary Care Physician] to write their Benzo, as Benzo prescription is also one of the things they look at. We would rather be careful than sorry.

A few months later, Dr. Ruan wrote to Dr. Couch about a supposed FBI raid of the clinic [14]:

> Just two weeks ago, you mentioned to me that we were supposed to be raided by FBI, which got you concerned, but thank god, it did not happen… Let's make a paper agreement between you and me, if you were out, or if I were out, the other party takes over the leaving party's business until the leaving party comes back, for whatever reason.

I knew some patients who obtained their medication at the pain clinic and when the clinic closed, they had to find a new doctor and be reevaluated before a prescription could be given. There were lots of patients searching for doctors right after the closure, causing a backlog of appointments. A friend with fibromyalgia, a very painful condition, waited a couple months before finding a doctor for treatment. Thus, these doctors not only contributed directly to patient addiction, but also left legitimate patients without the pain relief they needed.

On the contrary, I had a friend whose mother went to Dr. Couch with Chronic Obstructive Pulmonary Disease (COPD). He prescribed her OxyContin. She became addicted, and this friend was upset due to over-prescribing a narcotic. Physicians sometimes walk a fine line when deciding the appropriate treatment.

The raid on the Mobile clinic was part of a joint action by the FBI and DEA in four states to close down so called "pill factories" or "pill mills." According to an FBI report, the indictment stated [14]:

> Dr. Couch and Dr. Ruan conspired with each other and others to knowingly and willfully distribute and dispense Schedule II controlled substances outside the usual course of professional practice and not for a legitimate medical purpose. The Indictment identifies some of these Schedule II drugs as: oxycodone (brand name: OxyContin), oxycodone hydrochloride (brand name: Roxicodone), oxymorphone (brand name: Opana), hydromorphone (brand name: Dilaudid), morphine, fentanyl, and methadone.

The agent in charge of the investigation, Special Agent Clay Morris, stated [14]:

> Prescription drug abuse is the fastest growing drug problem in the country. It is particularly troubling to find doctors so complicit in this epidemic. We trust our doctors to heal our bodies, not poison our communities. We will not tolerate or accept this illegal behavior, and we will bring justice to those that condone this type of practice.

When the case went to trial, the prosecution proposed a conspiracy involving pharmaceutical companies Insys and Galena Biopharma Inc. These companies provided the instant release fentanyl drugs, Subsys and Abstral, respectively. According to Reuters [15],

> Prosecutors said the doctors received illegal kickbacks from Insys in exchange for prescribing Subsys and became among the top U.S. prescribers of the drug.

According to a Department of Justice report, both physicians were convicted of the following [16]:

> (1) RICO [Racketeering Influenced and Corrupted Organization] conspiracy;
> (2) Conspiracy to prescribe Schedule II and III Controlled Substances outside the usual course of professional practice;
> (3) Conspiracy to prescribe more than 40 grams of fentanyl outside the usual course of professional practice;
> (4) Conspiracy to commit healthcare fraud;
> (5) Conspiracy to commit mail and wire fraud;
> (6) Conspiracy to receive illegal kickbacks from IPM/CRX related to the workers compensation dispensary; and

(7) Conspiracy to receive illegal kickbacks from Insys Therapeutics in exchange for prescribing Subsys.

In addition, Dr. Ruan was convicted of both conspiracy and substantive money laundering counts. Each doctor was also convicted of several substantive illegal drug distribution counts related to prescriptions written to specific patients.

On May 25, 2017, Dr. Couch was sentenced to serve 240 months and Dr. Ruan 252 months in prison. They were also ordered to pay restitutions as follows: $6,282,023.00 to Medicare, $3,649,092.97 to Blue Cross/Blue Shield of Alabama, $2,285,170.70 to Tricare, and $1,695,929.00 to United Heath Group. In addition, they had to forfeit houses, beach condos, luxury cars, and $5,000,000 [16].

Four individuals associated with the clinic, directly or indirectly, pled guilty to charges and testified against Drs. Couch and Ruan. Two nurse practitioners working for Dr. Couch, Justin Palmer and Bridgette Parker, pled guilty to conspiring to prescribe Controlled Substances outside the usual course of professional practice and not for a legitimate medical purpose [16]. An employee for a dispensary management company, Christopher Manfuso, pled guilty to conspiring to pay illegal kickbacks to the doctors. In addition, a drug representative for Insys Therapeutics, Natalie Perhacs, pled guilty to conspiring to pay illegal kickbacks associated with the prescribing of Subsys.

Drs. Couch and Ruan appealed their convictions all the way to the Supreme Court. In 2022, the U.S. Supreme Court unanimously threw out the cases against them (and a third physician, Dr. Kahn, who was convicted on the same statute) stating that the justice department did not prove that the physicians *knowingly or intentionally* over-prescribed medications [17,18]. The cases returned to the Court of Appeals, which dropped several of the convictions but sustained enough convictions to justify maintaining the sentences as originally handed down [18,19].

Dr. Paul Volkman, Drugs Anyone?

Dr. Volkman graduated from the University of Chicago in 1974 with both an MD and a PhD in Pharmacology. After almost 30 years of practice in Illinois and Ohio, Dr. Volkman faced several malpractice lawsuits. After settling two malpractice lawsuits and being found guilty in two others, his Illinois insurance company had to pay $1.5 million. He subsequently could not obtain insurance. Thus, even though Volkman continued living in Chicago from 2003 to 2006, he worked at four pain clinics in Ohio. In 2003, Denise Huffman hired him to work at the TriState clinic, which had two branches in Portsmouth, Ohio. Dr. Volkman also worked at the Chilicothe Clinic and the Center Street Clinic, both in Chilicothe, Ohio. His choice of location seemed appropriate for his intentions.

Portsmouth is located in Scioto County, where opioid use, and abuse, exceeds that in most U.S. counties. In the state of Ohio, opioid prescriptions in 2010 were high at 67 pills per Ohio resident. But in Scioto County, that number reached 121 pills per resident [20]. Hospitals admitted five times the national average for opioid addiction. From 2000 to 2008, 117 people died of an overdose, 90% of them from prescription drugs. Sadly, 10% of babies born in Scioto County in 2010 were born addicted to drugs.

In less than a year working in Scioto County, the DEA received complaints of Dr. Volkman's prescription practices. The DEA's initial investigation suggested Dr. Volkman prescribed large quantities of controlled substances. In fact, many pharmacies in the area refused to fill prescriptions by Dr. Volkman; therefore, the Tri-State Clinic began dispensing its own prescriptions.

In 2004, Dr. Volkman purchased more oxycodone than any other health practitioner in the entire United States, and he purchased more than 97 times the national average. In Ohio, Volkman purchased more than 100 times the amount of oxycodone than the next highest practitioner, and in fact, purchased 96% of all oxycodone purchased by health practitioners in Ohio. He even purchased 3 times more oxycodone than any pharmacy in Portsmouth and 12 times the national average for pharmacies [20].

RX-Express Pharmacy in Russell, Kentucky, filled 85 – 95% of Volkman's prescriptions [20]. In the DEA investigation, pharmacists that did not fill prescriptions by Volkman indicated concerns about both the prescriptions and the patients. The prescriptions were often for high strength medications, high numbers of pills, and for more than one drug with serious interactions. The patients tended to be young, traveled long distances and often in groups, did not show signs of pain, and were not concerned about the cost – characteristics profiled by the DEA as indicating drug abusers.

Dr. Volkman predominately prescribed narcotics (98%) over other medications for his patients, with oxycodone by far his drug of choice. In addition, 85% of the prescriptions he wrote included two immediate release narcotics to the same patient on the same visit. In his presentation, Kresnak describes Dr. Volkman's prescribing practices as follows [20]:

Wrote for "the best and the most"

- Euphoria-inducing drugs:
    - Oxycodone
    - Hydrocodone
    - Benzodiazepines (Xanax or Valium)
    - Soma
- Large numbers of pills
- High doses of pills
- The more you came, the more you got

Through patient surveys, the DEA found that 18% of Volkman's patients had drug charges pending, and another 43% had been convicted of controlled substance charges.

The DEA provided much evidence for the prosecution of Dr. Volkman, including a letter from Dr. Volkman to Dr. Hurwitz, written in October of 2003. Much can be gained about the mindset of Dr. Volkman from this one letter; thus, it is transcribed from a copy of the Exhibit here [21]:

> Dear Dr. Hurwitz:
>
> Thanks for getting back to me. It is a pleasure to make contact with another professional of like mind and approach. I am new to the pain business, about 7 months, and have had to develop dosing schedules and strategies in total isolate (amidst scorn, vilification, and disapproval). It is reassuring to find that the approach that I have adopted appears similar to yours. Of course, you have been thrown in jail for your efforts, so I'm not sure why I am reassured.
>
> Paradoxically, due to Denise's [Huffman] former and continuing contact with area DEA and FBI

agents, our clinic has been well insulated from local corrupt officials and police. Denise once told me, when I asked if we were in any real danger, that the likes of Proctor and his cronies would never face anyone, and were frankly nervous about confrontations. In any case, she and her daughter and nieces all have guns in their purses.

As far as security, we have 3 to 4 large armed men in the clinic, and video and infrared cameras all about the building. Whenever the police hang out and take down patient license plates (to find out home addresses and send people over to steal prescriptions) our guys go out and ostentatiously take pictures of the cops, who drive away in a huff. I get the patients to buy a heavy safe for their pills. I have an apartment in town which has a security system and is about 3 minutes from the clinic. I always wear my seat belt so I won't get stopped and harassed by the Portsmouth police.

We are striving to be totally self contained, by having our own pharmacy so it really doesn't matter what the local pharmacists think of me and whether they fill any of my scripts. I think that is the only way for a pain clinic to operate.

Just 2 days ago a local drug mill (Iron Rock) where 200 to 300 people a day paid $250 to get in and get a handful of pain pill scripts (never seeing a doc), and where the local police got $50 for everyone entering was shut down by the feds. By the way, all of these bogus scripts were cheerfully filled by the local druggists, but not mine!

Regards, and good luck.

Paul Volkman, MD

Dr. Volkman practiced pain management for only seven months and yet developed his own treatments, suggesting he could have been an incompetent doctor trying to do good. However, his arrogance that he knows better than others, even though his type of practice got a colleague convicted, shows a narcissistic attitude, and maybe even psychopathic. He also played the victim in that pharmacists would not fill his prescriptions, but did so for a "local drug mill."

Kresnak also presented data of prescriptions for a single *new* patient [20]. The amount of medication prescribed is beyond imagination. Four medications were prescribed: oxycodone, 30 mg, 8 pills per day; Lortab (hydrocodone/acetaminophen), 10/325 mg, 8 pills per day; Xanax, 2 mg, 3 pills per day; and Soma (a muscle relaxant), 350 mg, 3 pills per day. Thus, this patient was prescribed 22 pills per day after seeing a doctor one time. I could not find information on the duration of that one visit.

Not surprising, at least 12 patients of Dr. Volkman died after receiving prescriptions for both opioids and sedatives. Eight of these patients died within three days of their last visit with Dr. Volkman. Eleven of these patients received increased doses of their prescriptions (one received an 800% increase), which may have contributed to their deaths.

Dr. Volkman, Denise Huffman, and Alice Huffman were all indicted.

On May 11, 2011, a U.S District Court jury found Dr. Volkman guilty of 12 counts of unlawful distribution of a controlled substance, including four counts that the illegal distribution resulted in death. The jury also convicted him of one count of conspiracy to distribute a controlled substance, four counts of operating or maintaining a drug premise, and one count of possession of a firearm in furtherance of a drug trafficking crime [22,23]. Denise Huffman and her daughter Alice plead guilty to operating and maintaining a drug involved premise, and then testified against Dr. Volkman.

On February 4, 2012, Dr. Volkman received a sentence of four consecutive life terms for illegally prescribing and dispensing pain pills outside the scope of a legitimate medical practice that resulted in the deaths of four people [22,23]. Sentences for the other crimes varied from 10 to 20 years and were to be served consecutively. The Court also ordered Volkman to forfeit $1.2 million obtained during the commission of these crimes. Denise Huffman was sentenced to serve 12.5 years in prison and Alice Huffman was sentenced to serve 5 years.

Dr. Volkman appealed his sentence to the Ohio Supreme Court [24]. He provided four arguments for his case: (1) the jury was denied the defendant's proposed jury instruction, "In other words, in order to find the defendant guilty, you must find that he used his prescription-writing power as a means to engage in the illicit drug-dealing and trafficking as conventionally understood." (2) the Government's expert witnesses improperly provided legal conclusions as to whether Volkman's actions had a "legitimate medical purpose," (3) there was insufficient evidence provided for conviction, and (4) the sentence was not reasonable. Volkman lost the appeal on all accounts.

Dr. Volkman continues to deny wrong-doing. In 2018, he wrote a letter to his victims and provided it to Doctors of Courage, an organization that helps health care practitioners charged with mis-prescribing controlled substances [25]. Several excerpts from this letter are provided:

> …Every one of my former patients, who had been threatened and coerced into testifying against me at my trial, described me as their trusted physician. Every patient was forced to testify that they had been prescribed “too much medicine”, and that they had become addicted to narcotics as a result of my treatment of them. Prior to initiating their treatment, I explained that EACH OF THEM would become DEPENDENT on opiates, but that this was NOT the same as DRUG ADDICTION, as none of them would experience a drug “high”, only relief of disabling chronic pain. These points are specifically included in the text of the OHIO INTRACTABLE PAIN TREATMENT ACT, controlling law for medical pain management in Ohio, where I practiced. Each patient related that the pain medicine I had prescribed enabled them to resume activities of daily living, while previously often virtually bedridden; some even resumed former jobs, played golf, had sex, cooked dinner!
>
> …Forrest Tennant, MD, a giant among America’s experts in pain management, testified at my trial that many of his patients passed away after two or three years; he considered it his role to provide palliative comfort care for essentially terminal patients. He prescribed the same combinations

and similar doses of the medications I had prescribed (since I had modeled my practice upon his published guidelines and practices). [Dr. Tennant's renowned Veract Pain Clinic has been shut down, and he has, shamefully, been raided for possible federal drug trafficking felonies!] During 2003 to 2006, some thirteen or fourteen of my patients out of about 500 died from their chronic conditions, actually far fewer than would have been expected from such a fragile population. Most were not autopsied, but the ones who were had evidence of severe heart disease, the presumptive cause of death. None of them had toxicology results which were consistent with drug overdose deaths. Despite these results, the government charged me with thirteen counts of "illegal distribution of controlled substances, resulting in death…

These horrifically misguided and unjust convictions, vacated by the Supreme Court, but then reimposed by the Sixth Circuit, naturally led many relatives of my former patients to blame me for their deaths, hate me, and scream for my execution, at my one hour sentencing hearing after my nightmarish, Stalin-esque, show trial.

I have nothing to apologize to them for. I was actually the victim in that grim scenario. These relatives, and most of the general public, have been duped by government propaganda blaming and scapegoating pain doctors for opioid overdose deaths, while over 90% of the deaths resulted from street heroin laced with deadly fentanyl, smuggled

> in by drug cartels. The government's WAR ON DRUGS, particularly its WAR ON PAIN DOCTORS, has made it almost impossible for some 100 million American chronic pain patients (according to statistics from the CDC and the Institute of Medicine) to legally obtain the pain relief they have an Eighth Amendment Right to. This has driven 50,000 people per year to overdose on ibuprofen and die from gastric bleeding, 50,000 to 100,000 to commit suicide, and 65,000 last year to opiate overdose death. About 900 may have died while taking physician-prescribed pain pills, but most of these deaths were likely from chronic conditions, as was the case with my patients.

The response from Doctors of Courage can be found on their website and follows [26]:

> This letter exemplifies the truth for ALL physicians charged for illegal distribution since the first, Dr. James Graves, in 2002. We have to spread this truth across the country. If you want to see the illegal treatment of doctors and patients end, you need to join doctorsofcourage.org HERE. Print this letter off and take it to your physician. We need to get medical associations involved as well. Physicians need to place this letter in their waiting room for all to see, and get on board with DoC.

Dr. Linda Cheek and Doctors of Courage

Dr. Linda Cheek founded Doctors of Courage after she was convicted of illegal drug prescribing. Doctors of Courage serves health care providers who treat pain, including those who prescribe high doses of narcotics. According to their website [26]:

> DoC's goal is to inform the public of the illegal action of the US Department of Justice to persecute medical professionals in order to confiscate their hard-earned assets. This is done through the illegal interpretation of the Controlled Substance Act, written to attack Drug Cartels, but then turned on doctors as a result of being ineffective in the War on Drugs. Since opioids are not the cause of drug abuse, the attacks on doctors is just as ineffective, but it has now become a method of job security and income for unethical US Attorneys and DEA agents.

The organization argues that treatment of pain is overregulated, preventing pain specialists from doing their job and providing patients a higher quality of life. The website has a slogan, "The Truth About the 'War on Drugs' becoming the 'War on Doctors'"

Linda Cheek worked as a secondary school science teacher in San Antonio Texas, before going to medical school at the University of Texas Health and Science Center at San Antonio at the age of 39 [27]. She graduated in 1992, completed one year of her family practice residency in Texas and the other two years in Roanoke, Virginia, where she continued her practice in family medicine. Dr. Cheek had specialty training in "Traditional Chinese Medicine, acupuncture, herbal medicine, Qi Gong, Clinical Issues in Primary Care, evidence-based wellness, clinical hypnosis, The Psychology of Health, Immunity and

Disease, numerous pain management courses, addiction and drug diversion courses and homeopathic courses, among others. [27]"

In 2008, Dr. Cheek was convicted of Medicare fraud and her license to practice medicine in Virginia was revoked. On February 12, 2009, the Virginia Medical Board reinstated her license [28]. However, on October 11, 2011 the DEA denied her a license to prescribe controlled substances (National Pain Report), stating numerous reasons including Dr. Cheek writing prescriptions for pain medicine using the DEA license of another physician working in her clinic. This physician denied giving Dr. Cheek permission to use her license [27]. In the DEA report of denial, Dr. Cheek, in reference to her illegal prescriptions, commented that if the DEA had timely issued her a new registration, "the complaint here would not have any substance" and that DEA's failure to grant her application demonstrates an "abject plan to create the scenario in which to charge [her] with committing a crime." [27]

In July, 2011, Dr. Cheek's medical license was again suspended, this time for writing prescriptions for controlled substances without a DEA license [28]. In May 2012, Dr. Cheek was charged with 86 counts of distributing controlled substances without holding a valid DEA license, 81 counts of using a DEA registration number issued to another person, five counts of using a DEA registration number which has been revoked or suspended and one count of maintaining a drug-involved premise. On February 21, 2013, a Federal Grand Jury found Dr. Cheek guilty of 172 felony counts of illegally dispensing pain medication. In October, 2013, she was sentenced to 33 months in prison and she had to pay $17,000 in fines. She spent 26 months in prison.

Dr. Cheek disagreed with the charges and became a political advocate for other health care providers facing similar charges. Before her trial ended, she stated the charges were a government attack on pain management physicians, “based on erroneous thought – that the pill causes addiction.” (Southwest Times) After her incarceration, Dr. Cheek wrote a book about drug addiction, **Target Pain Doc** [29]. On the Doctors of Courage website, she provides a pamphlet called *The Real Cause of Addiction* [30], and another called *Using the Law Illegally Against Doctors* [31]. The first pamphlet states the cause of addiction as follows [30]:

> **Addiction is a Disease**…but that is all conventional medicine has right. They have no clue as to the cause.
>
> **The Cause of Addiction**
> 1. Toxicity
> 2. Stress
> 3. Despair
>
> The principle of toxicity: What goes into the body must come out. If it doesn’t, eventually the toxins overload the cells and tissues, causing all chronic disease.
>
> Addiction is a lifestyle disorder, brought about by industrialization, processed foods (sugar, wheat, high fructose corn syrup, GMO, preservatives, etc.), acidifying diet, and conventional medicine such as anti-inflammatories, fever reducers, antibiotics, and vaccinations.

The pamphlet also describes the War on Drugs as a purely political ideology [30]:

> The motives behind the criminalization of opium and other controlled drugs was purely political to disenfranchise minorities as a Jim Crow method, and Vietnam anti-war protesters, as substantiated by John Erlichman.

Dr. Cheek also defended the actions of Drs. Couch and Ruan from Mobile, Alabama, and Dr. Volkman from Portsmouth, Ohio, as well as many other physicians. She wrote two books defending pain management specialists. In **Target: Pain Doc** [29], she writes about a fictional small town doctor pursued by the law for Medicare fraud and illegal drug distribution, a story similar to her own. In **Using the Law Illegally Against Doctors** [31], she argues that law enforcement is inappropriately interpreting the Controlled Substance Act to target health care providers.

Currently, Dr. Cheek operates a website on the holistic treatment of disease, including drug addiction. The Seven Pillars of Total Health [32] recommends cleansing the body of toxins and avoiding future toxins, restoring enzymes, and restoring your qi. The website recommends avoiding pharmaceuticals to treat illness. The following medical disclaimer is provided [32]:

> The information presented here is not conventional thought. It is not meant to take the place of your regular physician. No experiments have been done to establish the validity of this treatment. Keep in mind, conventional treatment suppresses expression of toxicity. With the release of toxins, healing can produce some slightly

> undesirable reactions. Given the chance, the body will always work toward healing.

Dr. Cheek has a third website, Linda Cheek, MD [27], which promotes Dr. Cheek for hire as a speaker. There are three topics promoted for speaking engagements: find healing, curing addiction, and defense witness.

While some patients may need larger than "normal" doses of narcotics to treat their pain, Dr. Cheek's arguments lose credibility to me primarily because of the following quote from one of the pamphlets she wrote on the Doctors of Courage website:

> US Attorneys and DEA agents with barely more than a high school education are now stating (without any proof) that the doctor's medical decision making is not "legitimate medical practice". But when asked on the witness stand, as in the trial of Dr. John Patrick Couch in Mobile, AL, to delineate what is "legitimate medical practice" in pain management, the DEA agent said he couldn't answer that question because "he wasn't a doctor." BINGO!!! ONLY THE DOCTOR CAN DETERMINE WHAT IS LEGITIMATE."

The last sentence to me delegitimizes anything Dr. Cheek wrote. Apparently, a doctor cannot be judged to be wrong because nobody else is intelligent enough to do so. When patients die of overdoses from prescriptions much higher than the "norm," then the government has the right and responsibility to investigate. In the case of Dr. Couch, three medical experts served as witnesses for the prosecution [33].

The Chapman Law Group assists Dr. Cheek in protecting doctors and others from charges for prescribing narcotics to treat pain [34]:

> For more than 30 years, we have been passionate about helping clients with matters involving health law, healthcare compliance consulting, professional licensing and regulatory compliance, DEA compliance, criminal defense, medical malpractice defense, correctional law, civil rights defense and business matters.

Clearly, their practice reaches beyond DEA compliance. On the home page of the website are links to healthcare professional blogs. The first blog listed is "10 Tricks the DOJ Uses Against Doctors — and How to Avoid Them." Several of these "10 tricks" were used in the prosecution of Dr. Volkman. For example, the data showing Dr. Volkman prescribed high volumes of narcotics compared to other practitioners is referred to by the Chapman Law Group as "statistical profiling." They claim that a total profile of the practitioner's prescriptions be evaluated to compare the number of narcotics to the number of other types of pain medication as some pain specialists simply have a higher volume of patients. The Law Group also questions the use of "red flag evidence," where an undercover agent goes to the clinic with a profile of doctor shopping, asking for medications by name, traveling a long distance to see the doctor, and showing up in groups. Undercover agents or informants used red flag evidence in the cases against Drs. Couch and Ruan, as well as Dr. Volkman.

Physicians, Pharmacists, Nurse Practitioners, ...

Physicians have company in the illegal distribution of opioids. The Department of Justice website [35] provides information about federal charges against those associated with illegal distribution of prescription opioids. One goal of the Department of Justice is to reduce opioid over-prescription as stated on the website:

> A leading component of the Department's Prescription Interdiction and Litigation Task Force, the Branch is pursuing both criminal and civil actions against wrongdoing entities and individuals up and down the prescription opioid supply chain.

An example of the actions of the Department of Justice occurred on April 17, 2019, when the Appalachian Regional Prescription Opioid (ARPO) Strike Force raided health care facilities in five southern states, resulting in the arrest of 60 people including 31 physicians, 8 nurse practitioners, 7 pharmacists, 1 dentist, 1 physician assistant, and 1 podiatrist [36]. The individuals arrested were associated with over 350,000 prescriptions for controlled substances, accounting for over 32 million pills.

The Department of Justice press release from April 17 showed that when health care providers stray from their oath, they often do it in more than one way [37].

For example, one of the physicians arrested in Alabama on that day allegedly "recruited prostitutes and other young women with whom he had sexual relationships to become patients at his clinic, while simultaneously allowing them and their associates to abuse illicit drugs at his house" [36]. A physician in Kentucky allegedly provided, "pre-signed, blank prescriptions to office staff who then used them to prescribe controlled substances when he was out of the office, and for directing staff at the clinic, including individuals not licensed to practice medicine, to perform medical services on patients." And the dentist arrested in Kentucky that day allegedly wrote "prescriptions for opioids that had no legitimate medical purpose and that were outside the usual course of professional practice, removing teeth unnecessarily, scheduling unnecessary follow-up appointments, and billing inappropriately for services." Several of the health care providers arrested were charged with health care fraud, and several provided prescriptions for opioids along with other dangerous drugs that are more deadly when taken together [36].

In the same press release, the Department of Justice provided information on previous activities in fighting illegal prescribing of controlled substances [36]. From June 2018 until the raid April 2019, 650 health care providers were excluded from Medicare, Medicaid and all other Federal health care programs for conduct related to opioid diversion and abuse. In addition, since July 2017, the DEA immediately suspended 31 controlled substance licenses and received 1,386 surrenders of licenses for violations of the Controlled Substance Act. The Chapman Law Group defended or is defending some of those charged with crimes.

Five months after the April 17 raid, the Appalachian Regional Prescription Opioid (ARPO) Strike Force struck again [37]. Thirteen individuals were arrested for illegal distribution of controlled substances and/or health care fraud.

In 2019, in Virginia Beach, Virginia, the U.S. District Court convicted dentist Gary Hartman for conspiracy to distribute prescription opioids and muscle relaxant pills without a legitimate medical purpose [36,38]. From 2014-2018, Dr. Hartman conspired to prescribe oxycodone and hydrocodone for his personal use and the use by co-conspirators. Specifically, Dr. Hartman would trade dental service to patients in exchange to providing the patient a prescription for opioids and having the patient return the pills to him. Dr. Hartman also worked with another dentist to provide each other prescriptions for opioids and muscle relaxants. Lastly, Dr. Hartman wrote prescriptions for friends, who then returned some of the pills to Dr. Hartman while keeping some for their own personal use. In all, 766 illegal prescriptions were written that included just under 40,000 oxycodone pills as well as other controlled substances. The U.S. District Court sentenced Dr. Hartman to 8.5 years in a federal prison.

A study by Katie Suda *et. al.* showed that dentists prescribe about 10% of the opioid prescriptions in the United States [39]. The study also indicated 25-50% of opioid prescriptions were for a greater dose than recommended by a 2016 Center for Disease Control guideline.

On the other side of the country, a U.S. District Court in Oakland, California, found Physician's Assistant David Lague guilty of 39 counts of distributing controlled substances without a legitimate medical purpose [40]. Medicare records show that Lague was the number one prescriber of opioid prescriptions in California in 2015 and 2016. He prescribed oxycodone twice to a patient who told Lague he planned on selling the drugs. The U.S. District Court sentenced Lague to ten years in a federal prison, and fined him $5,000. U.S. Attorney Tse commented about Lague's actions and the medical profession:

> The medical profession has made great strides in reforming prescribing practices, and the U.S. Drug Enforcement Administration has worked to decrease the total quantities of pills that

> pharmaceutical companies produce every year, all with the end of decreasing the numbers of powerful opioid pills getting into the hands of vulnerable members of our community.

I could go on with examples, but hopefully, you get the point. Clearly, some health care providers want more money and are willing to risk the lives of others to get it. Other health care providers, possibly including some who have been arrested, believe that prescription opioids are the only option to relieve a patient's pain and improve her quality of life. Many patients testified to that effect during the trial for Drs. Couch and Ruan. How can the Department of Justice determine who is trying to do what is best for the patient versus who is trying to do best for himself? And what about the cases where a physician makes an error?

Impact on Pain Treatment

Ben Rich and Lynn Webster studied thirty-five malpractice cases involving overdoses with opioid prescriptions, with twenty patients dying [41]. The purpose of the study was to determine how much patient error versus physician error contributed to the overdoses. Data showed that prescriptions for methadone to treat pain contributed most to death by overdose, with hydrocodone second. Physician error contributed to 75% of the deaths, and patient non-adherence or patch malfunction contributed to the other 25%. The physician errors included the following: starting the prescription at too high a dose, too rapidly increasing the dose to counter drug tolerance, inappropriately converting to a different opioid, and failing to screen patients for medical or mental comorbidities that can compromise opioid therapy. They concluded there is much complexity trying to relate the number of opioid deaths to prescription drugs being the cause:

> It is one thing to cite statistics supporting the observation that opioid- related deaths have risen drastically in recent years. It is quite another matter to obtain and interpret specifics of how the deaths occurred so as to inform clinical practice going forward.

Dr. Cheek proposes that overregulation of pain management has left many patients in unnecessary pain. Several medical doctors and biomedical scientists agree with her. According to Helen Edwards and Michael Bennett [42], opioids may be under-prescribed during palliative care. Anessa Foxwell *et. al.* argue that "Recent opioid legislation and increasing scrutiny from regulatory agencies have created multiple barriers for providers and patients to achieve pain control [43]. These challenges include prescription limitations and practitioner fear of litigation, which can result in the under-prescribing or refusal to prescribe opioids." Mark Barnes, *et. al.* wrote a review on how federal and state regulations written in the last decade affect the practice of prescription writing by respected physicians [44]. They argue that, "All of the legal interventions discussed above limit physician autonomy and interfere with professional judgment and discretion, thereby restricting this discretion and narrowing a physician's range of options for treating patients." Barnes, *et. al.* recommend putting more responsibility on hospitals and medical groups to regulate prescriptions from their facilities and dole out greater punishment to those health care providers who violate legitimate prescription practices. They also emphasize state laws that over-regulate prescription writing by physicians, such as occurred in New Jersey.

In 2017, in his final State of the State Address, New Jersey Governor Chris Christie spent most of his speech discussing the opioid crisis in New Jersey. Christie takes a strong stand on opioid prescriptions as they can lead to heroin addiction [45]:

> We know that the majority of heroin addicts first became addicted through the use of prescription opioids. According to the American Society of Addiction Medicine, four out of every five new heroin users got started by misusing prescription painkillers.

Thus, he calls for new laws regulating opioid prescriptions [45]:

> As a result, today I am directing Attorney General Porrino to use emergency rule-making and other regulatory reform to limit the supply of opioid-based pain medications that physicians, dentists and other licensed health care providers prescribe to patients presenting with acute pain.
>
> Presently in New Jersey, authorized health care providers can write initial prescriptions for opioid painkillers that provide up to a 30-day supply. This is dangerous, ill-advised and absolutely unnecessary. We know addiction to opioids can occur within days, we must work against potential addiction - and overdose - by limiting supply to five days that can be obtained at the outset of treatment. Prescribers would be required to consult the patient, assess their need and only then provide further authorization for additional quantities

Christie also called for investigations of prescription providers if needed [45]:

> If necessary, Attorney General Porrino should open an investigation of the prescribing practices of our medical community and their interaction with the industry manufacturing these drugs. Profit, by physicians or the pharmaceutical industry, must never be a rationale for contributing to the death of our citizens by overprescribing of these drugs.

Erin Krebs wrote of concerns over policies regulating opioid prescriptions that are not based on science [46]. No scientific studies support the five-day limit of opioids. In fact, there are concerns for a patient who has been taking long term opioids to treat chronic pain. Should this patient suddenly not be able to get the opioids normally prescribed, he could go into withdrawal, or turn to other sources (e.g. street drugs) to meet his needs.

Kelly Dineen and James DuBois wrote an excellent article reviewing the dilemma physicians now face when prescribing opioids [7]: *Between a Rock and a Hard Place: Can Physicians Prescribe Opioids to Treat Pain Adequately While Avoiding Legal Sanction?* The article starts with a quote from a study by Rosenblum *et. al.* that states concisely the dilemma faced by physicians treating pain [47]:

> Opioids play a unique role in society. They are widely feared compounds, which are associated with abuse, addiction and the dire consequences of diversion; they are also essential medications, the most effective drugs for the relief of pain and suffering.

Dineen and DuBois explain how increased prescriptions of opioids improved the quality of life for many patients, but unfortunately, an increase in drug abuse also occurred. If the drug abuse is associated with prescription drugs, then we need to determine who is at fault: the patient or the physician [7]. Dineen and DuBois argue the necessity of knowing the cause of the drug abuse, and if the physician is responsible, why? They then discuss pitfalls of the 4D Model for analyzing physician mis-prescription practices and provide a new mode, the 3C Model.

In the 3C Model, mis-prescribing is categorized as careless, corrupt, or compromised. The careless physician compromises his patients' health, but through reckless acts, not intention. A physician acting in this manner is incompetent and his ability to continue practicing medicine should be questioned. Corrupt physicians act to benefit themselves over their patients – they are the criminal physicians. Physicians who prescribe opioids for monetary gain are corrupt. They should be tried in a criminal court. Compromised physicians have a physical or mental impairment preventing them from practicing medicine skillfully and safely. Compromised physicians tend to fall in the category of those who abuse drugs or alcohol. As such, they tend to mis-prescribe opioids for themselves, not for patients.

Inequities in Prescription Medication

Although this chapter focuses on over prescribing or inappropriately prescribing pain medication, for many minority patients, pain medication is under prescribed. Michael Flores presented a study at the American Public Health Association Annual Meeting and Expo in 2019 [48]. He and his colleagues studied health records of over 22,201 adult patients of various races from the same low-income urban setting over 2.5 years for disparities in opioid prescriptions and opioid disorder diagnosis. They found that white patients receive opioid prescriptions more frequently than minority patients, and they had a higher risk of developing opioid use disorder. Thus, adults of racial and ethnic minorities are more likely to be undertreated for pain.

Minority children are also less likely to receive opioid prescriptions to treat their pain. Cornelius Groenewald *et. al.* studied racial differences in opioid prescribing in children 0-17 years old [49]. They used survey data from the 2003-2014 Medical Expenditure Panel Surveys. With 113,929 children participating, they showed that white and Native American children were more likely to receive opioid prescriptions compared to all other racial and ethnic groups. In contrast, white children received fewer prescriptions for non-opioid analgesics compared to all other racial and ethnic groups.

Their study also investigated relationships between health provider race and ethnicity and child race and ethnicity. White children had white health care providers 93.5% of the time, whereas black children had black providers only 34.3% of the time and Hispanic children had Hispanic providers only 42.7 percent of the time. In addition, less than 10% of black children had a Hispanic provider or Hispanic children have a black provider. Minority children were more likely to receive opioid prescriptions from a white provider compared to a minority provider. These data can be attributed to a shortage of Hispanic and black physicians.

Bridget Peteete reviewed twenty-eight studies on prescription drug misuse by racial and ethnic minorities in 2017, looking for potential risk factors that could be used to identify patients who might need to be monitored more carefully when treated with prescription drugs [50]. She found a high prevalence of prescription drug misuse among veterans, bisexual and gay young adults, and substance abusers compared to the general population. She also found prescription drug abusers tend to be young males, less educated, single, and with health or emotional issues.

The discrepancy in treating pain with prescription opioids is only one of many health care disparities between whites and minority groups in the United States. In a review by Taylor and Jonathan Santoro, biases in health care providers still exist to the extent that "opinions of white laypersons, medical students, and medical residents identified a continued belief that the black body is biologically 'different' from the whites."

Conclusion

Many patients complain of inadequate pain management, suggesting better training of physicians is required. Medical schools should increase the coverage of pain management in their curricula, and consideration should be given by the American Medical Association (AMA) to require residencies for those physicians who specialize in pain management.

But education will not stop those physicians out to make a dollar at the patient's expense. The Drug Enforcement Agency (DEA) has that responsibility, and it is not an easy one. Some patients require very high doses of opioids to get through a day, and if opioids improve the quality of life, that patient should get the prescriptions. Cooperation between the DEA and AMA might improve the ability to monitor and regulate the distribution of opioid prescriptions. However, one must be concerned that the DEA influence on patient prescriptions can parallel that of health insurance by taking away the authority of the physician to make the appropriate decision for his patient.

Unfortunately, the abuse by some physicians and patients makes it harder for other physicians to do their jobs and patients to receive the pain relief they deserve. That is how it works with evil people. To protect us from evil, legislators create laws that also limit our freedom.

Chapter 9

Eugenics

Gregor Mendel was a scientist, teacher, and a monk in the Augustinian order at the St. Thomas Monastery. Mendel is considered the "father of Genetics" because of his breakthrough work on inheritance of traits in peas. Previously, biologists believed that traits from parents mixed as they were passed on to offspring. However, Mendel found that the probability of inheriting a particular trait in peas was a constant dependent on the "parent" peas. He concluded that genes exist in pairs within an organism. However, the pairs do not have to be identical as there are two types of each gene, one dominant (G) and one recessive (g). As a gene gets passed on from the parent plant to the offspring, one gene comes from the male and one from the female. If a pea has one dominant gene for a trait, then it will have that trait. If it has two recessive genes, it will not have the trait.

For example, let's look at the "mating" of two peas, one with both dominant genes (GG) and the other with both recessive genes (gg). We will call these the parent genes. We will next follow the "matings" for two generations of offspring (F1 and F2).

Parents	GG	x	gg
One parent (GG) shows trait			
Generation 1 (F1)	Gg	x	Gg
Both offspring (Gg) show trait			

Generation 2 (F2)   GG   Gg   gG   gg

Three offspring (GG, Gg, gG) show trait; (gg) does not

In 1866, Mendel published his work *Experiments on Plant Hybrids*, introducing the world to inheritable traits and unknowingly changing the social norms of modern countries.

Charles Darwin, an English naturalist, biologist, and geologist, travelled on the HMS Beagle from 1831 to 1836. During this voyage, he studied animals, plants, rocks and fossils in Brazil, Argentina, and Chile, as well as in the Galapagos Islands. In 1839, he wrote a book on his studies called, **The Voyage of the Beagle**. Twenty years later, Charles Darwin wrote another book, **On the Origin of Species by Means of Natural Selection**, in which he proposed the theory of evolution based on natural selection. Those plants and animals best equipped with traits to survive do survive and generate offspring moreso than those with weaker traits. Therefore, over generations, the favorable traits get passed on to offspring whereas the less favorable traits decline.

Six years after Darwin published **On the Origin of Species by Means of Natural Selection**, his cousin Sir Francis Galton of England took Darwin's theory a step further and published an article in Macmillan's Magazine, in which he wrote the following:

> No one, I think, can doubt, from the facts and analogies I have brought forward, that, if talented men were mated with talented women, of the same mental and physical characters as themselves, generation after generation, we might produce a highly bred human race, with no more tendency to revert to meaner ancestral types than is shown by

> our long-established breeds of race-horses and foxhounds.

Galton followed the article with a book in 1869, **Hereditary Genius**, promoting the selective breeding of humans to improve the human race.

Improving the Human Race

The theories of Mendel and Darwin provided the framework for hypothesizing a means to improve the human race. In 1854, English philosopher Herbert Spencer coined the phrase "survival of the fittest." Some scientists, medical doctors, and politicians believed the population could improve by encouraging those with "fit" genes to reproduce while prohibiting those with "unfit" genes to do so. Francis Galton gave this movement a name, *eugenics*, in his book **Inquiries into Human Faculty and Its Development**, published in 1883. Francis Galton defined eugenics (derived from the Greek word eugenes, which means "well born") as "the study of the agencies under social control that may improve or repair the racial qualities of the future generations, either physically or mentally. [1]" Galton believed that survival of the fittest ceased to occur in humans because of social norms – namely, allowing the "unfit" to reproduce, and, according to some, at a rate greater than that of "fit" humans. To reverse the trend, promoting the "fit" to reproduce (positive eugenics) while forbidding the "unfit" to reproduce (negative eugenics) would force evolution in a direction to "improve" the human race. This would require separating the "fit" from the "unfit."

In 1876, Sir Charles Trevelyn developed a ranking system for people with mental deficiencies, which were believed to be highly correlated with poverty, crime, alcoholism, and sexual promiscuity. The most severe mental deficiency was idiocy, then imbecile, followed by the mildest feeble-minded. Idiocy refers to extreme folly or stupidity, mental retardation, or the state of a mental person with a mental age less than three. Imbecile indicates moderate stupidity, a person of mediocre mental retardation, or the state of a mental person with a mental age from three to seven. Feeble-minded means a deficiency in intelligence. All three terms are now considered offensive and no longer used in medicine.

Psychologist Henry Goddard, PhD, specialized in human intelligence and correlated intelligence to behavior and genes. In his 1920 book of lectures, **Human Efficiency and Levels of Intelligence** [2], Dr. Goddard defines intelligence as follows:

> …our thesis is that the chief determiner of human conduct is a unitary mental process which we call intelligence: that this process is conditioned by a nervous mechanism which is inborn: that the degree of efficiency to be attained by that nervous mechanism and the consequent grade of intelligence or mental level for each individual is determined by the kind of chromosomes that come together with the union of the germ cells: That it is but little affected by any later influences except such serious accidents as may destroy part of the mechanism.

In the early 20th century, Goddard coined the term moron as slightly more intelligent than feeble-minded, but like the other derogatory terms for intelligence, moron is no longer a medical term. However, eugenicists utilized these words in determining who *should not* and, for over eighty years, *could not* reproduce.

Eugenics in the United States

The eugenics movement started in the United States even before the term was defined by Galton. An early proponent, Dr. Isaac Kerlin specialized in mental deficiencies in children and served as the superintendent of the Pennsylvania Institute for Feeble Minded Children in Elwyn, Pa. from 1863-1893. In 1892, he addressed the Association of Medical Officers of American Institutions for Idiotic and Feeble-Minded Persons, promoting the surgical relief of "idiotic conditions." [3] During his speech, Kerlin posed the following question to the audience [3]:

> Whose state shall be the first to legitimize oophorectomy and orchitomia for the relief and cure of radical depravity?"

Indiana was the first state to legalize mandatory sterilization, but not until 1907. However, Kansas did not wait for a law. In 1893, Dr. F. Hoyt Pilcher, Superintendent of the Kansas State Asylum for Idiotic and Imbecile Children, implemented sterilization policies, as he personally performed 44 castrations and 14 hysterectomies [4]. The United States continued practicing eugenics for over 80 years through selective immigration laws and mandatory sterilization of those who were considered unfit.

Selective Immigration in the United States

A gift from France, the Statue of Liberty on Ellis Island symbolizes the openness of the United States to immigration (at least prior to 1882). The United States is a land of immigrants, as the famous words inscribed on the Statue say: "Give me your poor, your tired, your huddled masses." However, in 1875, the Supreme Court ruled that immigration was a federal responsibility [5]. Immigration laws passed in 1875 prohibited immigration of criminals and prostitutes, and prohibited the forced immigration of Asians (although they could immigrate to the United States of their own free will).

The Immigration Act of 1882 required immigrants pay a fifty cents tax to enter the United States. The Act also prohibited the entry of "idiots, lunatics, convicts, and persons likely to become a public charge."

Congress passed its first law prohibiting immigration based on ethnicity in 1882. The Chinese Exclusion Act suspended immigration of Chinese laborers for ten years, allowed for deportation of Chinese in the United States illegally, and barred Chinese from naturalization. When the Act expired, the Geary Act of 1902 extended the laws for another ten years, and in 1904, the law became a permanent ban until 1943.

In 1885 and 1887, Congress passed Alien Contract Labor Laws. The 1885 law, also called the Foran Act, prevented companies or individuals from bringing unskilled laborers to the United States to work for them. In 1887, mechanisms for enforcing the 1885 Act were passed.

In 1894, three Harvard graduates established the Immigration Restriction League, which promoted restricting the immigration of "undesirables." The league targeted immigrants from eastern and southern European countries. The League lasted until 1921, having successfully encouraged legislation requiring immigrants to pass a literacy test in 1917. Countering the Immigration Restriction League, Joseph Senner formed the Immigration Protective League in 1898. Senner wrote a letter to immigrant communities stating his thoughts on immigration, which some believe to be true today [6]:

> The nativistic authors of such entirely superfluous new laws pretend that their only object is to protect the American laboring man against foreign competition. This is, however, only a poor excuse for their real scheme, dictated by that hatred of the foreigner.

After World War I, immigration to the United States increased, especially from southern and eastern Europe. In 1924, the House of Representatives passed The Immigration Act (also called the National Origins Act), with Albert Johnson (Rep. Washington and author of the Act) stating the following in his argument [7]:

> It has become necessary that the United States cease to become an asylum,

The Immigration Act limited the annual number of immigrants from any particular country to two percent of the number of individuals from that country living in the United States in 1890, before a large immigration from southern and eastern European countries occurred. In addition, the Act prohibited immigration from Japan and Mexico, favored college educated immigrants and those with special skills, and limited the total number of annual immigrants to 150,000.

Compulsory Sterilization in the United States

For undesirable citizens and immigrants already in the country, compulsory sterilization became part of the solution. In 1902, Dr. David Jordan, the first President of Stanford University, published **The Blood of the Nation: A Study in the Decay of Races by the Survival of the Unfit** [8]. Jordan promoted maintaining a pure Anglo-Saxon Nordic race by banning interracial marriages and keeping the race strong by sterilizing those deemed unfit. He remained politically active in the eugenics movement for years.

In 1906, biologist Charles Davenport started the American Breeder's Association to "investigate and report on heredity in the human race, and emphasize the value of superior blood and the menace to society of inferior blood" [9]. Dr. Jordan became the first chair of one of the founding chapters. That same year, J.H. Kellogg provided the funding for the Race Betterment Foundation, which sponsored three national conferences on improving race and provided funding for Charles Davenport to start a registry of unfavorable inborn traits. In 1910, Davenport created the Eugenics Record Office in Cold Spring Harbor, New York. The Eugenics Record Office was run by Harry Laughlin and maintained data related to inborn physical, mental, and temperamental properties along family lines. Included in these records was the family line of Carrie Buck, whose case is described shortly.

Davenport studied animal and human heredity. He performed animal studies similar to Mendel's pea studies, and demonstrated Mendelian genetics of several animal traits. At the Eugenics Record Office, Davenport worked with Laughlin collecting family inheritance information through interviews of current living adults. Initially studying the hereditability of traits, he ultimately focused on inheritance of feeble-mindedness and other "negative" traits. His studies relied on recalled and subjective information from those his research team interviewed, questioning the validity of his conclusions, namely the hereditability of feeble-mindedness and several other "undesirable traits".

In 1908, Mary DeGarmo held the first Better Baby contest at the Louisiana State Fair [10]. Three years later, Dr. Florence Sherborn and Mary Watts organized a Better Baby contest in Iowa, and they spread to over forty states [10]. Babies were judged for shape of the head, width between the eyes, condition of tonsils, strength of spine, and ability to walk or crawl, similar to judging livestock.

Charles Davenport expanded his work to include director of the Biological Laboratory at Cold Spring Harbor and working at the Eugenics Record Office and at the Carnegie Institute of Washington's Station for Experimental Evolution [9]. He wrote to Mary Watts recommending heredity be given consideration when judging babies for fitness. A year later, Davenport sent her another letter stating, "A prize winner at two may be an epileptic at ten." When Watts shared this letter with Dr. Sherborn, they explored the possibility of analyzing the families for fitness, not just the baby. After several years of working with fair organizers and eugenicists, the Better Baby contest warped into the Fitter Families for Future Fireside contests [11].

Twenty families entered the first contest at the Kansas Free Fair in 1920. Families were judged for their eugenic qualities. If a family scored a B^+ or better, they received a medal engraved with, "Yea, I have a goodly heritage." Charles Davenport stayed involved with the contests, keeping eugenics alive at state fairs.

If undesirable traits are inherited, then selective breeding of humans could produce a better population of humans, similar to what happens with farm animals. Most people probably desire that we improve over generations. But what does improvement mean? Do we want to ultimately produce a human population of Jordan's (and Hitler's) Nordic race? Or do we want to produce a human population of artists, athletes, or scientists? And who should decide what we want?

Those questions need not be answered if one believes the studies of biologist Thomas Morgan, PhD [12]. Morgan studied heredity in fruit flies at the same time Davenport was studying other animals and humans. But Morgan's studies indicated different conclusions. Many traits in fruit flies were not inherited in accordance with Mendelian genetics - the inheritance was much more complicated. If this is true with simple traits in fruit flies, such as whether or not the legs had bristles, then one might question if something as complex as feeble-mindedness would follow basic Mendelian genetics.

Unfortunately, the United States government bought into Davenport's work promoting eugenics, and actively participated in eugenics through compulsory sterilizations. In 1907, Indiana passed the first law for mandatory sterilization of the unfit in the United States. Dr. Jordan substantially influenced the law, as he had served as President at Indiana University prior to Stanford. Dr. Jordan also provided critical input to the third state to pass a mandatory sterilization law, California, in 1909.

The Director of the Eugenics Records Office, Harry Laughlin, was concerned about the low number of sterilizations occurring in the United States, even after 12 states had passed laws legalizing them. Thus, Laughlin published a Model Eugenical Sterilization Law in 1914, proposing the authorization to sterilize people considered socially inadequate [13]. Specifically, the law considered those who were institutionalized or who could not sustain themselves in society inadequate, including feeble minded, insane, criminalistic, epileptic, inebriate, diseased, blind, deaf, deformed, and dependent (orphans, ne'er-do-wells, tramps, the homeless and paupers).

In 1924, Virginia passed its sterilization law for anybody who was feeble-minded, an imbecile, or epileptic. Before carrying out sterilizations, the state wanted the law to be challenged and hopefully heard by the U.S. Supreme Court to determine the legality of mandatory sterilizations. They used Carrie Buck to get that challenge.

The state of Virginia involuntarily institutionalized Emma Buck, Carrie's mother, for feeble-mindedness *evident in her sexual promiscuity.* With her mother institutionalized, Carrie Buck lived with an abusive adoptive family. They removed Carrie from school after the fifth grade even though she made the honor roll, and sold her services to neighbors. A nephew of her adoptive mother raped Carrie and got her pregnant at the age of 17. Her adoptive family had her institutionalized, as it was a common practice to consider unwed mothers feeble-minded. In 1924, Dr. Bell diagnosed Carrie as feeble-minded, with a mental age of only 9. Carrie's daughter Vivian was also considered "below average" and with a look that "was not quite normal." Thus, Dr. Bell wanted to sterilize Carrie against her wish. The case went to the Supreme Court in 1927 as Buck versus Bell [14]. The Supreme Court upheld the mandatory sterilization with the famous words of Supreme Court Judge Oliver Wendall Holmes:

> It is better for all the world, if instead of waiting to execute degenerate offspring for crime or to let them starve for their imbecility, society can prevent those who are manifestly unfit from continuing their kind...Three generations of imbeciles are enough.

This Supreme Court ruling increased sterilizations in the United States. Before 1927, about 3,000 mandatory sterilizations occurred. Following the Supreme Court ruling, 3,000 more mandatory sterilizations occurred in just four years. By 1940,

35,000 mandatory sterilizations were performed, 21,000 of them on women.

As for Carrie Buck, the state released Carrie on probation from the institution following her sterilization. Her daughter Vivian, like her mother, made the school honor roll, questioning the diagnoses of mother and daughter, and the mandatory sterilization. Carrie's younger sister had also been sterilized, unknowingly, during surgery for appendicitis.

In contrast to Carrie Buck, Ann Cooper Hewitt was a rich socialite and heiress. Ann's father Peter Cooper Hewitt died in 1921, leaving Ann two-thirds of his estate, and her mother one-third. But he also had a stipulation: if Ann died childless, her mother would inherit all the money. Shortly before her 21st birthday, when she would gain independence from her mother, Ann had an appendectomy. Unknown to Ann, her mother paid the surgeons $9,000 each to remove her fallopian tubes in addition to her appendix, stating that Ann was feeble-minded. The surgeons obliged, and Ann would die childless [15].

Ann Cooper Hewitt sued her mother, claiming her mother forced her to be sterilized to get the inheritance. She also sued the surgeons [16]. The cases went to trial in San Francisco in 1936. Mandatory sterilization loss favor in the United States after Germany initiated their own program in 1933 following Hitler's election to Chancellor. Thus, the defense team believed feeble-mindedness was insufficient justification for mandatory sterilization and changed tactics saying that Ann would make a poor mother because of her sexual promiscuity. Eventually, Ann dropped the lawsuit against her mother, accepting a settlement of $150,000. The judge dropped the cases against the surgeons because Ann was a minor and mandatory sterilization was legal in California. Her sterilization for feeble-mindedness, as that of

Carrie Buck, discredit the eugenics movement because these two women of normal intelligence were sterilized based on lies and misinformation. How many other Americans were sterilized under false pretenses?

The United States did not fully ban mandatory sterilizations until 1974, after approximately 65,000 sterilizations were performed in thirty-two states that passed mandatory sterilization laws. California exceeded all other states with 20,000 sterilizations, most of them before 1940. Even with the ban, a 1976 investigation by the Government Accountability Office showed that 25 to 50 percent of Native Americans were sterilized between 1970 – 1976, often without consent.

Eugenics Beyond the United States

The eugenics movement expanded far beyond the boundaries of the United States. Most modern countries participated in the movement to varying degrees, including Russia, Germany, Australia, Brazil, China, Japan, Sweden, and the United Kingdom. The Eugenics Education Society of Britain formed in 1907 and published a journal *Eugenics Review*. In 1912, the Eugenics Education Society of Britain organized the first of three International Eugenics Congress [17]. The Congress was held at the University of London and dedicated to Francis Galton, who died in 1911. Biologists, physicians, and politicians from Europe and America attended and presented their eugenic ideas. One of Charles Darwin's sons, Leonard Darwin, also presented at the Congress. Chitty wrote about the perspective of eugenicists at the Congress [18]:

> Western civilization was in danger of collapse, since we were preserving the weak and the "genetically undesirable" and allowing them to

> breed at an alarming rate. Indeed the pauper pedigrees presented at the Congress, featuring the heritability of undesirable traits among the dregs of society, proved conclusively that the poor and the feeble minded were highly fecund and would one day inherit the earth unless wise men intervened with a programme of genetic measures.
>
> On the final day of the Congress, one of Charles Darwin's sons gave the President's Farewell Address in which he "urged all present to return to their home countries determined to campaign for legislation designed "to stamp out 'feeblemindedness' from future generations."

Over 800 people attended the Congress. Dr. Schuster wrote a review for the Congress including the following commentary [18]:

> Dr. Hoffman's paper deals with a problem in eugenics of great importance to the newer nations whose population increases largely by immigration, namely the fecundity of the different elements thus introduced, compared one with another and with the native born. The data used were collected in the State census of Rhode Island in the year 1905. Whether judged by the percentage of married women who were childless or by the average number of children per family, the native born are found to be considerably less productive than the immigrants.
>
> Professor Niceforo has collected a large quantity of data in order to determine what differences, if any, exist between the upper and the lower social

> classes, with regard to their physical and mental powers… Individuals of the lower classes have, relatively to the members of the higher classes, poorer physical development, smaller cranial circumference, a lower degree of acuteness of sensation (as determined by the instruments used in experimental psychology), and of resistance to mental fatigue, a delay in the epoch when puberty manifests itself, a slowness of growth, a larger number of anomalies and cases of arrested development. Speaking generally, if men were grouped according to a scale of mental and physical development, the lower social classes would, on the whole, come near the lower end of the scale.
>
> Among the practical measures available for the prevention of procreation among these classes who form a burden on the community, sterilization has many advocates. Provided that it imposes no very serious risk or hardship on the person sterilized, it may be, even for them, a very desirable alternative to permanent confinement.

In 1921, the Second International Eugenics Congress was held in New York. Alexander Graham Bell served as the honorary president. Leonard Darwin and Dr. Davenport gave opening presentations. Following the Congress, a committee formed to help educate and promote eugenic ideas in America. This committee became the beginning of the American Eugenics Society.

Harry Laughlin organized The Third International Eugenics Congress in 1932 and held in New York. Over 250 exhibits for the Congress were displayed in the American Museum of Natural History for more than five weeks. Harry Laughlin prepared most of the exhibits, designed to "show the history, content, present researches and trends of eugenics, both as a pure and as an applied science…concerned primarily with racial and family-stock quality in the turn-over of population from generation to generation. [19]" Leonard Darwin served as the Chairman of the Congress, and Dr. Davenport served as Honorary President. According to Stillwell, the primary theme of the Congress was "mate selection and the differential birthrate resulting from the supposed 'outbreeding' of middle- and upper-class Americans by 'unfit' or 'degenerate' individuals." The Honorary Vice-President of the Congress gave a speech reminiscent of Nazi Germany's views on the role of women in society, namely that "fit" women should perform their duties to the nation and bear many children.

The Eugenics Society of Canada was invited to join the International Eugenics Congress and participated for the first time. This was the last Eugenics Congress, likely due to the rise of the Nazi party in Germany and the beginning of World War II.

The American Eugenics Society persisted until 1972, when it was reformatted into the Society for the Study of Social Biology. Founding members of the organization included Madison Grant, Harry H. Laughlin, Irving Fisher, Henry Fairfield Osborn, and Henry Crampton

Nazi Germany: From Eugenics to Racial Hygiene to Genocide

German medical advances in the late nineteenth and early twentieth centuries far exceeded that of most western countries.

Physicians in Germany focused on treating the country, not the individual. Thus, many physicians focused on bacteriology. But that soon changed, and physicians went from prominence to unscrupulous.

Nowhere in the world did eugenics progress as in Germany under Adolf Hitler. In his book **Mein Kampf,** Hitler referred to the American eugenics movement when proposing the concept of the superior Aryan race. But the concept of eugenics started in Germany long before Hitler took power.

German physician Alfred Ploetz wrote **The Excellence of Our Race and the Protection of the Weak** in 1895 [20]. His book introduced Germany to both eugenics and racial hygiene (*Rassenhygiene*). In 1904, Ploetz started a journal, *Archive for Social and Racial Hygiene*, which was the first journal in the world dedicated to eugenics research. Ploetz studied medicine under Swiss psychiatrist Auguste Forel, a proponent for eugenics [21]. Forel favored sterilization of the mentally ill, considering it a national sacrifice similar to "that of the soldier in times of war." Ploetz also studied in America under Charles Davenport. Ploetz wanted the Nordic races to grow in numbers in America under concern the Anglo-Saxons would be left behind unless they developed a policy that would change the relative proportions of the races.

In 1905, Ploetz founded the German Society of Racial Hygiene, which took eugenics one step further: declaring the Aryan race as superior and protecting the race from contamination by lesser races. The goal of the society, in Ploetz's words, was "for society to return to a healthy and blooming, strong and beautiful life." [22]

The following quote came from psychiatrist Ernst Rudin on the aims and purposes of the German Society for Racial Hygiene [22]:

> The formation of the German Society for Racial Hygiene goes back to 1905, when Alfred Ploetz, the founder of German racial hygiene and today an honorary member of the society, set out with a few friends (to whom I also belong) to create a wide dissemination of the ideas and principles of racial hygiene. Despite our continual attempts to call the public's attention to the fact that something had to be done for our race, and despite our revealing the degenerating quality of the Nordic race which had been going on since the beginning of this century, the immense danger of the declining birthrate and the unnatural nursing back to health of the genetically weak, sick and inferior, our ideas could not gain recognition at the upper echelons of government. Even when our spiritual movement succeeded in quietly and gradually winning over hearts and minds of our best Germans, the powers that be ensured that no racial hygiene measures could be taken. Only through the Fuhrer did our dream of over thirty years, that of applying racial hygiene to society, become a reality.

Contrasting the work of Ploetz, Dr. Wilhelm Schallmeyer promoted eugenics minus the racism. Schallmeyer focused primarily on the economic impact of maintaining life of those not deemed fit [23]. Schallmeyer wrote over 40 articles related to eugenics and a book, **Heredity and Selection in the Life Process of Nations** [24]. He was considered critical in the development of the German Racial Hygiene program, even by Ploetz. Schallmeyer disagreed with the concept of Aryan superiority, but he did believe in the importance of preserving the higher social classes of people at the expense of others.

Another key player in German Eugenics was Dr. Fritz Lenz, who studied medicine at the University of Freiberg where he met his mentor Alfred Ploetz. Lenz believed in the superiority of the Caucasians, and especially the Aryan race. While studying medicine, Lenz also met Eugen Fischer who founded the Society for Race Hygiene in Freiberg, and they co-wrote a book with Erwin Baur, **Principles of Human Heredity and Racial Hygiene** in 1921 [25]. This book became a popular German textbook on genetics and racial hygiene and was read by Adolf Hitler while incarcerated. Hitler utilized this book to scientifically justify his racial hygiene programs.

In 1907, Ploetz, Lenz, and psychiatrist Arthur Wollney formed a secret circle from within the Society of Racial Hygiene known as the Circle of Norda. By 1910, the circle morphed into the Nordic Circle and became a primary influence over German racial hygiene policies under Hitler.

German jurist Karl Binding and German psychiatrist Alfred Hoche published a widely discussed book, **Permitting the Destruction of Life Unworthy of Living** in 1920 [26]. Binding and Hoche argue for euthanizing the mentally ill and physically disabled as they carry bad genes and are a burden to the economy. A small part of the argument presented in the book follows:

> If one imagines a battlefield strewn with thousands of dead young men, or a mine in which firedamp explosions have trapped hundreds of industrious workers, and if, at the same time, one juxtaposes that image with our mental asylums,

> with their care for their living inmates – one is deeply shaken by the shocking discordance between the sacrifice of the finest examples of humanity on the largest scale, on the one hand, and by the greatest care that is devoted to lives that are not only absolutely worthless, but even of negative value, on the other hand.
>
> It can in no way be doubted that there are living human beings whose death would be a deliverance both for themselves and society, and especially for the state, which would be liberated from a burden that fulfills absolutely no purpose, other than that of serving as an example of the utmost selflessness.

Hitler expressed his belief in eugenics in 1925, writing, "He who is not healthy and worthy bodily or mentally is not allowed to perpetuate his malady in the body of his child....," [27]. At this time, however, Hitler did not believe unfit persons should be euthanized. That belief would change once he became Chancellor of Germany.

America continued to influence German political ideology when, in 1927, the Rockefeller Foundation funded the *Kaiser Wilhelm Institute for Anthropology, Human Heredity, and Eugenics* in Berlin. Eugen Fischer was Director of the institute.

When the Nazis came to power in 1933, the role of physicians in Germany changed dramatically. German physicians faced financial difficulties in the Weimar Republic following World War I. The government passed laws providing health insurance to Germans, taking away much of the private health care service. Following a strike by physicians, more laws were passed, limiting physicians to the number of patients they could see and opening clinics with salaried physicians. In response, in 1929, German physicians formed The National Socialist German Physicians' League, later to become the Nazi Physicians' League. In addition to battling for more economic freedoms, the National Socialist German Physicians' League promoted antisemitism.

Jewish physicians outnumbered other physicians, fueling resentment in an era when antisemitism did not need much fuel. After becoming the Nazi Physicians' League, the organization started removing Jewish physicians from the medical profession in March 1933. A month later, a law was passed forbidding Jewish physician civil servants from practicing medicine at universities and hospitals throughout Germany. Michael Kater writes, "Physicians became Nazified more thoroughly and much sooner than any other profession, and as Nazis they did more in service of the nefarious regime than any of their extraprofessional peers. [28]"

On July 14, 1933, the "Law for the Prevention of Genetically Diseased Offspring" required the compulsory sterilization of people with any of the following categories of disease: hereditary or congenital feeble-mindedness, schizophrenia, bipolar disease, hereditary epilepsy, Huntington's disease, chorea, hereditary blindness, hereditary deafness, malformation, and severe alcoholism. This Nazi law was based on Harry Laughlin's model sterilization law developed in 1914. For his contribution to the science of racial cleansing, Laughlin was awarded an honorary degree from the University of Heidelberg in 1936.

The book, **Permitting the Destruction of Life Unworthy of Life** [26], formed a blueprint for the organized mass murder that occurred during the Third Reich. In the book **Psychiatrists: the Men Behind Hitler**, the authors wrote [29]:

> The euthanasia program for the handicapped and mentally ill was organized exactly as prescribed in Alfred Hoche's and Karl Binding's notorious 1920 treatise…Hoche had also suggested in 1920 that a commission be appointed to order the euthanasia of 'dead-weight characters,' and Binding provided ideas for the removal of legal hindrances. Nearly twenty years later, the euthanasia program was established exactly as Hoche and Binding and others had imagined it a generation earlier when Germany was a republic rather than a dictatorship and when law, not hate, ruled. Theory in a republic had become an unspeakable reality in the twisted minds of totalitarian demagogues.

In 1929 Hitler said at a Nazi Party Conference in Nuremberg, "an average annual removal of 700,000-800,000 of the weakest of a million babies meant an increase in the power of the nation and not a weakening." In 1939, Germany advanced its eugenics program to include euthanasia of children. Hitler issued a decree commissioning doctors to perform "mercy killings" on those who were judged "incurably sick by medical examination." Hitler assigned his personal physician Dr. Karl Brandt and his Chancellery Phillip Bouhle to authorize physicians to euthanize children deemed unfit. Nurses and midwives assisted with the program. To minimize rebellion by the people, the euthanasia program started with infants and children under three years of age. But soon after, children up to the age of seventeen were euthanized [30]. Public Health authorities requested parents to submit their children to be institutionalized, even though the true plan was killing the children by overdose of drugs or starvation. According to the Holocaust Encyclopedia, "Conservative estimates suggest that at least 10,000 physically and mentally disabled German children perished as a result of the child 'euthanasia' program during the war years" [31].

In 1939, Hitler signed a secret authorization to expand the euthanasia program to include adults [30]. This secret program, coded as T4, was named for the street address in Berlin where the program coordinating office was located, Tiergartenstrasse 4 [31]. Hitler assigned Phillip Bouhler and Dr. Karl Brandt to head the program. To euthanize adults, six gassing units were established in Germany and Austria.

To select children and adults for euthanasia, questionnaires were sent to Public Health officials, hospitals, nursing homes, and mental institutions [31]. The forms specifically asked for identification of the patients unable to work; patients suffering from schizophrenia, epilepsy, dementia, encephalitis, and other chronic psychiatric or neurological disorders; those not of German or "related" blood; the criminally insane or those committed on criminal grounds; and those who had been confined to the institution in question for more than five years [31].

Teams of secret physicians evaluated the forms. T4 functionaries removed selected patients and transported them to one of the gassing installations for euthanasia. Upon arrival at the facility, the patients were taken to "showers" to be gassed with carbon monoxide. Their bodies were cremated and the ashes placed in an urn given to the surviving family members with a death certificate naming a false cause of death. Although the T4 program was secret, family members and clergy were suspicious and eventually the truth emerged. After public protest, Hitler terminated the Euthanasia Program in August 1941. By this time, the T4 program euthanized 70,723 adults [31].

Euthanasia of children continued after termination of the T4 program. In fact, adults continued to be euthanized, under even more secret measures by overdosing them with medication or starving them in lieu of sending them to the gas chambers. Historians estimate about 250,000 “unfit” people were euthanized in Nazi Germany.

When Hitler first became Chancellor of Germany, he planned on forcing Jews and gypsies out of Germany, and later, out of Germany occupied territories. When this failed, he turned to the Final Solution – eliminate the Jews and gypsies through euthanasia in gas chambers. Although this attempt at genocide ultimately failed, over six million European Jews were murdered.

The Evil of Eugenics and Racial Hygiene

Many physicians, scientists, and politicians participated in the eugenics movement worldwide and the racial hygiene movement in Nazi Germany. Are these people evil for their actions? Are the acts of eugenics and racial hygiene evil?

As always, there are few simple answers. Many eugenicists believed they could improve the human race by managing who could reproduce. There could be some truth in this, but does that make it right? Does the government have a right to control other people’s reproductive potential (take away their rights) for the betterment of society?

Furthermore, heredity is much more complicated than the Mendelian genetics that was known during the peak of the eugenics movement. In 2001, international scientists mapped the human genome. They expected to find between 100,000 and 150,000 genes to code for all the complexities of the human body, but only found 30,000. How can only 30,000 genes code for all human traits, from eye color to personality to intelligence level? Dr. Venter, one of the scientists on the Human Genome Project concluded [32]:

> We simply do not have enough genes for the idea of biological determinism to be right. The wonderful diversity of the human species is not hard-wired in our genetic code. Our environments are critical.

Thus, even today we do not have the genetic knowledge for selective breeding of humans to propagate good traits and eliminate the bad. Today, societies participate in eugenics by other means. Advancement of science has opened the door for selective abortions and genetic engineering during *in vitro* fertilization.

India and China favor male offspring over female, and for decades, couples often aborted female fetuses or even euthanized female babies. Now with the human genome project completed, we could abort a fetus whose genes indicate would have brown eyes when we wanted blue. Upon birth in the United States, there is a genetic screening for certain illnesses (this is why blood is taken from the baby's heel shortly after birth). However, the screening only includes diseases for which treatment is available and allows for starting treatment as soon as possible. Untreatable diseases were left off the mandatory screening.

A pregnant women can select to screen for genetic diseases during the fetal stage. For example, one can screen for Downs Syndrome, giving parents the option to abort the fetus. Currently, parents choose to abort in 60 to 90 percent of Downs Syndrome pregnancies, compared to only 18% of all pregnancies [33]. Unfortunately, these abortions come when Downs Syndrome people have shown increase production in society and longer life spans, as outlined in the Congressional Report [33]:

> •All people with Down syndrome are intrinsically and immeasurably valuable. In addition, people with Down syndrome are happy with their lives and make their families and communities better off.
> •Medical advancements have helped increase life expectancy of people with Down syndrome from about 10 years in the 1960s to over 50 years in 2020 according to JEC Republican estimates, but at the same time have led to expanded screening during pregnancy that opens the door to a rise in selective abortions.
> •JEC Republicans estimate that absent selective abortion, 80 percent more babies with Down syndrome would be born each year and that the Down syndrome population would be 217,000 people greater in 50 years, an increase which is greater than the current Down syndrome population.

Aversions to Downs Syndrome Children are often based on selfishness, as illustrated in a case from Indiana in 1982 [34]. A baby born with Downs Syndrome also had a fistula joining the esophagus to the trachea, preventing the baby from consuming food. Surgery successfully corrects the problem 90 percent of the time. The surgery was not performed and the baby died of starvation. The case went to a Civil Rights hearing, where the physician over the case, Dr. Walter Owens, stated that he advised the parents:

> that it would still be a mongoloid, a Down's syndrome child with all the problems that even the

> best of them have. That they did have another alternative which was to do nothing. In which case the child [would] probably live only a matter of several days and would die of pneumonia probably ... Some of these children ... are mere blobs.

After two other physicians recommended the baby be transferred to another hospital for surgery, Dr. Owens further advised the parents [34]:

> Obviously this has colored my thinking on the survival of such children. I believe there are things that are worse than having a child die. And one of them is that it might live.

The parents decided to forego surgery and allowed the baby to die.

Was Dr. Owens wrong in his advice to the parents? I presented this case in my Issues in Biomedical Sciences course every year, and overwhelmingly, the students agreed that the doctor gave more of an opinion than providing sound medical advice. Further, some students had siblings with Downs Syndrome and talked about how their siblings were happy and have a good quality of life. These students were upset at letting a child die.

Dr. Owens had a niece with a retarded child and may have believed that raising the child had negative impacts on his niece's life. However, that does not give him the right to pressure his patients to make a decision based on his personal observations. We often hear of doctors, "first, do no harm." Harm can be done actively, as in Nazi Germany, or passively, as in this case. Patients depend on doctors to give them the pros and cons based on medicine and science. Like the concept of eugenics overall, good intentions do not excuse bad deeds.

Today, we have the ability to take eugenics to a whole new level. Instead of waiting for evolution to improve the human race, we can engineer the genomes of future children. Genetic engineering provides many medical options, but also just as many abusive ones.

Genetic engineering refers to altering the DNA of an organism. We eat genetically engineered plants, often referred to as GMO (genetically modified organism) including corn, potatoes, soybeans, and many more. In fact, 92% of corn and 94% of soybean are genetically engineered in the United States.

In 2015, the FDA approved the first genetically modified animal source of food, salmon. Aqua Bounty spent over 30 years developing the AquAdvantage salmon, which have twice the growth rate of their non-modified species. But the FDA put many restrictions on growing and marketing the salmon including requiring labels as GMO [35,36]. With little interest in buying GMO salmon, Aqua Bounty lost millions of dollars and liquidated in December 2024. But the door has been opened, and more GMO meat products will be produced in the future.

Scientifically, genetic engineering in humans is possible, but ethically, should we go in that direction? During *in vitro* fertilization procedures, the genome of the blastocoele (undifferentiated cells between fertilization and fetal stage) can be determined and the parents can choose whether or not to implant. However, they cannot change the genome of the blastocoele to match their desired traits. We are closer to gene insertion, where a specific gene can be inserted into somatic cells (non-germ cells – sperm or ova) to treat a specific disease. Inserting into a germ cell would change the genome not only of the individual person, but of the offspring as well. Scientists extensively research this technique for disease prevention, such as Alzheimer's Disease or Parkinson's Disease. However, other scientists are studying this technique to "enhance" the human body. If successful, and made legal, parents could choose the properties of their children – intelligent, athletic, musical, or maybe all three.

Cloning is another type of genetic engineering and occurs on many levels from cloning DNA to cloning organisms. The DNA is located in the nucleus of most somatic cells of organisms. In organismal cloning, the nuclei in embryonic cells of an organism are removed and replaced with nuclei from another organism of the same species, thereby transferring the DNA. Scientists have cloned animals, but at this time, there is no evidence of a human clone. However, some claim it has occurred.

In 2002, Brigitte Bosselier announced the birth of the first human clone, named Eve. However, she provided no evidence that Eve existed, let alone was cloned [37]. However, Samuel Wood, a researcher at Stemagen Corporation in La Jolla, California, cloned himself from his skin cells in 2008 and formed embryos, demonstrating the possibility of human cloning [38].

Cloning of animals and humans, and to a lesser extent plants, is highly controversial among scientists, religious leaders, politicians, and others. However, humans (especially narcissists) have a tendency that when they can do something, they will do it. I would not be surprised to learn of a cloned human being and suspect the clone will be of the scientist who decides to do it, in true narcissist fashion.

Science continues to advance, possibly at a rate faster than we can establish the morality of certain experiments. An obvious example is the discovery and use of nuclear energy. Not only did this lead to the atomic bomb, of which there are enough globally to destroy the world several times, but we are generating nuclear waste without a means of safe disposal. A more recent example is development of the electric car to help decrease our reliance on fossil fuels and protect the environment. However, we again face issues with disposal of toxic material, in this case, batteries. Humans need to put more philosophical thoughts into their scientific practices.

Chapter 10

Mercy(?) Killings

I have owned twelve dogs and twenty cats in my life (so far), and the majority were euthanized prior to death. I preferred when my pets died at home, but if they showed signs of suffering, then we took them to our veterinarian to be euthanized. Euthanizing animals is accepted by society as a humane and merciful act when utilized to prevent suffering.

One of my students shadowed at a local veterinary office. Shadowing professionals allows students to learn what the career entails, without hands-on experience. The student reported to me something upsetting – the veterinarian he was shadowing euthanized a dog at the request of one of the legal owners, but the dog was not sick or suffering. The owners were getting a divorce, and the dog "belonged" to the husband. The wife, to hurt the husband, wanted the dog dead. This was not an act of mercy, but cruelty. In this case, I would consider the wife evil and the veterinarian negligent for euthanizing the dog without contacting the husband first.

Society has not yet decided whether humans can be morally euthanized if they are suffering. Some states in the United States and some countries in Europe have legalized physician assisted suicide and a couple of countries have legalized euthanasia. However, the majority of the western world opposes physicians directly or indirectly purposely causing the death of a patient. One argument against physician assisted suicide is that physicians would be taking on a God-like role by making a life-or-death decision for a person. Other concerns include who makes the decision for death if a patient is incompetent, and the slippery slope that the line at which a patient could be killed would continually decrease overtime.

Dr. Jack Kevorkian

Dr. Jack Kevorkian graduated from the University of Michigan Medical School in 1952 and completed his residency in pathology at the University of Michigan and Pontiac General Hospital. His only experience with living patients occurred during his residency and military service [1].

Kevorkian strongly promoted physicians assisting terminally ill patients to die with dignity. In 1985, Kevorkian published an article in the Journal of the National Medical Association advocating for the use of death-row inmates for experimentation during their last moments of life prior to lethal injection. In his conclusion, he states [2]:

> Postwar analysis of Nazi experimentations on human beings seems to have been so excruciating that it blinded the civilized world to a very important point in the formulation of the Nuremberg Code. Nowhere in the code is there any reference to experimentation (under anesthesia) on those who chose and desire it as an act of atonement when condemned to death by due process of peacetime jurisprudence. Those who oppose granting the condemned such a privilege fear that it would appear to endow the death penalty with undeserved beneficence and would vitiate their campaign for abolition of the penalty…

> In recent years, proponents have argued cogently, and in vain, that such experimentation, within the stipulated constraints, is just and right. The only circumstance in which a person may legitimately consent to a procedure implying certain death is as a necessary means to a very profound experimental end. In the United States where death rows are once again becoming over-populated, all condemned persons should be allowed to choose to submit to experimentation, or to organ donation, under strictly controlled anesthesia before ultimate death by lethal thiopental injection. All conflicting opinions aside, the decisive argument should be that of the condemned themselves.

After retiring from pathology, Dr. Kevorkian started his own business helping people die. His business card read as follows [3]:

> Jack Kevorkian, MD
> Bioethics and Obitiatry
> Special death counseling
> By appointment only

Obitiatry is a term created by Kevorkian and means the medical management of death.

In 1987, Kevorkian advertised in Detroit newspapers as a "physician consultant" for "death counseling". The Detroit Free Press Magazine published an article in 1990 about Dr. Kevorkian and his death machines [4]. The article opens with a description of the first death machine:

> The killing machine waits on the floor of his bedroom closet, next to a few cardboard cartons and a broken black-and-white television set.
>
> The frame is scrap aluminum, leftover from one of Dr. Jack Kevorkian's earlier experiments. The motor came from a toy car. Plug it into a wall socket, punch it into your arm, press a button and five minutes later, you're dead.

Near the end of the article is a description of his newer (1990) model [4]:

> Today's streamlined model, the third, stands 20 inches high and measures 14 inches across. It's a simple frame holding three glass bottles – one of saline solution, to keep the intravenous line open; one of pentothal, to induce coma; and one of potassium chloride to stop the heart.
>
> The plan calls for Kevorkian to insert the IV. Then it will be up to the patient to press a button, start the flow of pentothal and die.

The article also provides information on how you could obtain Kevorkian's help with dying [4]:

> Applications are being accepted. Oppressed by a fatal disease, a severe handicap, a crippling deformity? Write Box 261, Royal Oak, Mich., 48068-0261. Show him proper, compelling medical evidence that you should die, and Dr. Jack Kevorkian will help you kill yourself, free of charge.

Kevorkian created his death machines, the Mercitron (mercy killing machine) and the Thanatron (death machine) for use in the back of a Volkswagen van. Kevorkian would meet with his patients at the van and connect them to the machine, then let them die by own hands. The Mercitron used carbon monoxide gas applied from a gas cylinder to a face mask over the patient. The patient turned the gas on when ready. The Thanatron required an intravenous line, set up by Kevorkian. He started the line as a saline drip, and then the patient would press a button to start a series of two drug injections: thiopental (pentothal) for sixty seconds to put the patient into a coma, followed by potassium chloride to stop the heart [5].

According to an article in The Battle Creek Enquirer in 1989, Kevorkian believed the following [6]:

> Dr. Jack Kevorkian said he will offer the apparatus to patients who request it even though he could face criminal charges if the suicide device is used.
>
> "I have no fear about what I'm doing," Kevorkian, 61, said. "I'm here to help anybody who's in distress, or thinks he is. I couldn't let dogs suffer like that … how can I let a human?'
>
> "It's dignified, humane and painless, and the patient can do it in the comfort of their own home at any time they want."

Janet Adkins of Portland, Oregon flew to Detroit, Michigan in 1990. On June 4, she met with Kevorkian for assistance with her suicide. She wrote a suicide note, which included the following explanation [7]:

> I have Alzheimer's disease and do not want to let it progress any farther. I don't choose to put my family or myself through the agony of this terrible disease.

Janet Adkins entered the Volkswagen van and Kevorkian connected an intravenous line to her and started the saline drip. When Janet was ready, she pushed a button to start the thiopental infusion followed by potassium chloride. Minutes later, Janet died.

On June 8, Oakland County Circuit Court Judge Alice Gilbert ordered Dr. Kevorkian to stop using his suicide machine or assisting in suicide via any other means until prosecutors decide to try him for the death of Janet Adkins [8]. On December 3, Oakland County Prosecutor Richard Thompson charged Dr. Kevorkian with first-degree murder. On December 13, Judge McNally dismissed the case on the grounds that Michigan lacked laws pertaining to physician-assisted suicide [9]. Both the general public and the medical communities were divided on whether Kevorkian did wrong.

On October 23, 1991, Marjorie Wantz, a 58-year-old woman with pelvic pain, and Sherry Miller, a 43-year-old with multiple sclerosis, both from Michigan, met with Kevorkian at a cabin near Lake Orion, Michigan [10]. Wantz died from use of the Mercitron and Miller died from the Thanatron. In response, the Michigan Board of Medicine summarily revoked Kevorkian's license to practice medicine. After assisting in the suicide of three more people, the Michigan Legislature, on December 3, 1992, passed a law banning assisted suicide, to take effect on March 30, 1993. In the 4 months before March 30, ten more people committed suicide with the help of Kevorkian.

Between June 1990 and August 1995, Kevorkian assisted with the suicide of 25 patients, 2 by lethal injection and 23 by carbon monoxide poisoning. He claims to have assisted with the death of 130 patients by 1998 [10].

In September, 1998, Kevorkian went down the slippery slope, directly giving drugs to his patients. Without fear of reprimand, Kevorkian videotaped himself giving 52-year-old Thomas Youk, who suffered from amyotrophic lateral sclerosis (Lou Gehrig's disease), a lethal injection. The video was aired on 60 Minutes in November of that year [10].

In March 1999, the state of Michigan charged Dr. Kevorkian with second-degree murder in the death of Thomas Youk. Kevorkian's defense was based on Thomas Youk wanting to die; however, the judge instructed the jury that the victim's desire for death was irrelevant to the case [11]. The jury found Kevorkian guilty, and he was sentenced to 10 to 25 years in prison. With time off for good behavior, June 1, 2007, Dr. Kevorkian was released from prison after eight years and a promise never to assist with suicide again. He remained on probation for two more years [12]. On June 3, 2011, Dr. Kevorkian died a peaceful, non-suicidal death [13].

The American Medical Association defines physician-assisted suicide as "when a physician facilitates a patient's death by providing the necessary means and/or information to enable the patient to perform the life-ending act". The physician does not "kill" or euthanize the patient. In fact, euthanasia of people is illegal in the United States. As of August 2025, Physician-assisted suicide, also called death with dignity or medically-assisted dying, is legal in eleven states and the District of Columbia [14]. Another seventeen states are currently considering legislation to legalize physician-assisted suicide. As of July

2024, seven European countries allow physician-assisted suicide including Netherlands, Belgium, Luxembourg, Spain, Portugal, Switzerland, and Austria [15]. Three of these countries also allow euthanasia by physicians: Netherlands, Belgium, and Luxembourg. The Netherlands and Belgium even allow euthanasia of children [15]. France is the latest European country considering legalizing physician-assisted suicide [15].

Organizations exist that support physician-assisted suicide and those that oppose it. In 1980, Derek Humphrey founded the Hemlock Society in his garage in Santa Monica, California. The Hemlock Society was the prominent right-to-die organization of its time and promoted both physician-assisted suicide and voluntary euthanasia. Mr. Humphrey claims the organization was taken over by other right-to-die organizations in 2003 because the word "hemlock" was offensive to politicians and others [16]. The Hemlock Society gradually morphed into Compassion and Choices by 2005.

Compassion and Choices describes its mission as [17]:

> Compassion & Choices improves care, expands options, and empowers everyone to chart their end-of-life journey.

The visions of the organization are also listed on the website and provide better understanding of its purpose:

- Compassionate in our conviction that dying patients should be free of unwanted treatment, suffering, or outside interference.

- Respectful of the autonomy of individuals to decide what end-of-life options are best for them and their family.
- Courageous in our willingness to confront the toughest end-of-life health challenges, disrupt the broken status quo, and protect individuals' right to self-determination.
- Credible in all our education, advocacy and partnerships — our efforts are grounded in objective research and demonstrable facts.
- Resilient in our capacity to respond to opportunities and threats in the movement so that we can achieve our vision as quickly as possible.

Several disability groups oppose physician-assisted suicide as they fear it diminishes the importance and meaning of life of disabled people. Bob Kafka of American Disabled for Attendant Programs Today provides his view on how physician-assisted suicide is not a choice in today's world [18]:

> Seniors and people with disabilities who need assistance to do everyday tasks like dressing and bathing want the choice to get those services at home and to have control over how the services are delivered. They do not want to be forced into a nursing facility, nor see themselves and their spouse, and sometimes their children, forced to live in poverty to qualify for help with such basics. Unfortunately, that choice is not a reality for most. In states which have legalized assisted suicide, according to data from Oregon, over a third of those who request assistance to die do so because of "feelings of being a burden" and over 90% cite

"loss of autonomy" as a factor. If the only alternative to death that those in power offer people who require assistance is poverty and segregation in nursing facilities, then it makes no sense to talk about assisted suicide as a "choice". Society is failing to ensure that seniors and people with disabilities have access to consumer controlled long term services and support when they need them. The last thing we need is for those in power to make a public policy choice, during this time of vast budget cuts in Medicaid health and long term care, that an early death is the cost saving answer to these very real human needs.

Ari Ne'eman, President of Autistic Self Advocacy Network, describes how we are already sliding down the slippery slope [18]:

> There is significant evidence that proponents of physician assisted suicide have never intended to stop at the terminally ill. Legislation introduced in New Hampshire in 2014 and prior years includes a definition of terminal illness so broad as to allow for a lethal prescription in response to any condition which shortens lifespan without a known treatment – even if the individual in question might have years or even decades of life remaining. Or take the case of Tracy Latimer, a 12-year old Canadian girl with significant developmental disabilities who was murdered by her father. Following the father's conviction for second-degree murder, the executive director of the Toronto chapter of Dying with Dignity announced that their position was that he should serve no sentence, as "the Latimers had already

> lived under a sentence during the 12 years that Tracy was alive." Assisted suicide advocates have been adept at using an incremental strategy by focusing on people who are terminally ill, but their broader policy agenda is already well documented.

For clarification, Tracy Latimer suffered from severe cerebral palsy, causing mental and physical incapacities including inability to walk, talk, or feed herself. Tracy underwent four surgeries and needed another to repair her hip. She was in constant pain. Her mother described the care of Tracy as similar to caring for an infant. Her father, Robert Latimer, killed her in the cab of his pickup truck by piping carbon monoxide inside while he sat in the back. He called her death a mercy killing. This case also caught the attention of disability advocacy groups. A Canadian jury found Robert Latimer guilty of second degree murder, and the judge sentenced him to the mandatory minimum life sentence with no parole until 10 years, at which time Latimer received parole. To this day, Latimer believes he did what was best for Tracy [19]:

> It was pointless to torture our daughter any further. She had already had four operations. If she moved, her hip would go. She had rods in her back. She had been worked on enough.

Politicians in the United States still debate the legalization of physician-assisted suicide, but euthanasia is off the table for discussion, at least for now. However, one argument that needs attention is whether or not it *should be a duty* of a physician to assist in suicide if it is legalized. Patients want to believe that their doctor is doing what is best for him. If the doctor said it is time for a patient to die, that patient might take her advice. The power of influence doctors have over their patients (not necessarily purposefully) must not go unnoticed in these discussions.

Dr. Anna Pou

On August 25, 2005, category 1 (wind speeds of 80 mph) Hurricane Katrina struck Miami, Florida, where it killed two people [20]. Katrina then entered the Gulf of Mexico, where it strengthened to a category 5 hurricane. On Monday, August 29, Hurricane Katrina made landfall at Plaquemines Parish, Louisiana, approximately 45 miles southeast of New Orleans. It was a category 3 hurricane that spread out over much of the gulf coast and had a storm surge of 26 feet along the Mississippi coast. After passing Plaquemines Parish, Katrina crossed the Mississippi Sound and made a second landfall near the mouth of the Pearl River.

New Orleans first appeared to survive the onslaught. However, things changed in a matter of hours. Much of New Orleans is below sea level and levees keep water from Lakes Pontchartrain and Borgne from draining into the city. On the south side of New Orleans is the Mississippi River, which rises with storm surges. Katrina brought in a heavy storm surge and ten inches of rain, resulting in breaches in the levees at about 50 sites. When more than 80 percent of the city flooded, people were trapped at home, in the shelters, and in hospitals. The U.S. Coast Guard rescued hundreds of people from rooftops. About 30,000 people made their way to the Convention Center and Superdome, only to find a shortage of supplies and security. Over 1800 people died across seven states; most deaths occurred in New Orleans [20].

On Sunday, August 28, mayor Nagin of New Orleans ordered a city-wide evacuation, but tens of thousands of residents chose to stay home or shelter at the New Orleans Convention Center and the Superdome, as well as other smaller shelters. Most hospitals chose to shelter-in-place or were unable to evacuate patients due to transportation shortages. The Centers for Disease Control estimated 2500 hospital patients remained in New Orleans during the storm, and there was a high probability of power outages in the hospitals. The day before Nagin called for the evacuation, LifeCare, a small one-story specialized hospital in St. Bernard Parrish, evacuated its patients, most to Memorial Medical Center (Memorial), which housed a LifeCare campus on its 7th and 8th floors. After transferring the patients to Memorial, LifeCare administrators and physicians left the facility to ride out the storm elsewhere [21].

Tenet Healthcare Corporation of Texas owned Memorial Medical Center, located about three miles from the French Quarter and three feet below sea level. The complex was built in 1926 and opened as Baptist Hospital. Since its beginning, the hospital served as a shelter for the community during hurricanes. The same was true during hurricane Katrina, with about 1500 people taking refuge at the hospital, including family members of patients and employees. Many family members brought their pets to Memorial, with the pets housed in the Medical Records department on the first floor [21].

Dr. Anna Pou, head and neck surgeon on faculty at Louisiana State University School of Medicine, was on call at Memorial that Sunday, along with two residents under her guidance. Although the chair of her department recommended she stay home and assign her patients to other physicians, Dr. Pou sent her residents home and she reported to the hospital. On Sunday evening, 600 employees (25 of whom were physicians), 183 Memorial patients, and 55 LifeCare patients stayed at Memorial

[21]. The CDC warned of potential levee breaches that Sunday evening.

Early Monday morning, hurricane force winds knocked out windows at Memorial. Power was lost by 5:00 a.m. Emergency generators kicked in, providing minimal support, but without air conditioning, temperatures inside increased rapidly to 100°F. Outside Memorial, wind gusts reached 135 MPH and the streets flooded. Water soon entered the basement, threatening the generators on the first floor. By Monday night, however, it seemed to end as flood waters receded and employees could leave the hospital temporarily to check on their own property or to visit other hospitals [22]. New Orleans survived Katrina!

By Tuesday morning, physicians planned on releasing patients. But then the water started to rise again. Levees were breached, and the National Guard informed the hospital to expect fifteen feet of water [23]. Conditions deteriorated rapidly as the stench of sewage and body odor filled the air and temperatures inside exceeded 100°F. To make matters worse, looting occurred around the hospital and gun shots could be heard, with a shortage of security in the area.

Physicians had to make hard decisions. Fifteen feet of flood water would reach the generators and the hospital could lose all power. Memorial went into survival mode. Physicians had to prioritize treatments based on limited medical supplies. And with so many people taking refuge at the hospital, how do they allocate water and food supplies?

In a National Public Radio interview with author and physician Sheri Fink, Steve Inskeep asks Dr. Fink about discussions taking place in the hospital as the staff realized they would lose all power. Dr. Fink responded as follows [24]:

> That's when the very first tough ethical decision arose ... which was: Who are you going to save first if you know that you may be hours away from all power going out? ...
>
> They decided that the babies would be saved first, and the intensive care unit patients, whose lives really depended on electricity. They also decided at that point who would go last, and that was patients who had do not resuscitate orders.
>
> The doctors all agreed with this decision and, by the way, it was a small group of doctors who took this burden of making this decision on their shoulders.

The hospital physicians began triaging patients remaining in the medical center into three groups for evacuation priority. The first group, the most ill, was moved to the Emergency Room where National Guard trucks and then boats would pick them up at daylight. The second group was lined up on the parking garage rooftop to await helicopters. The third group would have to wait, and the four patients with Do Not Resuscitate (DNR) orders would be evacuated last. LifeCare patients were still in the LifeCare unit on the seventh floor, with no assigned physicians [25]. They were not included in the evacuation plan.

In the National Public Radio interview, Dr. Fink was asked about evacuating the DNR patients last [24]:

> As one doctor described it ... they would maybe have the least to lose compared with other patients. ...

> Doctors tend to be — and this is, some ethicists have explained this to me — doctors have kind of a utilitarian frame. If you ask them, "Who gets the organ transplant?" it's going to be the person who they think could benefit the most from it, and other considerations might not be taken into account, like justice or fairness. ... So you're switching, in a disaster, sometimes from looking at what doctors normally are supposed to do, which is treat each individual patient to the max, to this more population-based approach.

Surprisingly, the physician who first suggested holding back on patients with DNR orders did not understand what Do Not Resuscitate means. Medical chairman Dr. Deichmann believed that a DNR order meant that the patient had a certified terminal or irreversible condition [22]. In addition, neuroradiologist Dr. Armington assumed that patients who did not wish extraordinary measures to prolong their lives would want others to be saved before them [22].

Soon, the medical triage at Memorial resembled that of a military unit at war time with those who most likely would survive being evacuated first. Dr. Deichmann assigned two physicians to cover each of the fifteen patient wards, in lieu of caring for their private patients. The physicians would categorize patients for their transport destinations.

After determining the order of evacuation, the staff had to determine how to evacuate the patients. With flooded streets, patients could only be evacuated by boat or helicopter. Boats could pull up to the emergency room ambulance ramp to evacuate patients. Memorial had an old heliport that had not been used in years, but it would be put to use with questionable stability. Unfortunately, the elevator to the heliport was not wired to the emergency power. Only one elevator functioned at this time, and it went from the second floor to the rooftop level of the parking garage. However, a three-foot hole existed between the parking garage and heliport, providing a possible path to get patients to the heliport for transport [21,26].

The staff moved patients from the Intensive Care Unit (ICU) first, taking them to the Emergency Room or the heliport. A National Guard high water troop transfer truck arrived at the Emergency Room ramp and patients were loaded. As for the heliport, staff would need about 45 minutes to transport a patient there.

Anesthesiologist Dr. Primeaux waved in a Black Hawk helicopter, the first to land on the heliport. Dr. Primeaux was surprised, but very pleased, that the rusty heliport held. The pilot agreed to take a sick newborn. Dr. Primeaux rushed down the stairs, informed the incident command team, then got a baby from the neonatal ICU and took it up the stairs to the heliport. The pilot responded that the process took too long and there were too many people to save to repeat this procedure. The pilot was ordered not to return and he suggested that Memorial revise their evacuation procedures [22].

Meanwhile, a hospital in Baton Rouge, Louisiana, agreed to take all of Memorial's babies. Arrangements were made with the Coast Guard to provide helicopter transport in the afternoon. Staff prepared the babies for transport to the heliport. Some babies could be taken by elevator to the parking garage, carried to and through the three-foot hole to the heliport. However, the sickest babies had to be transported in isolettes, self-contained incubators with oxygen tanks, warming pads, and pumps to administer IV fluids. These incubators were too tall to travel in the back of a pickup truck through the parking garage to the hole in the wall to reach the heliport. Staff members and volunteers carried the incubators up three flights of stairs to the parking garage roof. In one of the incubators, the oxygen supply was not functioning properly and a nurse hand pumped oxygen into the baby's airways as she climbed the stairs with the men carrying the incubators. To their dismay, when they reached the heliport, no helicopters were waiting for them. Memorial administrators started contacting small helicopter companies who were also performing rescue missions in New Orleans. Many pilots refused to take neonates, but would take critically ill adults [22].

Neonatologist Dr. Gershanik had two sick babies whose incubators would not fit on a small helicopter that had just landed. As Dr. Gershanik ran to the helicopter pushing the incubator, he did the unthinkable. He removed one of the babies and took him in his arms along with an air bag to provide oxygen and sat in the seat next to the pilot. A nurse took the other baby and sat in the back of the helicopter. They were on their way to be rescued. But the pilot stopped for fuel, and had to wait until two Army helicopters refueled first. Dr. Gershanik expressed concern that having to wait could kill the babies, but the pilot explained that the Army helicopters were rescuing people from rooftops, who also might die if they get delayed. Dr. Gershanik realized Memorial was not alone in this disaster. All critically ill babies were ultimately evacuated by helicopter to Baton Rouge [22].

To transport adult patients to the heliport, staff and family moved patients from their rooms to the second floor to ride the elevator. When the patient reached the parking garage, they were taken by stretcher through the three-foot hole to be loaded on the helicopter for evacuation. Fifty patients were evacuated by night fall [21,26].

During the evacuations, Memorial lost its first patient of the Katrina crisis. Helen Breckenridge, 76 years old, suffered from complications of heart disease and diabetes. She had been under hospice care at another hospital, but her brother, a physician, had her transferred to Memorial for aggressive treatment. She was staying in the ICU on the 8th floor and needed to be taken by stairs to the first floor Emergency Room. During movement, air had to be supplied manually with the bag. When they arrived downstairs, the medical team was told to take her back up because she had a Do Not Resuscitate order. Shortly after, Breckenridge was back in her room and passed away [22]. She would not be the last.

During the middle of the night, the generator failed, ending any transport of patients by elevator. In addition, employees lost communication with the outside world as phones, computers, and other electronic devices became inoperable. Memorial still had 188 patients (including those from LifeCare). Seven LifeCare patients had been on ventilators, which also lost power, requiring nurses and volunteers to provide manual ventilation for these patients who could not breathe on their own – ideally at a rate of 10 per minute until power came back on, the patient was evacuated, or the patient died. Physicians now had to make those tough choices.

In an interview on 60 Minutes, Sheri Landry recalled, "It was like, where are the boats? Where are the helicopters? Why isn't anybody coming?"

In that same interview, Dr. Pou described the situation as follows:

> It was really shocking. The water was squirting out of the vents in the street, the gutters. And water

> started coming. And probably rose about a foot an hour.
>
> By late Tuesday, the hospital was flooded with ten feet of water and completely without power. Ventilators stopped, there were no telephones, limited food, and perhaps worst of all the 110 degree heat.
>
> I don't know if there's any way for me to describe to you how intense the heat was. It was relentless. It was suffocating. It made it extremely difficult to breathe. And with the heat came the terrible smell from all of the human waste and the fact that we didn't have water.

The Coast Guard agreed to transport more critically ill patients by helicopter, but staff and volunteers had to transport those patients on stretchers up several flights of stairs to reach the helipad. As hours passed, helicopters were diverted to other flooded areas of the city, leaving the hospital with little support. Although most of the LifeCare and ICU patients had been evacuated by then, many critical patients remained [21,26].

National Public Radio described the scene that morning as they were told from a family member of a LifeCare Patient [27]:

> Angela McManus' mother had been on the LifeCare floor for two weeks before Katrina hit. Wilda Faye McManus, 78, was battling a persistent infection due to complications from rectal cancer. Angela McManus says she was given a bed next to her mother and never left her side until Tuesday, the day after the hurricane. She

> says nurses told her that helicopters were coming for the seventh-floor patients and that McManus needed to get to the first floor and wait for evacuation boats.
>
> Once on the first floor, McManus said, she could hear gunshots outside the hospital. She saw looters sacking a corner drug store. Many sources confirm that at this point, there were 2,000 people - employees, patients and relatives - trapped in the hospital. According to McManus, "The sewer lines had all backed up, and we were down there in all that stifling heat and this odor was horrendous. People were trying to get into the hospital just to get to higher ground, and they weren't allowing that ... so they boarded the doors up, and we were just in there smothering all night long."

On Wednesday, Memorial received word that no government transportation would be coming that day. Unfortunately, due to lack of communications, employees did not know that Tenet Healthcare Corporation hired private transport to evacuate the medical center and other hospitals they owned in the area [25]. After hearing no help was coming, a group of staff met to determine how to treat the sickest patients who were least able to be evacuated. The group agreed to provide sedation, and according to Dr. Pou, the group made the decision that she would do so [28]. Dr. Pou and two nurses, Cheri Landry and Lori Budo, provided category 3 (last to be evacuated) and LifeCare patients midazolam (Versed) and/or morphine [27].

According to Ryan Bailey, Dr. Pou stated her intention was to "help the patients that were having pain and sedate the patients who were anxious" because she "knew they were going to be there another day, that they would go through at least another day of hell" [21,26].

Not all professionals at the medical center agreed with the decision. Dr. King stated, "I disagree 100%. The idea was stupidity itself" [29]. Several nurses expressed concern of this decision.

National Public Radio continued the description of events as told to them by Angela McManus [27].

> By Wednesday morning, Angela McManus learned her mother had not been evacuated as promised. She rushed back to the seventh floor and said her mother's condition had changed. "She was real lethargic," said McManus. "She would talk to me, then just doze back off. I was like, 'What's going on with her?' I was just sitting there talking to her and stroking her, and she was just sleeping and I'm like, 'Something is wrong'."
>
> McManus says nurses told her that her mother had been sedated. She grew concerned because she says her mother's pain had been manageable with Tylenol and an occasional painkiller. She stayed with her mom for hours and sang gospel hymns to comfort her. According to McManus, attempts were made to evacuate other patients from the seventh floor. She recalls seeing workers desperately trying to get one woman out of the hospital, only to see that the woman died in the process.

Angela McManus became seriously frightened for her mother when she overheard nurses saying a decision was made not to evacuate LifeCare's DNR patients. “DNR means ‘do not resuscitate.’ It does not mean do not rescue, do not take care of,” McManus said. She tried to rescind her mother's DNR order to no avail. On Wednesday evening, two full days after Katrina hit, Angela McManus says three New Orleans police officers approached her with guns drawn and told her she would have to leave. New Orleans police confirm that armed officers did evacuate non-essential staff from the hospital.

Confronted by police, McManus raced to her mother's bed. “I woke her up and I told her that I had to leave, and I told her that it was OK, to go on and be with Jesus, and she understood me because she cried,” McManus recalled. “First she screamed, then she cried. And I said, 'Momma, do you understand?' And she said, 'Yes.' And she asked me, she asked me to sing to her one more time. And I did it, and everyone was crying, and then I left. I had to leave her there. The police escorted me seven floors down.”

McManus says that when she left, only eight patients, including her mother, remained alive in LifeCare.

According to Dr. Fink, Dr. Pou was not alone in administering medications [22]:

> At some point on the day that the power failed, a few hours after that, some of the doctors told me that they walked around the hospital, had a look at the situation of the patients and felt that hastening death was the right choice.
>
> You know, one of them [Dr. Cook] had gone upstairs to the intensive care unit where there was just one patient left. Most of them had been taken out first, but she was very, very sick. She had this do not resuscitate order and had been held back. And he asked the nurse, "Give her enough morphine till she goes."
>
> The other doctor [John Thiele] actually was involved in injecting some of those patients on that Thursday, Sept. 1 of that year. ... And he also told me that the intent was to let these people die. He did describe to me having a moment after he did these acts where he did wonder whether it was the right thing. And he even hesitated just before he started injecting the patients and he asked the woman next to him, a nurse who was on the ethics committee, whether they could really do this.
>
> Now, one thing he didn't consider was staying there with the patients until they died.

Dr. Cook appears to have only been involved in the death of one patient, Jannie Burgess, in the intensive care unit (ICU), and the coroner ruled the cause of death as undetermined [30]. Dr. Thiele offered to help Dr. Pou with the drug administrations. Dr. Pou denied his offer, but he insisted that he help and not leave her alone with the task. He then participated in providing drugs to patients. Dr. Thiele describes his experience [29]:

> This was totally against every fiber in my body. We were abandoned by the government, we were abandoned by Tenet, and clearly nobody was going to take care of these people in their dying moments. I did what I would have wanted done to me if the roles were reversed.

After four days, all surviving staff, patients, and others at Memorial Medical Center were evacuated. Forty-five bodies were left behind for another ten days as the emergency conditions in New Orleans continued. Several of the deaths occurred naturally under unnatural conditions. However, autopsies showed the presence of morphine and midazolam in 29 of those 45 patients. Dr. Pou was believed to be responsible for nine.

Were the actions of these three physicians and two nurses meant to be long-term sedation or mercy killings?

What happened at Memorial Medical Center resembles a military medical facility near a war zone. When resources become scarce, instead of prioritizing treatment of those most in need, protocol is to do the greatest good for the greatest number. For example, if a soldier presents with severe shrapnel in much of the abdominal area and intensive bleeding, the time it would take to save that soldier could be used to save maybe 5 other soldiers with less severe injuries, but also life threatening. If morphine was plentiful, I would hope the army physicians would provide the patient that might be left to die with the pain killer to make the passing easier. Is this the same situation Dr.

Pou and her colleagues faced, without knowing that help was on the way?

Charles Foti, Jr., Louisiana attorney general, opened investigations into hospitals and nursing homes throughout New Orleans following hurricane Katrina. An attorney for LifeCare informed Foti of nine alleged cases of euthanasia at Memorial. Foti hired a coroner to perform autopsies on those nine patients, and all had morphine in their bodies. Foti then hired a forensic pathologist to review the toxicology reports on four of the nine patients. That pathologist ruled the four deaths homicide [21,26].

In July 2006, Dr. Pou, Cheri Landry and Lori Budo were arrested and charged with second-degree murder in the deaths of four patients at Memorial. The medical community showed mixed reactions, with some considering them heroes for staying behind during a hurricane and risking their own lives to care for others. Dr. Peter deBlieux, an emergency room physician from Charity Hospital in Louisiana, also stayed during Hurricane Katrina. He had this to say about the doctor and nurses [31]:

> We have people who are volunteering services and putting their lives on the line. It's going to make it less likely they're going to do that in the future.

The spokesperson for the Attorney General's office stated [31]:

> Where is the sympathy for the victims? Why is there not outrage for the people who would have not died if they had gotten out? These are not terminal people begging to be put out of their misery.

Landry and Budo were given immunity to testify against Dr. Pou in front of the grand jury. In July 2007, the grand jury refused to indite Dr. Pou, and all charges were dropped. Dr. Pou responded to the verdict as follows [32]:

> Today's events are not a triumph, but a moment of remembrance for those who lost their lives in the storm and a tribute to all of those who stayed at their posts and served people most in need.

Dr. Pou's attorney, Rick Simmons commented [32]:

> The certificates of death in these individual patients should read "abandoned by their government." That's what happened here.

The American Medical Association commended Dr. Pou, with the chair of the Board of Trustees stating [33]:

> We believe these physicians served as bright lights during New Orleans' darkest hour.

Following are some excerpts of a Newsweek conducted interview of Dr. Pou a month after the grand jury refused to indite her [34].

> Newsweek: "What was the hospital like? I know the hospital allowed those who worked there and their extended families to use it as a shelter."
>
> Dr. Pou: "It made the situation much, much more difficult because we had people who shouldn't have been there—physicians, nurses, employees, extended families, pets. That really comes into

play because instead of a few hundred people to take care of there were 2,000 people. That [policy] was one of those errs in human judgment."

Newsweek: "Tell me about conditions from Wednesday night until Thursday."

Dr. Pou: "By the time Wednesday evening came around, if you can imagine in our mind, there is a central area that is a sea of people. A lot of very sick patients in that central triage area. It's grossly backed up. Few patients had been evacuated. So there was just enough space to walk between the stretchers. It is extremely dark. We're having to care for patients by flashlight. There were patients that were moaning, patients that are crying. We're trying to cool them off. We had some dirty water we could use, some ice. We were sponging them down, giving them sips of bottled water, those who could drink. The heat was—there is no way to describe that heat. I was in it and I can't believe how hot it was. There are people fanning patients with cardboard, nurses everywhere, a few doctors and wall-to-wall patients. Patients are so frightened and we're saying prayers with them. We kind of looked around at each other and said, 'You know there's not a whole lot we can really do for those people.' We're waiting [for help]. The people in that area could have [been evacuated] by boat but no boats were coming. I would do what I could with the nurses: changing diapers, cooling patients down with fanning. It wasn't like, 'I'm a doctor, you're a nurse.' We were all human beings trying to help another human being, whatever it took.

Newsweek: “What happened Thursday?”

Dr. Pou: “On Thursday morning we were told nobody was coming and we had to fend for ourselves. Everybody was kind of like at a loss here. What is plan B? Or plan C?”

Newsweek: “Tell me about the decision to administer painkillers to the nine people on the seventh floor?”

Dr. Pou: “There were patients, all of us knew, still remaining in the LifeCare unit. They were category three [in the reverse triage system]. We all believed eventually everybody was going to leave the hospital. We just didn't know when or what was the time frame. So we knew that patients were going to be there for a long time. We knew they were going to be there another day. That they would go through at least another day of hell. Basically it was decided to give the patients sedation.”

Newsweek: “What was your intention when you administered injections to the nine patients in the acute unit?”

Dr. Pou: “The intention was to help the patients that were having pain and sedate the patients who were anxious. That was it. Reverse triage meant the sickest would be the last to be triaged. We

> didn't know how much longer they would be there. I take care of patients with cancer, so if I was a murderer, it would really be an interesting combination, very incongruous."
>
> Newsweek: "Did you consider when giving the injections that they might hasten these patients' deaths?"
>
> Dr. Pou: "I guess the thought crossed my mind. Any time you give medicine it crosses your mind. There's always a risk of hastening death. There is a risk with every single thing we do in medicine. Every time you give antibiotics there's a risk."

Today, Dr. Pou is on faculty at the Louisiana State University School of Medicine. She performs clinical work and performs medical research in the Cancer Center. She is a member of several professional organizations in her field of expertise [35]. Dr. Pou survived Katrina and its aftermath.

Nurse Charles Cullen

Charles Cullen worked as a nurse for sixteen years at several nursing homes and hospitals in Pennsylvania and New Jersey. He was fired from five nursing jobs and resigned from two others [36]. He last worked at Somerset Medical Center in New Jersey. On December 12, 2003, Charles Cullen was arrested at Somerset for one count of murder and one count of attempted murder. Two days later during an intense interrogation, Cullen admitted to the charges, adding that he killed 30-40 patients in what he called, "mercy killings" over the past 16 years.

Cullen struggled during childhood, losing his father when he was an infant and his mother when he was 17. He claims to have attempted suicide at least 20 times, the first at age 9 when he created a drink from his chemistry set. After his mother died, he dropped out of high school and joined the U.S. Navy. He attempted suicide while in the navy, thereby receiving a medical discharge in 1984. Cullen went to nursing school and got his first job in 1987 at St. Barnabas Medical Center in Livingston, New Jersey. He also got married that year, and later had two daughters.

Cullen committed his first murder at St. Barnabas in 1988 by giving a patient medication to which Cullen knew the patient had an allergy. During his later confession, he admitted murdering 11 patients at St. Barnabas. He quit his job in 1993 and took another at Warren Hospital in Phillipsburg, New Jersey. He confessed to killing three elderly women at Warren Hospital by giving them overdoses of digoxin (medication used to treat a weak heart and irregular heart rhythms) [37].

In 1993, Cullen's wife filed for a divorce. Shortly after, she filed a domestic violence complaint and requested a restraining order. According to a police report filed for Mrs. Cullen [38]:

> Ms. Cullen stated that her husband is an alcoholic, had done bizarre things such as placing pets in trash cans, has placed lighter fluid in other persons' drinks in the past and recently requested funeral rates from the Finnegan Funeral Home. Ms. Cullen said her husband is a nurse and has access to drugs and [she] is in fear of her safety as well as the safety of her children. She said her husband has never acted this strange before.

1993 was a bad year for Cullen, as he attempted suicide three times. After the first attempt (six days after a judge denied his wife's request for a restraining order), Cullen was admitted to the behavioral health unit at Muhlenberg Hospital Center in Bethlehem, where he stayed six days. Although his mental health diagnosis is not given, his wife diagnosed him as a borderline personality (has difficulty managing emotions and thoughts, subject to impulsive and reckless behaviors). Three months later, after killing one of the Warren Hospital victims, Cullen self-checked himself into an addiction and mental health treatment center, the Carrier Clinic [38].

Cullen almost got caught killing his third patient at Warren Hospital, but evidence was scarce. He quit his job there in December and moved to Hunterdon Medical Center in Flemington, New Jersey. And he killed there, and he moved on, and he killed…

Charles Cullen was charged with murder in both New Jersey and Pennsylvania. In April 2004, Cullen plead guilty in a New Jersey court to 13 counts of murder and 2 counts of attempted murder. In May, he plead guilty to 3 more counts of murder. Six months later, Cullen plead guilty to 6 counts of murder and 3 counts of attempted murder. The courts found Cullen guilty of more murders, totaling 29 in the two states, and 6 attempted murders, all patients [36].

In New Jersey, Cullen received 11 consecutive life sentences. He avoided the death penalty by making a deal with prosecutors [36]. One of the family members of a victim, John Shanagher, made a statement during the sentencing hearing [36]:

> [my family] will never feel safe in a hospital again. We will never feel we can trust the medical profession again.

This statement elucidates the dangers when people in trusted professions (health care, clergy, law enforcement, etc.) violate their oaths and do harm to others.

During Cullen's sentencing trial in Pennsylvania, he challenged the judge and requested the judge to remove himself from the case. Judge William Platt refused, and Cullen continued stating, "your honor, you need to step down," for 25 minutes. Deputies placed a mesh hood (spit hood) over Cullen's head to muffle his voice as family members of victims spoke to the court. In a June 2004 interview prior to this hearing, District Attorney James Martin asked Cullen about whether there were two reasons as to why he murdered the patients – a need to do it and a desire to let seriously ill patients die with dignity. Cullen responded [38],

> Yeah. I would say it was a combination. I worked in critical care, so I worked around people constantly in very serious and grave condition all the time. There was a certain amount, I guess, of a feeling of a compulsion to do it, yes. I also believe that I also selected people that were terminal or progressively getting worse, too. I think both played a factor in when I would attempt or caused someone's death.

Essex County Chief Assistant Prosecutor Norman Menz summed up his thoughts of Cullen's actions [37]:

> There is evil in this world and sometimes there is no explanation for this evil. In the many hours I

have spent with Mr. Cullen, I have gotten no explanation. There is a deeper evil involved.

Murder or Mercy?

Many more health care providers killed patients in the name of mercy, but the three examples above illustrate sharp contrasts in motivation. Dr. Kevorkian considered himself an advocate for those who wanted to die. It became his cause and he took drastic steps to promote physician-assisted suicide, and even euthanasia, albeit to a lesser extent. Some may consider Dr. Kevorkian's assisted suicide a form of civil disobedience. His actions, whether right or wrong, started a conversation and changes in laws including both laws banning physicians from assisting in suicide and laws allowing it in different states. Some of his ideas, such as using death row prisoners for end-of-life experiments, violate the research principle against taking advantage of vulnerable individuals, which prison inmates are considered. Whether he really desired to allow death row inmates to die with dignity or he has a fascination with death and controlling it, we will never know.

Dr Pou never planned on euthanizing patients, but extreme circumstances forced her and some of her colleagues to make tough decisions a physician never had to make before (except possibly during war). Her actions, in my opinion, were merciful in that she wanted to avoid future suffering of her patients, not knowing that help was on the way. However, her actions may have given her lifelong suffering with ethical debates about her decision. Her claim that she was only strongly sedating patients is questionable.

Nurse Cullen was quite simply a murderer and tried to hide it in the guise of mercy. In fact, Nurse Cullen could have been covered in the next chapter on serial killers. His actions demonstrate some weaknesses in how students are selected for professional schools, and how medical institutions monitor their employees. His license should have been revoked long before he killed thirty or more patients. Over the past 20 years, U.S. politicians have discussed how to keep guns out of the hands of mentally ill people. But just as important is how to keep certain mentally ill people from having jobs where they can control a patient's life. How do we draw the line where a person is capable of handling a health care job (or a gun) versus when they become a danger to others, and also maintain individual rights?

Chapter 11

Medical Serial Killers

The Federal Bureau of Investigation (FBI) defines serial killings as, "a series of three or more killings, not less than one of which was committed within the United States, having common characteristics such as to suggest the reasonable possibility that the crimes were committed by the same actor or actors. [1]" There are enough serial killers who are health care professionals that they get their own name, "medical serial killers," and are employed at various levels of healthcare including doctors, nurses, and various assistants. Some of these professionals consider themselves "Angels of Mercy" as they kill patients to relieve the patient of suffering. Other times they kill for personal gain.

Figure 1 shows the number of identified serial killers in the United States and all other countries (where data is kept) by decade since 1900 [2]. The number of serial killers increased considerably starting in the 1960's, then started to decline again in the 2000's. Note that mass murderers fall in a different category as they kill more than one person on the same day.

Studies on serial killer occupations vary in outcomes. According to psychiatrist Herbert Kinnell, "medicine has arguably thrown up more serial killers than all the other professions put together, with nursing a close second. [3]" Kinnell also argues that the medical profession attracts some individuals infatuated with the power of life and death [3]. Psychiatrist Eindra Khin presented data at the annual meeting of the American Academy of

Psychiatry and Law suggesting that the number of healthcare killings in the United States is decreasing, possibly due to increased use of electronic health records, but increasing in other countries [4]. Internationally, the number of health care serial killings were 10 in the 1970's, 21 in the 1980's, 23 in the 1990's, and 40 from 2000 to 2006 [4]. Comparing these numbers to the total number of serial killers since the 1970's (Figure 1), health care professionals comprised less than 3% of all serial killers

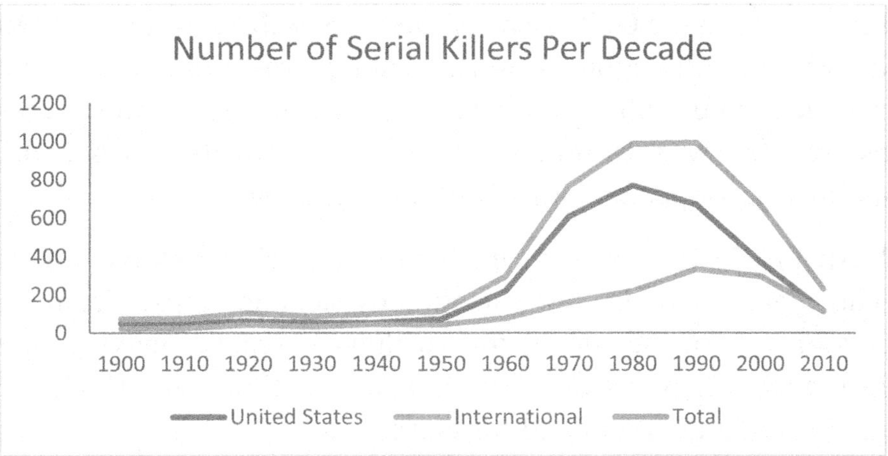

In another study, Beatrice Yorker and colleagues reviewed case histories of 90 health care professionals prosecuted for serial killings from 1970 to 2006 in 20 countries, 40% of which occurred in the United States [5]. Fifty-four of the professionals were convicted: 45 for serial murder, 4 for attempted murder, and 5 plead guilty to lesser charges. Nurses made up the majority of healthcare professionals prosecuted (86%), followed by physicians (12%), then allied health professionals (2%). These findings differ from those of Kinnell, who stated physicians made up the majority of health care serial killers [3]. Either way, the data are disturbing.

Yorker also found that the 54 convicted serial killers murdered at least 317 patients, with over 2,000 patient deaths likely [5]. Intravenous injection (opioids, potassium chloride, or insulin) was the primary method of killing, followed by suffocation, poisoning, and then tampering with equipment [5]. Most health care murders occur in hospitals, followed by long-term care facilities [6].

Dr. H.H. Holmes

Many historians consider H.H. Holmes the first documented serial killer in the United States. Herman Webster Mudgett, born in 1861 in Gilmanton, New Hampshire, was an intelligent but troubled child with aspirations for medicine. He expressed his aspirations by performing surgeries on animals. As a child, authorities suspected him of killing a classmate.

Herman Mudgett went to medical school at the University of Michigan. He funded his education by stealing corpses from the school and using them for insurance scams, making life insurance claims from $10,000 to $20,000. In 1884, he graduated in the middle of his class.

In 1885, Herman Mudgett changed his name to Henry H. Holmes and moved to Chicago, IL, where he worked in a pharmacy. In Chicago, he started his career as a swindler and murderer, killing people to steal their property. Eventually, he became owner of the pharmacy near Jackson Park.

Holmes had a three-story house built near the pharmacy. The ground floor consisted of store fronts and the second floor had living space for apartments. Holmes never had the third floor completed, unknown to some of his swindled victims. Newspaper articles at the time claimed the house had trap doors and chutes that went to the basement where he trapped his victims then destroyed the bodies, and they called the house

Holmes' "Murder Castle." These stories are now considered exaggerated. Most of the building burned down in 1894 and the presence of trap doors has not been verified while chutes were common in homes and apartment buildings for laundry.

In 1893, Chicago hosted the World's Fair in Jackson Park. With the location close to Holmes' house, he rented out rooms during the fair. During this time, Holmes killed at least 9 people, but some claim he killed up to 200, including female tourists he seduced at his hotel [7,8]. As with the claim of trap doors and chutes, 200 murders is now considered an exaggeration [7,8].

While in prison awaiting trial for murder, Holmes wrote a book [9], **Self Portrait of a Serial Killer, H.H. Holmes in His Own Words**, in which he described how he swindled a lot of insurance companies, but did not kill anybody. His book contradicts police and news reports of the time.

According to the St. Paul Daily Globe, police had evidence that murders occurred in the castle [10]:

> More conclusive evidence than has heretofore been collected by the police in the Holmes case was revealed by the investigation tonight. Inspector Fitzpatrick believes it to be certain that at least three murders were committed in the "castle" of Holmes…

The police also had evidence that two bodies were buried in lime and quicklime in the basement of the "castle. [10]"

In 1891, H.H. Holmes likely killed Julia Connor, a women with whom he had an affair, and her 6-year-old daughter. Julia may have assisted Holmes with insurance fraud. Ironically, she became the subject of a fraud herself. Holmes had another accomplice, Mrs. Patrick Quinlan, pretend to be Julia Connor to collect an insurance claim [10]. Holmes claimed he killed Julia accidentally while giving her an abortion [8].

Holmes showed a liking toward killing those he involved in his swindles, including Emeline Cigrand and Minnie Williams. He also likely killed Minnie's sister Annie and her brother Horace, both who disappeared about the same time as Minnie [10]. Holmes claimed that Minnie killed her sister Annie by hitting her with a stool, and that he disposed of the body in a lake [11]. The skeletons of Emeline and Annie may have been saved by Holmes, who supposedly hired an articulator to piece the bones together [10]:

> The police have found a man… He told the police that he had mounted three skeletons for Holmes, and that the skeletons were from bodies taken from the house of Holmes on Sixty-third street. One of these he said was the body of a man, and the other two were women. … The police declare that they have almost positive proof that the first female body was that of Emeline Cigrand and the second that of Annie Williams.

Police caught Holmes after he murdered his swindling partner Benjamin Pitezel, and possibly Benjamin's children [12]. Police found the bodies of Pitezel's three daughters in the cellar of a house rented by Holmes in Toronto, and the cause of death was ruled a homicide by hypodermic morphine injection. Police believed Holmes killed Benjamin Pitezel in Philadelphia and his son in Detroit [11].

Holmes used the murder of Benjamin Pitezel in a fraud scheme. Shortly after purchasing a $10,000 life insurance policy with Philadelphia Mutual Life Association, Pitezel rented a house in Philadelphia under the name of B.F. Perry. Pitezel died by an explosion in that house, and his body burned so severely that he was buried under the name B.F. Perry. Holmes claimed

Benjamin Pitezel was still alive living in San Salvador, but in his book, he describes Pitezel's death as a suicide, which he supposedly covered up so the family could collect the insurance money. In fact, Holmes and Pitezel's wife went to Philadelphia to identify the body and claim the insurance money. The insurance company paid out $7500, but had the case investigated as to whether the death was accidental [11,13].

Authorities arrested Holmes in November 1894, and he confessed to killing 27 people. He was convicted of murder in 1895 and hung to death for his crimes in Philadelphia, Pennsylvania, on May 7, 1896 [14].

In the last section of his book, Holmes writes his confession to 27 murders, describing the manner of each. He also writes [9]:

> I could not help the fact that I was a murderer, no more than the poet can help the inspiration to sing.

Most of his book was dedicated to proving his innocence by providing alternative stories related to all the other deaths for which he was accused and/or convicted of murder. Holmes wrote the stories with incredible details, including dates and locations. His knowledge of the details suggests maybe he knew too much to not be guilty.

Dr. Martin MacNeill

Dr. Martin MacNeill, who also had a law degree, and his wife Michele had eight children. Michele gave birth to four children in five years, and the couple adopted four younger children, three of whom came from Ukraine [15]. Dr. MacNeill was the clinical director at the Utah State Developmental Center, a residential facility for people with developmental disabilities, from October, 2000 until September 21, 2007 [16].

Dr. MacNeill was having an affair with Gypsy Jillian Willis, a registered nurse, for several years. His wife suspected the affair and confronted MacNeill. He responded to her that she should improve her appearance, and he encouraged her to have a face lift [15]. He also claimed he was dying of cancer [17]. MacNeill found a plastic surgeon for Michele, and she had the surgery on April 3, 2007. MacNeill also directed the surgeon to prescribe certain medications for his wife, including Valium, Ambien, Phenergan, Percocet, and Lortab, all central nervous system depressants that should not be taken together [15].

MacNeill told his children he would take care of their mother the night following surgery, making sure she took her medications as scheduled. Michele could not see due to eye patches placed by the surgeon. One of the children, Alexis Somers, was in her first year of medical school and could not be with her mother the day of the surgery. However, spring break started the next day, and Alexis checked on her mother and found her over-sedated. Alexis confronted her father, who admitted he gave Michele too much medicine. Alexis told her father she would take over responsibility for making sure her mother had the right medicine. Apparently, Michele suspected MacNeill might try something after her surgery and told Alexis, "Give me each of the pills so I can feel it with my finger so if he tries to give me something else, I'll know what he's giving me." Alexis also recalls her mother telling her a few days after surgery, "If anything happens to me, make sure it wasn't your dad. [15]"

Alexis returned to school after spring break, and visit her mother on April 11, 2007. Michele's body was found by her six-year-old daughter Ada, whom MacNeill picked up from school and then sent her in the house to find her mother, which she did – dead in the bathtub. MacNeill would later be convicted of her murder [15].

Alexis immediately suspected her father killed her mother and wanted to determine how much medication her mother had taken. However, MacNeill discarded her mother's remaining medication keeping that information hidden [15].

MacNeill forbade Michele's family from attending the funeral. Alexis and her sister Rachel offered to move in with MacNeill to help with the younger children, but he denied their help and hired a live-in nanny – Gypsy Jillian Willis. Alexis knew about MacNeill's affair with Gypsy, and made her father choose between his children or Gypsy living at home. He chose Gypsy. Alexis recalls her conversation with her father [15]:

> He said, "If you fight me, I'm going to get you thrown out of medical school… I'm going to take you down." I was told that I needed to leave the home because I wasn't nice to Gypsy.

MacNeill kicked Alexis and Rachel out of the house, "without our keys or shoes. [15]" Three months later, he proposed to Gypsy.

The coroner ruled Michele died of natural causes, and the police closed the case. But Alexis and Rachel disagreed and, out of concern for their younger siblings, fought to have her death investigated [18]. Alexis also had personal reasons to suspect her father. One month after her mother's death, Alexis accused MacNeill of sexual assault. She filed a complaint resulting in his arrest on October 2, 2007 [16]. In 2014, MacNeill was convicted and sentenced to one to fifteen years in the Utah State Prison for sexual assault [19].

After much effort, Alexis and Rachel finally convinced Utah County Attorney's office investigator Doug Witney to investigate the death of their mother. Doug Witney found much more than just evidence for a murder [18].

Martin MacNeill lived a life of lies. Martin served in the US Army for less than 2 years in the mid 1970's, having been discharged for a psychiatric disorder. Over the next 30 years, while working as a physician, he collected more than $100,000 disability from the VA [18].

In 1977, Martin wrote phony checks in Orange County, California. He was sentenced to 180 days in jail and felony probation for 3 years [18].

While on probation, Martin MacNeill used another student's college transcripts to apply to a Mexican medical school. The school accepted Martin and he studied there for one semester. He then used the same transcripts to get into Western University of Health Sciences College of Osteopathic Medicine of the Pacific, and graduated with a doctorate of osteopathic medicine (DO) in 1984. Dr. MacNeill practiced medicine 30 years [18].

Two years after Michele's death, MacNeill and Gypsy were convicted of fraud for identity theft of one of MacNeill's daughters adopted from Ukraine, 16-year-old Giselle. MacNeill asked Alexis to accompany Giselle to Ukraine to visit her biological sister for the summer. However, he kept Giselle's U.S. passport so Giselle could not return. Gypsy then applied for a social security card using Giselle's name, and MacNeill had Giselle's birth certificate changed to Jillian MacNeill, another new name for Gypsy. In fact, MacNeill and Gypsy both accumulated numerous aliases under which they carried out fraud. While Gypsy used Giselle's name, Giselle's cousin flew to Ukraine and brought her home, unveiling the stolen identity. On January 15, 2009, MacNeill was arrested at the Salt Lake City Airport and Gypsy at their home. MacNeill received a four-year sentence and Gypsy two years [18].

With MacNeill in jail, Witney expanded his investigation into the death of MacNeill's wife. The toxicology report showed high levels of the pain killers Oxycodone and Lortab, the sedative Valium, and the sleep medicine Ambien. The coroner changed the cause of death from natural to undetermined. The police re-opened the investigation into Michele's death [18].

After being released from prison for his fraud charges in 2012, MacNeill faced new charges for murdering his wife [20]. During the trial, several of his daughters testified against him, and all his family except for his one son, Damian, believed him guilty. While in law school, Damian killed himself in January, 2010 [20]. (Foy) Also at MacNeill's trial, Anna Walthall, a mistress of his from 2005, testified that MacNeill described how he could murder someone and make it look like a heart attack [21]. She also testified MacNeill claimed he killed his brother and tried to kill his mother, and he killed handicapped people at work out of mercy.

In November 2013, the jury found MacNeill guilty of first degree murder and obstruction of justice [20]. He was sentenced from fifteen years to life in prison for the murder conviction, and one to fifteen years for the obstruction of justice conviction. In April 2017, MacNeill committed suicide while in prison.

Dr. Joseph Michael Swango

Michael Swango graduated valedictorian of his high school in 1972 and joined the U.S. Marines. He left the Marines in 1976 and went to college at Quincy University in Illinois, where he graduated *summa cum laude* (graduated with a grade point average of 3.90 or higher). Michael then went to Southern Illinois University School of Medicine. While in medical school, Swango worked as an ambulance attendant. Most medical schools require or recommend that students not work while enrolled due to the intensity of the educational program.

Michael Swango was a bright student, but demonstrated peculiar behavior, including fascination with dying patients. Many classmates and instructors expressed concern of Swango's behavior, including suspicions of falsifying patient records. Several of Swango's patients went into cardiopulmonary arrest (coded), with at least five dying. A month before he was to graduate from medical school, Swango falsified records during his OB/GYN rotation and got caught. The disciplinary committee met, and a unanimous vote was needed to expel Swango from medical school. However, one dissenter allowed him to remain and make up his OB/GYN rotation and other work with an additional year of medical school [22].

In 1983, after graduating from medical school, Dr. Swango started his surgery internship at Ohio State University Medical Center. Patients there started dying under suspicious circumstances. Nurses became concerned and reported to the hospital administrators, but an investigation had inconclusive findings. However, the investigation did show poor work ethics by Swango, and the medical center dismissed him after his one-year internship, rescinding an offer for a neurosurgery residency [22].

Swango next became an emergency medical technician in Quincy Illinois. His colleagues with the Adams County Ambulance Corps started noticing trends of becoming ill when Swango prepared the coffee or provided food or drinks. Investigators found arsenic in the coffee. Swango was convicted of aggravated battery for poisoning six of his colleagues and he served two years in prison, being released in August 1987. The Ohio State Medical Board revoked Swango's license [23]:

> **02/12/1986:** BOARD ORDER – REVOCATION BASED ON CONVICTION IN THE ADAMS COUNTY, ILLINOIS, CIRCUIT COURT, OF

> SIX FELONY COUNTS OF AGGRAVATED BATTERY. DOCTOR PLACED POISONOUS SUBSTANCE IN FOOD OR BEVERAGES WHICH WERE CONSUMED BY INDIVIDUALS WHO NEITHER CONSENTED TO NOR KNEW OF THE ADULTERATION. ORDER MAILED 2/27/86; ORDER EFFECTIVE 2/12/86.

Swango's high school teachers and friends expressed shock and disbelief. Many claimed he was incapable of committing such a crime [24].

Even as early as high school, Swango showed a fascination with death. One classmate describes the response of Swango when the class watched movies of the holocaust [24]:

> Most of us were grossed out, but Mike had no problem watching it—the holocaust films.

Paramedics who worked with Swango also described his fascination with death. One paramedic testified that Swango kept three scrapbooks filled with news articles of violent accidents and murders. The director of the ambulance service described Swango's response to a televised news report on a McDonald's mass shooting that killed 21 people. Swango turned up the volume and stared at the screen commenting that, "every time I have a good idea, someone beats me to it." Swango testified that he was joking when he made that comment [24].

After his conviction, Ohio State University Hospital investigated a suspicious death of a patient under Swango's care. No charges were filed [25].

In 1989, Swango worked as a counselor at the Virginia state career development center in Newport News. He was fired after getting caught working on a scrapbook of disasters during work hours. He next found a job in the same city, working as a laboratory technician until 1991, when he resigned to seek a job as a physician. During the time he worked in the laboratory, several employees suffered from stomach pains.

Swango changed his name to Daniel Jackson Adams prior to applying for several residencies across the country using falsified information, including altering a court document on his conviction for assault and battery that the case was in regard to a bar brawl [26]. In March 1992, he finally found a hospital that would accept him at the University of South Dakota in Sioux Falls, which was associated with the Veterans Affair Medical Center. The medical center found out the truth of Swango's conviction and fired him in December 1992 [26].

Swango next applied for and was accepted into a residency program at Stony Brook Health Sciences Center in conjunction with the Veterans Affairs Medical Center in Northport, NY. This residency lead to more patient deaths under Swango's care. In July and October 1993, Swango falsely informed physicians and staff that patients had agreed to a Do Not Resuscitate (DNR) order prior to killing them. Swango murdered three patients and tried to murder a fourth through intravenous administration of medications [26]. In late October, the director of the residency program found the true basis of Swango's assault and battery, and immediately suspended him from patient care and dismissed him from the residency program.

Remarkably, Swango still had more killing to do under the guise of a physician. Swango applied for employment with the Zimbabwe Association of Church Hospitals using falsified information. He was hired in November 1994 as a physician working at Mnene Missionary Hospital in Zimbabwe, Africa. He tried to murder at least two patients before being fired in July 1995 [25].

Swango managed to lie his way into yet another job working as a physician in Saudi Arabia. However, his lies caught up to him and he was arrested in the Chicago O'Hare Airport on his way to Saudi Arabia in 1997. He was charged with providing false statements and inappropriate use of controlled substances. Following a more intense investigation of the deaths in Ohio and New York, Swango confessed to four murders. He was sentenced to three consecutive life terms in prison [23]. Swango is suspected of having poisoned and killed 35-50 people in the United States and Africa [25].

Dr. Harold Shipman

Dr. Harold Shipman graduated from Leeds Medical School in England in 1970. Dr. Shipman killed between 215 to 260 of his patients over 27 years, mostly when he made house calls, but also during surgeries. He used excessive opioids to kill his patients [6].

In 1974, Dr. Shipman became a general practitioner in Todmorden, and by 1976, he started taking pethidine (opioid) which he supplied himself through false prescriptions he wrote. He was caught, fined, and had to undergo psychiatric treatment. The General Medicine Council issued him a warning.

In 1977, Shipman moved to Hyde where he joined a group practice. In 1992, he went solo and his murders increased. He would falsify death certificates to cover his tracks. Another general practitioner in Hyde reported suspicion of Shipman's practice. His last murder was of 81 year old Kathleen Grundy, the former mayoress of Hyde. He then forged her will stating [27]:

> I give all my estate, money and house to my doctor. My family are not in need, and I want to

reward him for all the care he has given me and the people of Hyde.

Police identified the will as a forgery and initiated an investigation of Shipman. Police accused Shipman of killing 215 patients, and possibly another 45. He was charged and convicted of killing 15 women, for which he was sentenced to life in prison. Shipman hanged himself in jail on January 13, 2004 [27].

Nurse's Aid Donald Harvey

Donald Harvey, born in Cincinnati, Ohio in 1952, dropped out of high school and then got his GED. He was depressed, suicidal and became interested in the occult. Harvey would place a skull next to a burning candle, then read a list of names waiting for the flame to flicker. Whatever name was read when the flame flickered was the next to die [8].

His dying grandfather was in Marymount Hospital, Kentucky, and Harvey visited him regularly. At the age of 18, he was hired as an orderly at Marymount Hospital. Two weeks later, Harvey killed his first victim, a stroke patient, by smothering him with a pillow. Harvey killed at least 12 other patients at Marymount in the next two years, usually elderly or severely sick patients whose deaths were ruled accidental. Like most medical serial killers, Harvey claimed he killed out of mercy for the patients [29].

Harvey continued in health care, working as a nurse's aid in two Cincinnati hospitals: VA Memorial Hospital and Drake Memorial Hospital. Harvey continued murdering patients at both. He claims he killed 80 people, not all patients. Harvey chose poisons as his weapon, including cyanide, arsenic and rat poison, sometimes administering the poison through the feeding tubes of patients [28].

Not until 1987, after 17 years of killing, was he caught. Dr. Lee Lehman of the Hamilton County Coroner's Office performed an autopsy on the body of John Powell, who supposedly died of complications following a motorcycle accident. Dr. Lehman noticed the smell of almonds emanating from the opened stomach, making him suspicious of cyanide poisoning. Indeed, the toxicology screen showed cyanide in the stomach contents and blood, and Dr. Lehman ruled the death a homicide [29]. Donald Harvey confessed to the murder of John Powell.

When WCPO news anchor Pat Minarcin reported on the murder of John Powell, he questioned during his broadcast whether a nurse's aid who worked in hospitals for 17 years and was just caught having killed one patient might have killed more patients previously. Tips started coming in to the newsroom, including from nurses who claim they filed complaints and suspicions with supervisors, but were told not to discuss it [30]. Anchor Minarcin asked Harvey's attorney William Whalen if his client killed more than one person. The attorney then asked Harvey, who answered yes. Donald Harvey then confessed to killing 24 people at Drake Memorial Hospital and 13 at Marymount Hospital. According to his attorney, Harvey confessed to killing 57 patients. In Ohio, Harvey was sentenced to four consecutive 20 years to life sentences. In Kentucky, Harvey was sentenced to eight life terms plus 20 years for 12 deaths at Marymount Hospital, and another three life sentences plus three terms of 7 to 25 years for the murder of three people not associated with the hospital.

In 2011, while incarcerated at the Toledo Correctional Institution, another inmate beat Donald Harvey to death [30].

Dr. Christopher Duntsch

Dr. Christopher Duntsch did not set out to kill his patients, but his gross incompetence resulted in the loss of many lives.

Christopher Duntsch went to high school in Memphis Tennessee, and his goal was to play college football at Colorado or Colorado State. However, when he got his football scholarship, it was for Millsaps College in Mississippi, an NCAA Division III school. After a year at Millsaps, Christopher transferred to Colorado State and made it on the football team as a walk-on (non-scholarship). Christopher only stayed at Colorado State for one year, transferring to Memphis State University (now the University of Memphis). He hoped to play football, but his multiple transfers made him ineligible. Christopher Duntsch now focused on his second goal, becoming a doctor [31,32].

In 1995, Christopher Duntsch was accepted to the University of Tennessee at Memphis College of Medicine dual MD/PhD program, a rigorous program during which a student typically spends an extra three years in medical school doing original research to earn a PhD. His research focused on brain cancer and potential treatments with stem cells. Christopher completed both degrees by 2002 [32].

Dr. Duntsch did his surgical residency at the University of Tennessee in Memphis, where he also continued research. The research lead to a patent with DiscGenics on a technique for growing stem cells. Dr. Duntsch had a promising future ahead [32].

Unfortunately, Dr. Duntsch had a drug and alcohol problem, and frequently performed surgeries under the influence. Not surprising, Duntsch made frequent errors during surgery. He refused to take a drug test, but enrolled in an impaired physician program and completed his residency [32].

In 2011, the Minimally Invasive Spine Institute in North Dallas hired Duntsch, who received a contract with Baylor Regional Medical Center at Plano (Baylor Plano) to have surgical rights at the hospital. Duntsch performed one operation at the Spine Institute, after which he left for a vacation to Las Vegas with nobody assigned to look after his patient. Duntsch was fired from the Spine Institute, but still had surgical privileges at Baylor Plano where over two years he performed 38 operations that left 31 patients paralyzed and 2 dead. He resigned from the Medical Center.

Despite his lack of surgical prowess, Dallas Medical Center temporarily gave Duntsch surgical privileges in July, 2012, when he was able to provide them with a letter from Baylor Plano stating the following [31]:

> On behalf of the Medical Executive Committee of the Medical Staff of Baylor Regional Medical Center at Plano, I am authorized to notify you of the following:
>
> All investigations with respect to any areas of concern regarding Christopher D. Duntsch, M.D. have been closed.
>
> As of this date, there have been no summary or administrative restrictions or suspensions of Dr. Duntsch's Medical Staff membership or clinical privileges during the time he has practiced at Baylor Reg. Medical Center at Plano.

During his first week at Dallas Medical Center, Duntsch performed two surgeries. His first surgery was to remove a disc and fuse two vertebrae on Floella Brown. She suffered a brainstem (lower part of the brain responsible for life sustaining functions) stroke after Duntsch accidentally cut her vertebral artery by removing the wrong piece of bone. The day following surgery she was pronounced brain-dead [31]. His second surgery occurred the same day Ms. Brown died. Duntsch operated on

Mary Efurd to have two vertebrae fused. When she woke from anesthesia, Ms. Efurd was partly paralyzed and in extreme pain [31]. A CT scan showed that a nerve root coming off the spinal cord was severed, screw holes were placed inappropriately in the vertebrae, and a screw was placed into another nerve root (the screw should have been placed in the bone). Before completing his first week of work, Dallas Medical Center removed Duntsch's surgical privileges [31].

However, Duntsch was able to get new privileges at yet another hospital, University General Hospital [31]. Here, he performed a cervical vertebrae fusion surgery in May 2013, that lead to a formal complaint by a colleague. During the surgery, Duntsch found a mass in the patient's neck and had to stop the surgery. However, he left behind a sponge causing an infection. The doctor who filed the complaint had watched Duntsch perform surgery at Baylor Plano and decided to check on the patient at University General. In his complaint, the doctor called Duntsch a sociopath who would "continue to maim and kill innocent patients," if he was not stopped. Once again, Duntsch lost his surgical privilege at a hospital. This time, however, following a temporary suspension of his medical license, in December 2013, his license was revoked by the medical board [31].

In July, 2015, Duntsch was arrested and charged with five counts of aggravated assault and one count of injury to an elderly person [33]. He was convicted only of the latter charge and sentenced to life in prison [34].

Not only did Duntsch fail as a health care provider, but the medical system failed by allowing him to continue practicing medicine. This case reminds me of how the Catholic church handles child molestation cases – send the priest to another parish. In this case, the doctor moved to another hospital. The health care system needs to develop plans to prevent this type of malpractice, and supervisors should be required to report to a national oversight committee or database complaints about a health care provider.

What type of person would continue practicing surgery after injuring so many patients? Several articles refer to Dr. Duntsch as egotistical or narcissistic. Dr. Duntsch wrote an e-mail to his assistant that shed some light on his frame of mind [33]:

> anyone close to me thinks that I likely am something between God, Einstein and the antichrist. … you cannot understand that I really am building an empire.
>
> What I am being is what I am, one of a kind, a [expletive] stone cold killer.

A friend of mine was the prehealth advisor at the University of Memphis and remembers Christopher Duntsch. I asked him if there were any signs Christopher would become this cold, egotistical, and incompetent as a physician. He said no, that Christopher was an excellent student and was expected to go far in medicine.

Nurse Genene Jones

Genene was born in 1950 and adopted by Dick and Gladys Jones who lived near San Antonio, Texas. The family had three other adopted children. The family started out well-to-do but then faced hard financial times. Her younger brother Travis died at the age of 14 when he made a pipe bomb that blew up on him. Genene was 16. A year later, her father died of cancer. Genene married shortly after graduating high school. She then went to beauty school and, after graduation, worked at the Methodist Hospital Beauty Parlor. She had one child, then got a divorce after 4 years of marriage. Her older brother died of cancer shortly after her divorce. Genene then quit work at the beauty salon and trained to be a licensed vocational nurse (LVN) [35].

In 1977, Genene Jones started her first nursing job at San Antonio's Methodist Hospital, but she was fired after just eight months. She got another job at a community hospital, and was quickly fired again [35]. The Pediatric Intensive Care Unit of Medical Center Hospital in Bexar County Texas, a teaching hospital associated with the University of Texas Medical School, hired Genene Jones in October of 1978. She worked the 3:00 – 11:00 pm shift, later to be called the "death shift. [36]"

In December 1980, Medical Center Hospital admitted 10 month old Christopher Hogeda with a congenital heart defect, pneumonia, and diarrhea. In May of 1981, he developed hepatitis and an infection that spread throughout his body. He started having cardiac arrhythmias and died of cardiac arrest on May 21, 1981. Jones later talked about his death [37]:

> He was my boy.
>
> I would bathe the children, and I would sing to them while I bathed them. If that sounds insane, tough shit. If you can't die with dignity, why live with dignity? We talked to them even after death. We're not God. We don't know when the spirit leaves the body.

Medical Center Hospital admitted 6-month-old Jose Flores to the pediatric ward on October 6, 1981 with fever, vomiting, diarrhea, and dehydration. Jose developed seizures and was taken to the radiology department for a brain scan. While there, he went into cardiac arrest, but doctors revived him and took him to the pediatric ICU. Jose was bleeding and the doctors did not know why. He went into cardiac arrest again, and this time, he died on October 10 [37].

Dr. Robotham, medical director of the pediatric ICU, recognized a pattern of excessive bleeding of patients. Physicians had been diagnosing the patients with disseminated intravascular coagulation (decreased ability to form blood clots caused by a severe infection), but the frequency exceeded that to be expected in a hospital. Another possibility was excessive use of the blood thinner heparin, which was used in intravenous lines to prevent blood clotting [37].

Dr. Robotham informed the acting chair of pediatrics, Dr. Franks, of possible heparin misuse. Dr. Robotham also ordered new procedures in the handling of heparin, including having two nurses present whenever heparin was drawn from the bottle and the nurses had to initial the bottle. In addition, Dr. Robotham reviewed all patient records with disseminated intravascular coagulation indicated to determine if the cause of bleeding could have been heparin overdose. Dr. Robotham found no evidence of wrong-doing [37].

Investigations after the fact indicated that ten children died in the pediatric ICU of "unexpected events" in 1981. Doctors and nurses in the pediatric ICU recall other peculiar cases in 1981 and early 1982. In response to increased patient death from illnesses they usually survive, nurse Toni Grosshaupt stated, "it began to happen more frequently until it happened every day or two." About Jones, Grosshaupt stated [37]:

> I'd leave a patient I thought was stable. She'd come on, and I'd find out the patient had a bad spell—had seizures or codes. That happened consistently.

Another nurse, Pat Alberti, spoke about coming in to work after Jones' shift [37]:

> I struggled with it for eight hours, and the kid was still alive. Day shift had it for eight hours, and the

> kid was alive. [Genene] came in for three hours, and the kid was dead.

Susan Maldonado, a nurse hired in 1980, commented about Jones [37]:

> I thought, "All these children died in the same period of time, and it just so happened that Genene was taking care of them." If the kid was sick, why didn't he die on me? Why didn't he die on somebody else?

Maldonado was concerned enough that she reviewed the records of patients who died. She made a complaint to Pat Belko, head nurse of the pediatric ICU. Belko knew the nurses in ICU talked about Genene. She brought Maldonado's concerns to Dr. Robotham. An internal investigation was born [37].

With no clear evidence, Jones kept working in the pediatric ICU. But on January 6, 1982, that changed. Four-week-old Rolando Santos had pneumonia and respiratory problems when admitted to the pediatric ICU on December 27, 1981. A few days later, Rolando had seizures, but brain scans gave no indication of why. He then went into cardiac arrest but was revived. On January 1, Rolando bled from old needle puncture sites, causing a drop in his blood pressure. Dr. Copeland ordered a test for heparin, and it came back positive. On January 10, Rolando started bleeding again. This time, Dr. Copeland gave Rolando protamine, a drug that reverses the effects of heparin, and the bleeding quickly stopped. On January 12, Dr. Copeland ordered Rolando be transferred out of the pediatric ICU. Rolando was released from the hospital four days later. Many in pediatrics suspected foul play [37].

Hospital administrators could not prove who caused the heparin overdose, but Jones was high on the suspect list. More patients mysteriously died, concerning their attending physicians, with some physicians threatening to take their pediatric patients to another hospital. Physicians relayed their concerns to the director of county hospital district, B.H. Corum, who called a meeting of medical school administrators and the hospital's malpractice attorney. The group decided to continue with an internal investigation, but not to fire Jones nor to notify the district attorney as there was not enough evidence [37].

Dr. Conn, a visiting professor from Canada who came to San Antonio for research on drowning victims, lead the internal investigation. Dr. Conn formed a team of six investigators which included medical directors and head nurses from three external pediatric ICUs. The review found suspicion of "a nurse," but no evidence of foul play. They recommended Dr. Robotham and Pat Belko be relieved of their duties. They also recommended that all LVNs in the ICU be replaced by registered nurses (RNs), thus relieving Jones of her duties. But it also meant six other LVNs would be moved from the ICU. Chair of the surgery department, Dr. Arthur McFee commented [37]:

> It was a case of having to use a huge stick because it was impossible to single out one. If we had just gone out and fired her, we would have had a substantial suit.

The system fails us again. Jones was not stopped, just moved. She quickly got another job with Dr. Holland, a third-year pediatrics resident about to start her own practice in Kerrville. Dr. Holland heard rumors about Jones before offering her the job. In fact, a colleague and friend of Dr. Holland recommended she reconsider hiring Jones. Dr. Holland also asked Pat Belko if she should be concerned about hiring Jones, and Pat supported the hiring, but warned Dr. Holland that Jones is assertive. The hospital gave Jones a good letter of recommendation, and even offered her another job, outside of pediatrics. Dr. Robotham, on the other hand, approached Dr. Holland and recommended that she reconsider the hiring, sharing with her some of his suspicions, to no avail.

Before Dr. Holland's clinic opened, Jones inadvertently let her nursing license expire, thus working for Dr. Holland approximately three months without a valid license.

On August 24, 1982, Dr. Holland's second patient at her new clinic was 14 months old Chelsea McClellan, who, according to her parents, had a bad cold. When the parents entered Dr. Holland's office, Chelsea started knocking items off the desk. Genene Jones offered to take Chelsea out of the office so the adults could talk. While with Jones, Chelsea stopped breathing and had seizures. Jones called Dr. Holland to come out, and Dr. Holland saw Jones providing air to Chelsea with a bag respirator and an IV line in Chelsea's scalp. An ambulance was called and Chelsea was taken to Sid Petersen emergency room. Chelsea was breathing on her own and kept in the ICU for nine days. With no indications of what could have caused her respiratory distress or seizures, she was released. The parents expressed gratitude to Jones and Dr. Holland, "We worshipped the ground these people walked on. [37]"

On August 27, Mr. and Mrs. Benites brought their one-month-old baby Brandy Lee to Dr. Holland's office because Brandy had blood in her stool and diarrhea. Dr. Holland and Jones took Brandy to an exam room as the parents stayed in the waiting room. Shortly after, Dr. Holland informed the parents that Brandy had stopped breathing. An ambulance was called to take Brandy to the local hospital, but after waiting half an hour, the parents wanted to transfer Brandy to a San Antonio hospital, which would also be by ambulance. Jones traveled in the back of the ambulance with Brandy, and Dr. Holland followed in her car. The Emergency Medical Technician (EMT) tells what happened next. Jones began pleading to the baby, "Please, baby, don't die! C'mon! C'mon!" The EMT was concerned that Jones was "getting out of control. She turned the whole situation into 'the world's falling in.' The baby was in bad shape, but calm, collected movements are better than going crazy." Brandy nearly died and the ambulance pulled over so that Dr. Holland and Jones could revive her. Doctors could not determine what caused Brandy's emergency [37].

After two other babies suffered similar incidences in the care of Dr. Holland and Jones, on September 17, Petti McClellan brought her son Cameron, who had the flu, and Chelsea accompanied them. As to why Chelsea came, Petti McClellan provides a different reason than Jones and Dr. Holland.

Petti stated, "Kathy Holland said she wanted me to bring Chelsea in. I said, 'Chelsea's fine.' She said she wanted to check on her. She just insisted I bring her in. [37]"

Jones stated, Petti "had called in and told Gwen [that Chelsea] was having increased blue spells. She also stated she'd had two seizures that lasted three minutes each that day. She felt the blue spells were getting worse. [37]"

Dr. Holland stated, "The call was not to set up an appointment to see Cameron. She told me that Chelsea had a cold that day, that she was having some funny little spells again—frequently. I didn't ask her just out of the blue to bring Chelsea in. [37]"

Dr. Holland ordered two routine vaccines for Chelsea. Jones prepared the syringes and then Petti McClellan brought Chelsea into the exam room. After the first shot, Petti thought Chelsea was having respiratory problems and, concerned that Chelsea might be having a seizure, asked Jones to stop the injections. However, Jones insisted she must give the second injection and did so, into her right thigh. Chelsea immediately stopped breathing and had a seizure. Chelsea was once again rushed from Dr. Holland's office to the local hospital by ambulance, and after stabilization transferred via ambulance to the hospital in San Antonio. Chelsea never made it to San Antonio. She went into cardiac arrest, and while Jones and an EMT were trying to revive Chelsea, they stopped at a small community hospital to no avail. Chelsea died. The McClellans once again gave thanks to Dr. Holland and Jones, this time in the form of a newspaper ad [37].

Three hours after Chelsea died, five-month-old Jacob Evans had an appointment with Dr. Holland. Jacob's mother Lydia was concerned of his crying spells, but he seemed healthy otherwise. When Jacob arrived at Dr. Holland's office, the doctor had not yet returned from the ambulance trip with Chelsea, but Jones was there. Jones took Jacob into an exam room to run blood tests and she wanted to start an intravenous line. Lydia questioned Jones' reasoning. According to Lydia, "She said it was in case he went into a seizure while they were running tests, so they could give him medication right away. I said, 'Jacob's never had any seizures.'" Jones then asked Lydia to leave the room. Shortly after Lydia heard Jacob scream a few times and then stopped mid-scream. Jacob had difficulty breathing and was taken by ambulance to the local hospital. Lydia describes what happened when Jones informed her in the hospital that Jacob had recovered fully: "She was shaking all over, and she was sweating very badly. The expression in her eyes [was] like she was almost in a utopia, like she was really enjoying it....Her eyes looked very excited—a happy excited. She looked like she was enjoying every minute of it. She looked like she was in her element." The doctors had no explanation for why he had an episode [37].

It looked like Jones was back to her old tricks. At the local hospital, Sid Peterson started an investigation into Dr. Holland's office. On September 24, the hospital notified the Texas Rangers of the suspicious cardiac and respiratory arrests occurring with Dr. Holland's patients and a legal investigation began.

One of the concerns raised by the hospital regarded a missing bottle of Anectine (succinyl choline, a muscle relaxant which causes paralysis). Jones found the missing bottle on September 26, but Dr. Holland noticed the bottle was full but had several needle holes in the rubber top where drug had been pulled out into a syringe. When she confronted Jones about this and demanded an explanation to give to the investigators, Jones responded: "I don't think we should explain them at all. I think we should just throw it out. We thought the bottle was lost. We should say we never found it."

As for the bottle being full, most of the fluid was saline. Dr. Holland also discovered that an extra bottle of Anectine had been ordered for the office, but not by her, and nobody signed the invoice. Dr. Holland told three of her colleagues, "If that child died because of something Jones did, I'll never be able to forgive myself. [37]"

The Kerr County grand jury indicted Genene Jones on one count of murder and seven counts of injury to a child, all victims having been patients of Dr. Holland. Jones was arrested on May 25, 1983 [38]. Prosecutors presented the motive as Jones wanting to show the need for an intensive care unit for children at a local hospital [39]. Jones was convicted on February 15, 1984, of all charges, with the jury taking only three hours to deliberate [40]. She was sentenced to 99 years in prison [41].

Some of the comments during the trial were heart-wrenching, including this testimony from the mother of Chelsea describing how her daughter responded to an injection given by Jones, she "went limp like a rag doll." Jones injected Chelsea with succinyl choline, the muscle relaxant [40].

After the conviction, Chelsea's grandmother commented on how Jones should have been stopped before Chelsea died [40]:

> It had to be stopped. It could have gone on forever. If it had been stopped in San Antonio, it would have never gone to Kerrville. They were warned about Genene. They didn't pay heed.

In 1983, a special grand jury in Bexar County, Texas, began investigating the deaths of infants who died at the Medical Center Hospital from 1978 to 1982, under suspicion that some of the infants may have died from injections that caused cardiac arrest. Bexar County district attorney Samuel Millsap started the investigation because of information implicating, "the possibility that multiple infant homicides had occurred at the Medical Center Hospital. [42]"

Texas had a mandatory prison release law and Genene Jones was scheduled to be released from prison in 2018. However, San Antonio prosecutors filed five separate murder charges for deaths of children in 1981 [43]. It took a long time, but on January 16, 2020, Jones plead guilty to killing an eleven-month-old baby in 1981. The other four charges were dismissed in the plea agreement. The judge made his feelings on the case clear, stating [44]:

> [The plea deal] doesn't come close to what you did to these families and the tragedies that you caused. You took God's most precious gift, babies, defenseless babies. I'm going to follow this agreement that you agreed with your attorney and state. But I truly believe that your ultimate judgement is in the next life.

Can Medical Serial Killing be Stopped?

Ideally, society could identify children likely to become criminals at an early age and provide ethical behavioral changing techniques to keep criminality from happening. But we cannot. Possible causes of evil were described in earlier chapters, and until we can identify the cause, we cannot develop a treatment. But some children and young adults show signs of low levels of empathy and high levels of cruelty indicating a potential for criminal behavior. What can we do, if anything, to prevent potential evil actions by such individuals without violating their rights?

Those who murder and feel no remorse likely have brain abnormalities in the frontal lobe of their brain (see Chapter 6). A lack of impulse control and feelings of empathy indicates sociopathy or psychopathy. Approximately 30% of prison inmates suffer from psychopathy. Other mental disorders also increase the likelihood of criminal activity.

Bucknall *et. al.*, compared the prevalence of narcissism, psychopathy, and Machiavellianism in healthcare workers to the prevalence in the general population. The definitions of these disorders as used by the authors follow [45]:

> Narcissism is defined as a "normal" personality trait at low levels and is characterized by the expression of pride, egotism and vanity.
>
> Psychopathy manifests as antisocial behaviour expressing callousness, remorselessness and selfishness, with little empathy and a high level of impulsivity. It has been proposed that 2 variants of psychopathy exist: primary and secondary. Primary psychopathy is believed to be a genetic predisposition to certain behavioural patterns,

> which include cool, calm, meticulous forward planning and emotional dissociation. Secondary psychopathy, however, is an adaptation to environmental factors and involves impulsivity and emotional instability, which can place others at risk.
>
> Those expressing Machiavellian traits are deceptive, self-interested and succeed through the exploitation of others.

Health care workers participating in the study were recruited through poster advertisements in a teaching hospital, and included surgeons, hospital physicians, general practitioners, acute medical physicians, nurses and technicians with limited patient contact. General population participants were recruited through on-line advertisement on social websites. Previously validated tests were used to assess each of the three disorders. Analysis over all health care professionals showed a lower level in all three personality disorders compared to the general public. However, 60 surgeons in the study showed an average higher level of narcissism and primary psychopathy compared to other medical professionals (but still lower than the general population), and 49 nurses in the study showed an average higher level of secondary psychopathy compared to other health professionals. The primary outcome of this study for our purpose is that health care professionals exhibit fewer personality disorders than the general public, although the sample size in the study was rather small [45].

Lloyd and Gartrell studied medical school students and found a prevalence of certain mental disorders. Medical students showed more psychiatric symptoms than the general population, including anxiety, depression, obsessive compulsive behavior, and interpersonal sensitivity [46]. But did they have these symptoms before entering medical school, or were they caused by the rigors of medical school?

A study in Iran indicated that 44% of medical students, interns and general practitioners showed indications of potential mental disorders, with the students showing the highest probabilities [47]. The study also suggested, "medical students and doctors had high levels of indifference and cynicism."

Working with pre-health students, I have seen my share of students who had issues suggesting they were not suited for a health career. I also had annual meetings with pre-health advisors from other schools around the country, and we shared the types of incidences we faced, without sharing names or the year an incident occurred. The stories I share below come from these sources.

A pre-med student stalked a patient who was in the hospital following a car accident in which the patient's young daughter died. He even went to the daughter's grave site and took a picture of the site with flowers and brought it to the mother. The hospital banned the student, with his picture provided to all security to keep an eye out for him.

During a research project, a pre-med student abused animals at the medical school to which he was applying. Needless to say, he did not get accepted, at least to that medical school, that year.

Several advisors, myself included, knew of students caught cheating on one or more exams. Many of these students went on to health professional schools.

As a pre-health advisor and teacher, I knew a lot about my students. However, I could not share that information with professional schools because of FERPA (Family Education Rights and Privacy Act). If a student cheated in my course, I could not even tell my colleagues in the same department to watch out for such behavior in their courses.

When students apply for professional schools, they usually complete an extensive application (for medical school, it often exceeds 10 pages) that includes disclosure of any misconducts associated with the schools or legal convictions. For professional schools to find out if applicants had cases of misconduct (academic or otherwise) during their undergraduate education, the applicant must sign a waiver for the undergraduate school to release any such records to the medical school. Many professional schools do this, but not all. Further, the student's undergraduate school does not always release the information even if a waiver is signed.

Professional schools generally accept students conditionally in that a background check is completed before the student can start school. Not all legal issues keep students out of professional schools. Repetitive driving under the influence and convictions of drug abuse/distribution are examples of red flags for the schools. But a one-time charge for underage drinking their freshman year can be overlooked, as long as the student was honest about it in their application. Dishonesty in an application can result in dismissal from a professional school up to the moment of graduation from that school.

Clearly, students get accepted to and graduate from health professional schools who are or become criminals. It happens in all professions. Unfortunately, when it happens in health professions, those people are in a unique position to cause bodily harm. Congress passed FERPA to protect the rights of adult students, as it should. Thus, professional schools must develop better methods of finding the right students to serve in health care.

Chapter 12

Human Experimentation on Vulnerable Populations

I swear by Apollo the physician, and Asclepius, and Hygieia and Panacea and all the gods and goddesses as my witnesses, that, according to my ability and judgement, I will keep this Oath and this contract: To hold him who taught me this art equally dear to me as my parents, to be a partner in life with him, and to fulfill his needs when required; to look upon his offspring as equals to my own siblings, and to teach them this art, if they shall wish to learn it, without fee or contract; and that by the set rules, lectures, and every other mode of instruction, I will impart a knowledge of the art to my own sons, and those of my teachers, and to students bound by this contract and having sworn this Oath to the law of medicine, but to no others. I will use those dietary regimens which will benefit my patients according to my greatest ability and judgement, and I will do no harm or injustice to them. I will not give a lethal drug to anyone if I am asked, nor will I advise such a plan; and similarly I will not give a woman a pessary to cause an abortion. In purity and according to divine law will I carry out my life and my art.

I will not use the knife, even upon those suffering from stones, but I will leave this to those who are trained in this craft.

Into whatever homes I go, I will enter them for the benefit of the sick, avoiding any voluntary act of impropriety or corruption, including the seduction of women or men, whether they are free men or slaves.

Whatever I see or hear in the lives of my patients, whether in connection with my professional practice or not, which ought not to be spoken of outside, I will keep secret, as considering all such things to be private.

So long as I maintain this Oath faithfully and without corruption, may it be granted to me to partake of life fully and the practice of my art, gaining the respect of all men for all time. However, should I transgress this Oath and violate it, may the opposite be my fate.

Hippocratic Oath, Translated by Michael North, National Library of Medicine, 2002 [1].

Medicine existed since ancient times, after all, it is the 4th Century BCE Hippocratic Oath that is recited at most medical school graduations in the United States. Although Hippocrates is often considered the father of western medicine, whether he wrote the oath is debatable. Nevertheless, Hippocrates promoted many doctrines utilized in western medicine today, including basing medicine on scientific studies; basing medicine on nature and not religion; practicing patient-oriented medicine; taking the least harmful approach to treating an ailment; and treating medicine as a craft that physicians practice on patients and should, at the least, "do no harm".

Anthropologists found evidence of surgeries occurring before Hippocrates. For example, they discovered skulls with holes bored into them for what is believed medical purposes, in a process called trepanation. The purpose of the surgery is unknown, but it may have provided a passage by which evil spirits could escape (treating mentally ill), or it may have decreased the pressure on the brain relieving epilepsy or headaches. Whatever the purpose, trepanation occurred across numerous civilizations 7,000 years ago.

We have advanced considerably since trepanation, with many mistakes made along the way. For example, people used bloodletting to treat various ailments for thousands of years. Bloodletting began in ancient Egypt and continues today. Ancient physicians including Hippocrates and Galen believed that the body was filled with four humours: yellow bile, black bile, phlegm, and blood. Health depended on a balance between the humours, and illness was often associated with too much blood. Thus, removal of some blood could bring the humours back to balance and return the person to health.

From the middle ages to the 1800's. Physicians and barbers performed bloodletting, with leeches or scalpels, to remove "bad blood" and return balance. However, more patients died from blood loss during bloodletting than patients cured. The white rags used during the procedure became quite bloody, and the barber-surgeon would place the rags around the poles outside their shop to advertise their craft, thus the design of the barbershop pole. Doctors use leeches and other forms of blood-letting in medicine today, but sparingly and for very specific ailments.

Before the 1800's, what today would be considered minor infections could result in death; injuries to limbs often required amputations; and exploratory surgeries often were the only means to diagnose internal illnesses. But medicine became more science-based. We now have antibiotics, antiviral medicines, prosthetic devices and numerous methods for viewing the internal body such as ultrasounds, x-rays, CAT scans, MRIs, PET scans, and more. How has medicine advanced so rapidly over the last 200 years? Through scientific research, including research on animals and humans.

Biomedical Research

Biomedical research studies the human body in terms of normal function, diseases and other ailments that alter normal function, and potential treatments. The research is hypothesis-driven, meaning that researchers propose a hypothesis of what is expected to be the outcome of an experiment, perform the experiment and analyze the data, then make a conclusion based on the data, often generating more questions. Based on these results, the same or different researchers try to answer the new questions. Using this *scientific method* of experimentation, the data drive the direction of research. In addition, researchers publish their studies in open, peer reviewed journals. Exceptions include research by pharmaceutical companies and the military. Most biomedical research is performed on cells, tissue, or animals for safety and other ethical reasons, using humans only when necessary or when studies are deemed "harmless." However, some scientists have strayed and used humans in unethical ways.

Biomedical researchers include medical doctors (MD), doctors of osteopathic medicine (DO), and doctors of philosophy (Ph.D.) in the sciences and social sciences. Medical doctors and doctors of osteopathic medicine are trained in the art of healing based on science. While most MDs and DOs practice in clinical settings, some also perform biomedical research, especially those employed at university-associated training hospitals. In contrast, scientists with a Ph.D. primarily perform research at universities, government institutions (CDC, NIH, military, NASA, etc.), and private companies (pharmaceutical, biotechnology, etc.). A small percentage of scientists earn both an MD and a Ph.D., and, therefore, are best trained for human research.

In their book, **Against Their Will**, Hornblum, Newman, and Dober describe how the eugenics movement that began in the late 19th century shaped the way for human experiments on institutionalized people, including orphans, prisoners, and mentally ill [2]. With the dehumanizing concept imbedded in eugenics, using the "unfit" as experimental subjects was a good use for people who could otherwise not contribute to society. Even after Hitler showed the world the horrors of eugenics, and Americans wrote the Nuremberg Code to try Nazi doctors for their atrocious experiments on humans (see Chapter 14), the United States carried out some horrendous experiments of their own.

Medical Research on Slaves 1845-1849

Dr. James Marion Sims is considered the father of gynecology as one of the first physicians to take on women health issues. He was quite successful, being elected the President of the American Medical Association in 1876, cofounder of the American Gynecology Society, and President of the American Gynecology Society in 1880. However, his notoriety has been questioned because of his use of slave women and children in human experiments. Sims was a slave owner, and he wrote in his autobiography [3], "There was never a time that I could not, at any day, have had a subject for operation."

While working in Montgomery, Alabama, Dr. Sims developed a surgical procedure to correct vesicovaginal fistula that often occurred following childbirth. A vesicovaginal fistula is caused by tears of the uterus and bladder forming a path by which urine seeps out of the body through the vagina. The result of such fistulae includes urinary incontinence and pain. At least three enslaved women were used in experiments to develop the surgical technique. His first subject, 18-year-old Lucy, gave birth to a child resulting in a vesicovaginal fistula. Sims tried to fix the fistula through a surgical technique without using anesthesia. During the surgery, Lucy contracted blood poisoning and almost died. Sims described the effect of surgery on Lucy as follows [3]:

> Lucy's agony was extreme…I thought she was going to die… It took Lucy two or three months to recover entirely from the effects of the operation,"

After four years of experiments, Sims finally found a successful technique following 30 surgeries including one on a 17-year-old slave. The experimental subjects did not have to give approval, and if damage occurred during the process, there were no guarantees of further treatment. A statue of Dr. Sims was erected in New York City's Bryant Park and then relocated to Central Park [4]. In 2018, the statue was removed following protests against celebrating a doctor who utilized slaves for experimentation.

Sims also used enslaved children in experiments, including testing surgical techniques to treat neonatal tetanus. He never found a suitable surgery.

The End of Poliomyelitis

The first evidence of poliomyelitis (polio) came in an Egyptian stele, dating from 1500 – 1300 BC, depicting a priest with a shortened leg and deformed foot consistent with polio. In the 1800's, Scottish novelist and poet Sir Walter Scott described his crippling childhood disease which resembled polio. Dr. John

Badham of Scotland described the first outbreak in 1835. German physician Jacob von Heine identified the target of the disease as the spinal cord in 1840.

In the latter 19th century and the first half of the 20th century, breakouts became more frequent in western Europe and the United States. The first polio outbreak in the United States occurred in Vermont in 1894, resulting in 18 deaths and 123 cases of permanent paralysis [5]. Dr. Charles Caverly studied the cases and concluded that polio could become an epidemic, most victims were children, and some patients experienced only mild symptoms and quickly recovered [6].

In 1905, Swedish physician Ivar Wickman first identified the contagious nature of polio. He also confirmed Caverly's finding that many infections were asymptomatic. In 1908, Austrian physicians Karl Landsteiner and Erwin Popper isolated a particle from the spinal cord of a polio casualty and demonstrated the particle is smaller than bacteria, and injection of the particle into two monkeys (but not mice, guinea pigs, or rabbits) caused paralysis of one monkey and death of the other [6]. Later, three forms of the poliovirus would be identified [5,6].

In 1915, the United States suffered from only 691 polio-related deaths [6]. Those numbers would change quickly as spring and summer heat hit New York. In 1916, a densely populated center of Brooklyn called Pigtown suffered an outbreak of polio, with the death of 6,000 (mostly children), and permanent physical disability of another 21,000 people. Polio hit New York City hard as well, infecting 9,000 children, with 2,343 of them dying [6].

Polioviruses are enterovirus, meaning they are transmitted primarily through the intestines, usually through ingestion of the feces of an infected person. Less frequently, it can be transmitted by coughs and sneezes. Most infected people (72%) are asymptomatic; that is, there are no indications the person has an infection, but they can still transmit the virus to others. Another 25% of infected people show symptoms like the flu (sore throat, fever, exhaustion, nausea, headache, and stomach pain), but the symptoms usually go away on their own in a few days. The remaining 3% of infected individuals have symptoms caused by the virus infecting the spinal cord. These neurological symptoms include tingling, numbness, muscle weakness, and paralysis, giving polio its other name: infantile paralysis [7].

Because the paralysis could include the muscles needed for breathing, those with serious cases of polio often died. The respiratory muscles work by creating pressure gradients across the chest wall that expand the lungs to allow air in, and when the muscles relax, the lungs recoil inward moving air out. In 1929, scientist Philip Drinker and physician Charles McKhann developed a breathing machine, initially named the Drinker respirator and later called the iron lung. The iron lung was a metal tank approximately 6 feet long and 2.5 feet wide, with a pump that would change the pressure inside. The pump performed the function of the respiratory muscles. The patient would lay inside the tank, with the head on the outside. As the tank pressure changes, the lungs would expand and recoil causing air to move in and out of the lungs. Most patients only needed to stay in an iron lung for one to two weeks [7,8].

The iron lung was a lifesaving device, but it cost as much as a house in the 1930s. About 1,000 iron lungs were in use in hospital polio wards by 1939. Marshall Barr, who was infected with polio at the age of 7, described his experience of having to use an iron lung for breathing when he was in his 20's [8]:

> The bit that you lay on actually pulls out like a tray. You would lay on it and get pushed inside. The mechanics are underneath the machine so

> you're laying on the pump. And of course you get the vibration underneath. Like: … breathing, bump; breathing, bump … It was not quite like a smooth breath.
>
> You can eat in the iron lung because your head is outside but the rest of your body is inside, although since you are flat on your back you really need to be careful when you swallow; you have to swallow in rhythm with the machine because it's pulling your diaphragm in and then pushing it out again. You just wait until it's breathing out and then you swallow. Coughing was a bit more difficult because you don't cough in rhythm with the iron lung. It was something you had to work round. But that was just sort of a down side. You cannot turn over or anything. The iron lung had port holes on the side which came in useful for physiotherapy. They had a rubber seal so you could open them on the down breath and put a hand in, to do physiotherapy or anything inside.

After leaving the hospital, Marshall Barr found that he needed the iron lung to breathe on a regular basis. Thus, he had one at his home and used it at night, being able to function without it during the day [8].

In another example of the harshness of polio, a nurse at Pittsburg Municipal Hospital describes her experience working in the polio ward [9]:

> In all my career there has been no experience like Municipal Hospital before the Salk vaccine. One year the ambulances literally lined up outside the place. . .One of our resident physicians never went to bed for nights on end, except for stretching out

> on a cot in his clothes. We nurses could never get home on time, either. To leave the place you had to pass a certain number of rooms, and you'd hear a child crying for someone to read his mail to him or for a drink of water or why can't she move, and you couldn't be cruel enough just to pass by. It was an atmosphere of grief, terror, and helpless rage. It was horrible. I remember a high school boy weeping because he was completely paralyzed and couldn't move a hand to kill himself.

During the late 1940's and early 1950's, polio paralyzed an average of 35,000 Americans a year, mostly children. Outbreaks occurred mostly during summer months. Polio cases in the United States peaked in 1952, with 57,628 cases of which 3,145 died. When Jonas Salk announced the discovery of a polio vaccine on March 26, 1953, people were relieved and eager to get vaccinated. The vaccine was approved for use in the United States in 1955. Following an aggressive vaccine program, the United States eradicated polio by 1979 [7]. Most of the world followed in a massive polio eradication program, leaving only two countries still inflicted by the wild polio virus: Afghanistan and Pakistan [10]. Although rare, some new strains of poliovirus formed from one type of vaccine, called vaccine-derived polio virus. Today, polio is making a comeback, especially in third world countries.

Eradicating polio from the United States and much of the world (as well as other diseases such as smallpox) show the value of vaccines. However, development of the polio vaccine has a dark history.

In the 1930's, physicians Maurice Brodie and William Park discovered they could inactivate poliovirus obtained from monkeys with formalin (a chemical often used in preserving animal tissue), producing a potential vaccine that successfully immunized 20 monkeys and increased the production of poliovirus antibodies in 12 children. The vaccine seemed effective, and the study was expanded to 2300 children and a few adults. None of these participants developed paralysis [11]. More field studies were carried out on 7000 children and hospital workers, and again, none developed paralysis although some had mild symptoms [12]. However, in future monkey trials, the vaccine proved ineffective against injections of virus into the brain. Later, it would be determined that there were three different strains of poliovirus, and the Brodie and Park vaccine probably protected against only one strain and was dismissed [9].

At the same time as Brodie and Park's studies, John Kolmer tested a live virus with activity attenuated by ricinolate, a soap-like substance. Kolmer administered his vaccine to thousands of children, resulting in 12 cases of paralytic polio, 6 of which were fatal. His vaccine was deemed a failure [5].

In 1938, President Franklin D. Roosevelt created the National Foundation for Infantile Paralysis (now known as the March of Dimes). Mothers would march to raise money for research, and some of that money went to Jonas Salk.

When scientists test the effects of drugs on a person, the scientifically correct method is to have half the subjects receive drug and the other half placebo (usually the substance in which the drug is mixed with for administration), unless more than one dose of drug is tested. However, if there is high probability the drug will have the desired effect and be safe, providing half the experimental subjects with non-active placebo would be morally wrong. Salk wanted to do the morally correct experiment, as expressed in his letter to the National Foundation for Infantile Paralysis [9]:

> . . .if we are aware of the fact that the presence of antibody is effective in preventing the experimental disease in animals and in man, then what moral justification can there be for intentionally injecting children with salt solution or some other placebo for the purpose of determining whether or not a procedure that produces antibody formation is effective. . .I would feel that every child who is injected with a placebo and becomes paralyzed will do so at my hands. . .The use of a placebo control, I am afraid, is a fetish of orthodoxy and would serve to create a "beautiful epidemiologic" experiment but would make the humanitarian shudder and would make Hippocrates turn over in his grave.

Jonas Salk, like Brodie and Park, used formalin to inactivate the poliovirus. Inactivating the virus, as opposed to attenuating its activity, was not common, and scientists were skeptical. Salk first tested his vaccine on tissue cultures and in monkeys to test for safety and efficacy. Salk performed the first human trials of his polio vaccine on institutionalized children: 43 children from the D.T. Watson Home for Crippled Children whose residents had already been affected by polio, and 161 healthy children from the Polk State School for feeble-minded children. Once the vaccine was determined safe and efficacious in these vulnerable children, the vaccine was tested on thousands of children in the Pittsburg area and the results were positive. However, Salk and other scientists believed more tests should be run prior to mass immunization with the vaccine. The National Foundation for Infantile Paralysis funded a field trial of over a million children across the United States; this time the children were not chosen from vulnerable populations. The vaccine had an effective rate of greater than 50% (that is, if 100 people were exposed to the polio virus, fewer than 50 would become infected) [5,9].

With a successful trial run, the vaccine was approved for public use. Unfortunately, 266 people (94 of whom were children) caught polio from the vaccine, with 11 people dying and 75% of the cases paralytic. Vaccine distribution was terminated until the source of infections was identified - one supplier of the vaccine failed to completely inactivate the virus. Once this source was removed from the supply chain, vaccinations continued with no further cases of vaccine-induced infections [5].

With mass immunization using the Salk vaccine in 1955, polio cases fell rapidly in the United States. However, by 1958 the number of cases started to increase, and by 1959, the cases had doubled. The Salk vaccine would require regular booster shots.

As Salk's vaccine was used for mass immunizations, Dr. Sabin from the University of Cincinnati worked on a different type of polio vaccine, a live virus with attenuated activity. Sabin also targeted all three variants of the polio virus. Sabin's vaccine provided longer protection than the Salk vaccine, and it could be taken orally (generally on a sugar cube or in a drink) whereas the Salk vaccine had to be injected. By 1960, the Sabin vaccine was approved for use, and by 1968 completely replaced the Salk vaccine for use in the United States [13].

Scientists continued working to improve the polio vaccine. The United States (and most of the world) primarily used the oral attenuated-virus vaccines, even though some cases of vaccine-induced polio occurred. The last case of a wild-virus infection occurred in 1979. However, 8-10 vaccine-induced polio infections occur every year. In 2000, the United States returned to an improved version of the Salk dead virus vaccine to decrease the likelihood of vaccine-induced polio [5,13]. By 2015, one of the three variants of the polio virus was completely eradicated.

Salk's polio vaccine began the eradication of polio, but he used vulnerable children to test the safety of the vaccine before testing the vaccine on protected children (parental permission required). Does the end justify the means? Is it okay to sacrifice a few to save the multitude? And who chooses who can be put at risk? The United States does not use vulnerable children anymore for scientific experiments. But what if another epidemic struck the United States or a pandemic struck the world?

It did—COVID-19. The virus responsible for COVID-19 is more transmissible and deadly than the polio virus. Yet scientists discovered effective vaccines in less than a year, without testing vulnerable children. Because the disease affects adults more than children, the vaccines were tested on adults first, and only after Federal Drug Administration and Center for Disease Control approval in adults were tests on children volunteers, with parental consent, used in studies. But what if a new deadly childhood disease arises? Will the world wait for volunteers?

The Tuskegee Syphilis Experiment

Syphilis is a sexually transmitted disease that was prominent for centuries and took the life of many historical figures of various walks of life, including the author Oscar Wilde, gangster Al Capone, composer Franz Schubert, and philosopher Friedrich Nietzsche. Many other historical people suffered from syphilis infections but did not die from it, including dictator Vladimir Lenin, Pope Julius II, and U.S. President Abraham Lincoln.

Syphilis is caused by a bacterial infection, usually through sexual contact but it also crosses the placenta putting fetuses at risk for congenital syphilis. Early names for syphilis included the Great Pox, venereal disease, and French disease. The more common theory on the origin of syphilis in the western world suggests Columbus and several crew members caught syphilis in Haiti and brought it back to Europe in 1493. Within two years, syphilis became an epidemic, possibly promoted by the French overthrow of Naples, Italy. Syphilis then spread to India in 1497 and China in 1505 causing epidemics in those countries as well, and it is now a world-wide disease [14,15].

Syphilis is characterized by three stages with a latent period between stages two and three. Primary syphilis generally occurs about three weeks after infection with development of sores called chancre at the site of infection. The sores tend to be firm, round and painless. Secondary syphilis occurs from four to eight weeks after the appearance of chancre, with appearance of a skin rash or lesions of mucous membranes, swollen lymph nodes, and a fever. Mucous membranes are continuous with the skin but extend into open cavities of the body such as the eyes, nose, mouth, anus, and vagina. A latent period follows secondary syphilis, during which there are no symptoms, but damage to organs may be occurring. After a latent period, a person may return to symptoms of secondary syphilis, progress to tertiary syphilis, or remain asymptomatic. Tertiary syphilis causes severe damage to organs such as the brain, heart, and blood vessels. The symptoms vary based on which organ(s) is affected. For example, neurosyphilis can cause difficulty moving, numbness in the extremities, confusion, dementia, headaches, and even personality changes. Tertiary syphilis can ultimately lead to death if not treated [16].

In 1905, Schaudinn and Hoffman identified the bacterium causing syphilis as Treponema pallidum, and by 1910 Wasserman devised a serological test for syphilis. Early treatments for syphilis included heavy metals (such as mercury and bismuth) and arsenic compounds. These treatments caused pain and other adverse reactions, and were not very effective. In 1943, Mahoney successfully treated four cases of syphilis with penicillin, and penicillin (or another suitable antibiotic) has been the drug of choice since [14,15]. Although antibiotics can cure syphilis, they cannot reverse damage already done to the body. Thus, early treatment is critical.

In 1908, the United States Congress established a Division of Venereal Disease in the Public Health Service. Many states followed with their own venereal disease programs to monitor and control the spread of diseases. A study in 1926 showed that the incidence of syphilis was 4.05 cases per 1,000 people, with the incidence in Caucasians at 4.0 per 1,000 and that of African-Americans at 7.2 per 1,000.

In 1929, Surgeon General Hugh Cumming requested financial support from the Julius Rosenwald fund to study the control of venereal disease in the south. The Julius Rosenwald fund promoted African-American welfare, and provided funding to the U.S. Public Health Services (PHS) to develop health programs for African-Americans in the south. In 1930, the U.S. PHS chose Macon County, Alabama, as one of five sites in which to study the prevalence of syphilis. The Tuskegee Institute provided local support. Other sites were also in southern rural areas with majority African-American populations [17,18].

Macon County had 27,000 residents at the time, 82% African-American. The area was poor and rural, with many people living in shacks with dirt floors, no indoor plumbing, and poor sanitation. Health care was not readily available. When public health officials arrived in the county to test people for "bad blood," volunteers agreed to participate without knowledge that the officials were testing them for syphilis. The study showed that 36% of the African-American population had syphilis. The PHS wanted to expand the study, but the Julius Rosenwald fund refused to fund any further study due to concern the study would unfairly associate syphilis with African-Americans in the south [17].

Before describing the Tuskegee Syphilis study in detail, the status of African-Americans living in the rural south and the way white Americans saw them needs to be addressed.

The Rockefeller Foundation and Carnegie Corporation joined forces to increase industrial productivity in the south by educating African-Americans and improving the health of laborers. At one of their conferences in the late 19th century, an organizer stated the following in relation to the goal of recruiting cheap African-American labor [17]:

> He will willingly fill the more menial positions, and do the heavy work, at less wages, than the American white man or any foreign race which has yet to come to our shores. This will permit the southern white laborer to perform the more expert labor, and to leave the fields, the mines, and the simpler trades for the Negro. The union of white labor, well organized, will raise the wages beyond a reasonable point, and then the battle will be fought, and the Negro will be put in at a less wage, and the labor union will either have to come down in wages or Negro labor will be employed...this will not be a clearly defined issue until your competition in the markets of the world force you to compete with cheap labor in other countries...

A series of conferences from 1898 to 1914, sponsored by the Rockefeller Foundation and Carnegie Corporation, campaigned to increase the health and education of African-American workers. The Julius Rosenwald Fund joined forces as the three organizations embarked on their mission [17]:

> The labor unit of the South is still the negro, emancipated, but ignorant, unambitious and less trained than when a slave. In his present condition he renders difficult, if not impossible, the changes requisite to intensive and diversified agriculture and retards the development of all industries in which he is employed. As a race he is less skilled than during slavery. The industrial development of the South demands that the negro be either improved or gotten rid of...The problem is not political but purely, industrial...His colleges of law, of medicine, of theology and of literature, science and art should be turned into schools for industrial training. Hampton Institute and Tuskegee should be duplicated in every Southern State...

Contrary to the views of southern politicians, Andrew Carnegie further supported development of the African American worker when he said [17]:

> ...our colored people...constitute one of the most valuable assets of the Republic, viewed from an economic standpoint. It is certain we must grow more cotton to meet the demands of the world, or endanger our practical monopoly of that indispensable article. Either the efforts of Europe will be successful to grow in other parts [India], even at greater cost for a time, or the world will

> learn to substitute something else for it. We cannot afford to lose the Negro. We have urgent need of all and of more. Let us therefore turn our efforts to making the best of him.

Clearly at the time, the foundations (and politicians) had little desire to promote the health and education of African Americans other than for economic gain. This sets the scene for the beginning of the United States Public Health Services long-term experiment on the progression of syphilis in Macon County Alabama.

When results of the 1931 study showed a very high prevalence of syphilis in African Americans, the PHS decided further study was needed to better understand the progression of syphilis. However, the data of the 1931 experiment were unreliable because the test used for syphilis (Wasserman test) was not accurate, regularly resulting in false positive (test positive but no disease) and false negative (test negative but disease present) results [17].

Even with poor testing for syphilis, the PHS decided to choose a site of study that showed a high prevalence of syphilis in the population, supposedly the entire population of the region and not focusing on the African American population. However, other areas in the south had higher prevalence rates of syphilis than Macon County suggesting the site was chosen for another reason.

According to Barrett, Macon County, Alabama, was chosen for the following reason [19]:

> In the late 20's various of the foundations began their studies of health conditions in the south which were to eventuate in the development of local health units. One of the most striking

> findings in the early surveys of disease prevalence was the high rate of syphilis among the majority of the Negro groups studied. In one of the study areas (Macon County, Ala., home of Tuskegee Institute) initial efforts at control of syphilis were followed by further moves on the part of the United States Public Health Service to bring diagnosis and treatment to the population. With the finding of high prevalence of syphilis in the survey and with certain other factors apparent in the community it became evident that it might be possible to institute in this region a prospective—in contrast to a retrospective—study of the results of untreated syphilis in the Negro male. Such a study was needed to assist in the planning and execution of the national venereal disease control program which was then being planned for a later time.

In 1932, the PHS decided to monitor syphilis in the African-American male over the period of six to eight months. The experiment compared the health of a control group and a syphilis group. The Public Health Service recruited the Tuskegee Institute to assist in the study as their African-American personnel could help build trust with the test subjects.

Volunteers for the study were recruited by flyers advertising free blood tests and free treatments for "bad blood", a term used to describe several ailments including syphilis [20]. Participants were also offered free meals, free physicals, and free burials, the last being of particular importance as most participants were poor sharecroppers who struggled to bury dead family members. The flyer had no mention of syphilis [19].

The PHS recruited 600 African American men, 399 of whom had syphilis. Of the 201 volunteers without syphilis at the beginning of the study, all contracted syphilis by the end of the study, which was in 1973, not six to eight months. The PHS withheld the diagnosis of syphilis from the participants.

The Tuskegee experiment had many ethical concerns, but a common misconception about the experiment is that the participants were given syphilis by the experimenters – this is not true. However, by not telling participants they had syphilis, the participants often spread it to their wives, who then passed it to their fetus.

Although this study started before World War II, it continued well after the war ended and the Nuremberg Code was written. Disgustingly, the United States did not follow the same guidelines by which they tried the Nazi doctors. Violations of the code include the following.

Code: *The voluntary informed consent of the human subject is absolutely essential.*

Although subjects consented to the experiment, it was not "informed consent," and if anything, it was misinformed consent. Participants were not informed of the true purpose of the experiment, nor of the duration (which kept being extended).

Code: *The experiment should be so conducted as to avoid all unnecessary physical and mental suffering and injury and no experiment should be conducted where there is an a priori reason to believe that death or disabling injury will occur.*

These two items of the Nuremberg Code equal the adage *First, do no harm.* The syphilis study exposed subjects and their families to potential harm. By not informing subjects who were infected with syphilis, the men did not refrain from sexual behavior thereby infecting their wives who later infected their fetuses. In addition, eleven years into the study, penicillin was approved for treatment of syphilis, and all infected test subjects should have been offered the treatment, especially as they were in more dangerous stages of the disease. No treatment was officially given, but the head nurse of the experiment secretly provided some test subjects with penicillin. Several subjects died of syphilis during the course of the experiment.

Code: *The degree of risk to be taken should never exceed that determined by the humanitarian importance of the problem to be solved by the experiment.*

To study the natural progression of a disease without providing treatment, let alone knowledge to the subject that he had the disease, is inhumane and a risk far exceeding that of the purpose of the study.

Code: *During the course of the experiment the scientist in charge must be prepared to terminate the experiment at any stage, if he has probable cause to believe, in the exercise of the good faith, superior skill and careful judgment required of him that a continuation of the experiment is likely to result in injury, disability, or death to the experimental subject.*

Irreversible damage to organs and death occurred in many subjects and the PHS had an obligation to terminate these experiments and provide appropriate treatment. However, the study did not end until news agencies broke the story forcing the Assistant Secretary for Health and Scientific Affairs to convene an ad hoc committee to review the study. The committee recommended termination [20].

Was the Tuskegee Syphilis Study a racist study? Absolutely. Would the PHS try to conduct this experiment in an affluent primarily white neighborhood, or even in a poor white neighborhood?

In 1974, then President Richard Nixon signed the National Research Act to protect human experimental subjects. This Act formed the National Commission for the Protection of Human Subjects of Biomedical and Behavioral Research, which was charged with identifying the ethical principles of conducting research on human subjects. After four years, the Commission published the Belmont Report [21].

The Belmont Report addresses four topics the Commission was charged to address: "(i) the boundaries between biomedical and behavioral research and the accepted and routine practice of medicine, (ii) the role of assessment of risk-benefit criteria in the determination of the appropriateness of research involving human subjects, (iii) appropriate guidelines for the selection of human subjects for participation in such research and (iv) the nature and definition of informed consent in various research settings. [21]" Of particular importance to the Tuskegee Experiment are the sections on Basic Ethical Principles and Applications.

Basic Ethical Principles address the concepts of respect for persons, beneficence, and justice; three principles absent from the Tuskegee Experiment. People should be treated as autonomous, that is, capable of making decisions affecting their lives and goals. People incapable of making such decisions due to mental or physical illness or other aspects of life affecting their autonomy (such as being in the military or in prison) should be protected from having unjust decisions made for them. Beneficence, the act of providing care and kindness to an individual, is an obligation during human experimentation. Human subjects also need to be treated justly, that is, without prejudice (except of a scientific nature such as studying sickle cell disease in African Americans). Clearly, the Tuskegee Syphilis Experiments disregarded basic ethical principles [21].

Applications refer to applying ethical principles to the conduct of research including informed consent, risk/benefit assessment, and the selection of subjects of research. Informed consent requires appropriate communication to the subject regarding the purpose of the study, potential benefits or harms from participating in the study, and the consent must be voluntary. Assessment of risks and benefits focuses on whether the research in humans is necessary to address the scientific question (for example, has testing already occurred in animals suggesting a potential benefit to humans), if the research may harm the individual physically, mentally, or even economically, and if the benefits such as a potential cure outweigh the risks. Selection of subjects protects against targeting certain populations of people as subjects for non-scientific reasons [21].

Ironically, much of what is present in the Belmont Report was also in the Nuremberg Code, and had the United States simply adopted research principles they put on Germans, then the Tuskegee Experiments would have been terminated 30 years earlier, although that is still ten years too late. The experiments were unethical from the beginning.

The Monster's Study of Stuttering 1939

University of Iowa student Wendell Johnson was a stutterer. He started stuttering at about 5 years of age, and he hypothesized that his stuttering was learned – as he stumbled in his speech, anxiety of speaking grew, which caused him to stumble more until he became a stutterer. If stuttering was a learned behavior, possibly caused by the anxiety of speaking, then stuttering could be unlearned. Johnson continued at the University to earn a master's degree in psychology and became a speech pathologist [22].

As an assistant professor, Johnson designed an experiment to test his hypothesis, and his graduate student, Mary Tudor, would carry out the experiment as part of her Master's degree dissertation with the assistance of other graduate students. They utilized children from the Soldiers and Sailors Orphans' Home in Davenport, Iowa. Their goal was to make normal children anxious about speaking to see if they later develop a stutter. In addition, they tried to convince stuttering students they spoke fine to see if the stutter would disappear [22,23].

Twenty-two children from five to fifteen years of age were chosen for the study, ten of whom stuttered. The children, or experimental subjects, were told they would receive speech therapy, not that they were participants in an experiment [22,24].

For the stuttering children, they were randomly divided into two groups. The test group (IA) was told their stuttering was just a phase and would go away with time [24]:

> Do you like to speak? Do you like to read aloud? You should do more speaking and reading aloud. Many children have this same kind of trouble that you seem to be having. But it really isn't trouble; it's just a certain stage that children grow through. In no time, you'll outgrow it [the stuttering] and you will be able to speak even much better than you are speaking now.... Pay no attention to what others say about your speaking ability for undoubtedly they do not realize that this is only a phase in speech development that you are undergoing.

The control group (IB) was honestly told about their stuttering. They were to answer about their stuttering and how it affects their interactions with others.

Of the five stuttering children who were told they spoke normally (IA), two showed a slight increase in fluency, two a slight decrease, and one did not change. Thus, the data were inconclusive toward Johnson's hypothesis.

The normal speaking children were also randomly divided into two groups. The control group (IIB) was honestly told they spoke fine [24]:

> Do you enjoy speaking? You speak very well. Your speech is of very good quality. Speak whenever you have an opportunity. You have the earmarks of a fine speaker.

In contrast, the experimental group (IIA) was told the following [24]:

> The staff has come to the conclusion that you have a great deal of trouble with your speech. The types of interruptions which you have are very undesirable. These interruptions indicate stuttering. You have many of the symptoms of a child who is beginning to stutter. You must try to stop yourself immediately. Use your will power. Make up your mind that you are going to speak without a single interruption. It's absolutely necessary that you do this. Do anything to keep from stuttering. Try harder to speak fluently and evenly. If you have any interruptions, stop and begin again. Take a deep breath whenever you feel you are going to stutter. Don't ever speak unless you can do it right. You can see how [the name of a child in the institution who stuttered rather severely] stutters, don't you! Well, he undoubtedly started the same way you are starting. Watch your speech every minute and try to do

> something to improve it. Whatever you do, speak fluently and avoid any interruptions whatsoever in your speech.

In addition, Tudor provided the following instructions to the teachers and caretakers at the orphanage who interacted with these children [24]:

> The staff has come to the conclusion that these children show definite symptoms of stuttering. The types of interruptions they are having very frequently turn into stuttering. We have handled a number of cases very similar to these children. You should impress upon them the value of good speech, and that in order to have good speech one has to speak fluently. Watch their speech all the time very carefully and stop them when they have interruptions; stop them and have them say it over. Don't allow them to speak unless they can say it right. They should be made very conscious of their speech, and also they should be given opportunities to talk so that their mistakes can be pointed out to them. It is very important to watch for any changes in the child's personality, in his attitude toward his school work, in his attitude toward his playmates, etc.

Tudor, herself, reinforced the "stutterer" label at least once a month for several months through meetings with the children, their teachers, and other caretakers at the orphanage [22,25].

At the end of the study, Tudor wrote the following about the six children who spoke normally but were told they stuttered [24]:

> All of the subjects ... showed similar types of speech behavior during the experimental period. A

> decrease in verbal output of all six subjects; that is they were reluctant to speak and spoke only when they were urged to. Second, their rate of speaking was decreased. They spoke more slowly and with greater exactness. They had a tendency to weigh each word before they said it. Third, the length of response was shorter. The two younger subjects responded with one word whenever possible. Fourth, they were more self-conscious. They appeared shy and embarrassed in many situations. Fifth, they accepted the fact that there was something definitely wrong with their speech. Sixth, every subject reacted to his speech interruptions in some manner. Some hung their heads: others gasped and covered their mouths with their hands; others laughed with embarrassment. In every case the children's behavior changed noticeably.

Tudor claimed the results of her study supported the hypothesis that children labeled "stutterers" would learn to stutter. However, only two of the children in the study showed a decrease in speech fluency. In fact, two other children showed increases in speech fluency and the other two showed no change. Thus, any label effect on the children was only 33% effective, and with such a small number of experimental subjects, I find it difficult to support or dispute the hypothesis.

Whether the hypothesis was supported or negated is irrelevant to the problems with the design of the study. Johnson chose orphans as subjects because of their accessibility. According to the NY Times article, Johnson expected to induce stuttering in the children, similarly to how he became a stutterer [22]. In contradiction, Silverman writes in his article that Johnson formulated the theory that labeling a child as a stutterer could change that child's speech fluency only after Tudor completed her study [25]. But without a hypothesis, why would Johnson had even proposed such a study?

The study had negative outcomes in that the labeling of the normal speaking students as stutterers had severe negative impacts (beyond stuttering) that significantly affected their quality of life. These changes in behavior continued long after the study concluded [25].

As reported in Tudor's dissertation, all six of the children's performance in school decreased, and several showed other signs of diminished confidence. Tudor described some of her meetings with the students in her dissertation [22,24].

> Second meeting with 5-year-old Norma Jean Pugh: "It was very difficult to get her to speak, although she spoke very freely the month before."
>
> 15-year-old Hazel Potter: "much more conscious of herself, and she talked less."
>
> 11-year-old Clarence Fifer: "He stopped and told me he was going to have trouble on words before he said them," When Tudor asked him why, he said the sound "wouldn't come out. Feels like it's stuck in there."
>
> One boy refused to recite in class.

Another graduate student wrote that 9-year-old Betty Romp, "practically refuses to talk," and he indicated that she "held hand or arm over eyes most of the time." [22]

The sixth student in the study group, 12-year-old Mary Korlaske, seemed to be most affected by the study at the time. She became withdrawn. When Tudor met with Korlaske, she asked if Korlaske's best friend knew about her stuttering. When

Korlaske responded, "no," Tudor inquired why, and Korlaske sadly responded, "I hardly ever talk to her." Korlaske ran away from the orphanage two years later. In a 2003 NY Times article, Korlaske claims, "It just ruined my life." [22]

Mary Tudor was herself changed by the study, after which she volunteered at the orphanage several times working with the study participants to undo the damage the study had done. She wrote in a letter to Johnson, "I believe that in time they . . . will recover, but we certainly made a definite impression on them." [22]

Johnson remained a leader in stuttering research, and some useful information was gained from the study on Iowa orphans, also called "The Monster's Study." As stated in the 2003 NY Times article [22]:

> Johnson's findings about the nature of stuttering, once it has begun, remains the accepted wisdom to this day. The disorder does respond to conditioning, and once established, stuttering can have a ruinous momentum. Often, the worse someone stutters, the more he fears speaking, and the worse his speech becomes.

However, speech pathologists often recommend parents to point out their children's errors in speech [25]. Should they stop doing this? I had a lisp when I was a child, and I had speech therapy. I was constantly corrected at school when I spoke with the lisp, and I rarely showed the lisp after grade school – my therapy was effective. But a stutter is different, and there is more of a negative stigma attached to it. I was not teased for having a lisp, but classmates who stuttered were teased.

Johnson and Tudor's research on the orphans was never published other than in Tudor's dissertation. The young participants never knew they were part of a study until 2001, when the San Jose Mercury News published a series of articles about the study. The now grown children learned of their role in the experiments that negatively impacted their lives sixty years after it happened. A lawsuit was filed for the living children and the estates of those who had passed on. The State of Iowa settled the lawsuit with a payment of approximately $1 million, and the University of Iowa issued apologies to all living participants and made a general apology.

Willowbrook Vaccine Trials 1955-1970

Following frequent hepatitis outbreaks, including one during World War II affecting 50,000 American troops, scientists searched for a vaccine. Hepatitis affects the liver and causes jaundice (yellowing of the eyes and/or skin), weight loss, fever, and liver tenderness, and it can be deadly. There are currently three known hepatitis viruses: A, B, and C.

In 1941, the World War II outbreak of hepatitis and other diseases lead the U.S. Surgeon General to form the Board for the Investigation and Control of Influenza and Other Epidemic Diseases in the Army, subsequently called the Armed Forces Epidemiological Board until its termination in 1992.

In 1938, New York legislators allocated money to build Willowbrook State School on Staten Island, to house and care for up to 4,000 mentally challenged adults and children. The facility was completed in 1942. However, the U.S. government took over the institution to care for disabled veterans from World War II. After the war, then governor of New York Thomas Dewey argued that the facility was needed for its original purpose, and described those needs to the director of the Veterans Administration, General Omar Bradley [26]:

> Every year in the State of New York, thousands of children come into this world who are mentally and physically defective and feeble minded, who never can become members of society. They require constant care, both medically and physically, and in many cases, for social, psychological and economic reasons, few parents can afford to place them in private institutions. Even if such institutions existed in sufficient quantity, the result is that the state must take responsibility for the care of these children and do so with a high degree of tenderness and attention.

Thus, after World War II, Willowbrook once again was purposed to house the thousands of children in New York who were "mentally and physically defective and feeble minded, who never can become members of society." In 1947, the first residents moved in to Willowbrook, with occupancy reaching 6,000 residents by 1965. In addition to overcrowding, residents were underfed, sexually assaulted, and left unattended for hours. Senator Robert F. Kennedy visited Willowbrook that year and described the conditions of the residents as, "living in filth and dirt, their clothing in rags, in rooms less comfortable and cheerful than the cages in which we put animals in a zoo." He then stated that Willowbrook "borders on a snake pit." [27] A physician from Willowbrook described it as "a human warehouse."

In 1972, journalist Geraldo Rivera did a documentary on Willowbrook in which he described Willowbrook and nearby Letchworth Village as "mere depositories for the retarded." [27] A European expert on mental retardation stated, "My God, you don't care for your mentally retarded children as well as we in Europe care for our cattle."

But Willowbrook residents had more troubles than the filth and abuse. Dr. Saul Krugman, faculty at New York University and Bellevue Hospital, and his colleague Joan Giles, also from New York University, initiated studies on hepatitis at Willowbrook in 1955, with some funding from the Armed Forces Epidemiological Board. The studies continued through 1972, with over 700 students from Willowbrook serving as experimental subjects. The only method of hepatitis diagnosis at the time of Krugman's studies was based on symptoms. (Shortly after the studies, blood enzyme assays became available for diagnosis.) During the time of his studies, scientists believed hepatitis was a virus exhibiting two different transmission patterns: infection hepatitis transmitted via feces and medically-induced hepatitis transmitted via contaminated needles, syringes, or other medical devices [28].

Krugman first carried out an epidemiology study to determine the prevalence of hepatitis at Willowbrook and its progression. Data showed that hepatitis was acquired at Willowbrook and not prior to admission to Willowbrook. Transmission of hepatitis occurred from student-to-student as opposed to an infected food supply. Approximately 30 to 50 percent of students staying at Willowbrook would eventually catch hepatitis (some reports state that 100 percent of residents caught hepatitis). However, the strain of hepatitis at Willowbrook was relatively mild and no deaths occurred [28].

Krugman next tried to find a mechanism of protection from hepatitis infection. He tested the effects of gamma globulins (a type of antibody) on its ability to protect children from infection with hepatitis using a challenge study, which involves exposing healthy individuals to the infectious agent. Krugman used more than 50 children, aged 5-10, in the study. The children were divided into two groups, a test group that was given gamma globulins and a control group that did not receive the gamma globulins. Both groups of students then received either injections of blood serum from infected patients or chocolate milk with infected feces mixed in it. The children given the gamma globulin were protected against hepatitis infection for 39 weeks; the other children had no such protection [28].

According to Robinson *et al.,* Krugman's research at Willowbrook was ethically sound. Parents consented for their children to participate, and the residents of Willowbrook were benefitted by a reduction of hepatitis infections to 15-20% of what it was prior to the immunization. But others argue that Willowbrook residents were chosen based on their disabilities, purposeful exposure of healthy children to a virus was not necessary, and the consent parents gave was coerced and parents were not fully informed. A nickname for Willowbrook was "Pediatric Tuskegee" [28].

Cancer Research at the Jewish Chronic Disease Hospital in Staten Island

For over a decade, Physician Chester Southam studied the immunology of cancer at the Memorial Sloan-Kettering Cancer Center. He published results of several experiments involving hundreds of human subjects regarding the ability of cancer cells to grow in healthy people (subjects were from the Ohio State Penitentiary) versus cancer patients (from the Cancer Center). In these experiments, subjects received injections of living cancer cells that produced nodules. For the cancer patients, the type of cancer cells injected differed from the type of cancer from which they suffered. The robust immune system of healthy people attacked the cancer cells rather quickly and the nodules regressed. In contrast, it took four to six weeks for the weakened immune system of cancer patients to reject the cancer cells. These experiments left a critical question unanswered: was the slower rejection in cancer patients because of the cancer or because of poorer general health secondary to the cancer [29,30]?

In 1963, Dr. Southam asked physicians working at the Jewish Chronic Disease Hospital in Staten Island, NY, to participate in a study injecting live cancer cells into terminally ill patients and comparing their rate of rejection to that of the cancer patients. Southam first asked Dr. Kagan, who not only refused to participate, but resigned from the hospital once the experiments were approved. Two other physicians also resigned from the hospital. Dr. Mandel agreed to work with Dr. Southam and injected 22 terminally ill patients with live cancer cells. These 22 patients rejected the cancer cells in the same time frame as healthy subjects, nullifying their hypothesis [29,30].

The cancer patients and prisoners used in the studies signed consent forms, and they were fully informed of the type of cells they would receive – cancer cells. In contrast, the terminally ill patients gave oral consent, and Dr. Southam purposely left out the word "cancer" from the description of the cells. In addition, some of the patients legally could not consent. William Hyman, a lawyer and director of the Jewish Chronic Disease Hospital, stated the following [29]:

> The 22 patients were between 43 and 83 years old. All were sick. They were not informed that they were being given live cancer cells, and did not consent to it. Some of the patients were not even capable of giving consent: several were senile, some spoke only Yiddish, and one was deaf.
>
> Four of the participants in the experiment are now dead.

Southam defended his research and his decision not to use the word "cancer" in describing the cells because it would cause unnecessary concern to the research subjects. He also said informed consent was rarely obtained in human experimentation, despite the existence of the Nuremberg Code [29].

Following concern of potential research misconduct, the Division of Professional Conduct of the New York Department of Education reviewed the case and presented it to the Regents of the State of New York. The Regents found both Dr. Southam and Dr. Mandel guilty of unprofessional conduct and of fraud and deceit in the practice of medicine [30]. The Regents voted to suspend the licenses of both physicians, but instead, both were put on probation [29].

The research conducted at Jewish Chronic Disease Hospital was published in the Annals of the NY Academy of Sciences. In 1968, Dr. Southam was elected president of the American Association for Cancer Research [29]. Does professionalism really matter, or is it all about results?

Was Evil Present in Human Experimentation?

The cases presented in this chapter are horrific in the disregard for the human subjects. Does this indicate that the researchers were evil?

As science lead to rapid advances in modern medicine, many physicians became scientists, especially physicians associated with universities. The concept of "publish or perish" is well known to any university professor who seeks tenure. University administrations expect and require their faculty to bring money into the university, and the way to do this is to obtain research grants, usually through federal organizations, but also non-profits such as the American Heart Association. Grants generally not only support direct research costs, but also subsidize the investigators' salaries and provide funds for "indirect costs," which are established by agreements between the universities and funding agencies. Indirect costs supposedly cover electricity, space, etc., but the administrators can determine how that money is spent.

Tenure is a two-bladed sword. Some people oppose tenure as it makes firing a professor more difficult. On the other side, to be denied tenure can be the end of a professor's career. Obtaining tenure at research universities requires obtaining external funding (grants) for research and publishing articles in peer-reviewed journals. The pressure is high. Physicians who serve on the faculty at medical schools are under this same pressure to obtain tenure, although if they are denied tenure, they can get jobs as clinicians.

Another factor contributing to the desire of researchers to become leaders in their fields is ego, and that is plentiful in university faculty. Prestige comes with the number of publications, type of publications, and number of invited presentations. The likelihood of getting funded increases the more one received funding as this generally leads to more prestige. Often, the non-tenured faculty have greater difficulty receiving funding as they have not yet developed the research history required to demonstrate their competency.

Thus, faculty are pressured to complete experiments and publish their results in the most prestigious journals possible. This is not an excuse, but a fact. Bioethicist and physician Henry Beecher wrote in 1966 his description of unethical experiments [31]:

> During ten years of study of these matters it has become apparent that thoughtlessness and carelessness, not a willful disregard of the patient's rights, account for most of the cases encountered. Nonetheless, it is evident that in

> many of the examples presented, the investigators have risked the health or the life of their subjects.
>
> …Taking into account the sound and increasing emphasis of recent years that experimentation in man must precede general application of new procedures in therapy, plus the great sums of money available, there is no reason to fear that these requirements and these resources may be greater than the supply of responsible investigators. All this heightens the problems under discussion.
>
> Medical schools and university hospitals are increasingly dominated by investigators. Every young man knows that he will never be promoted to a tenure post, to a professorship in a major medical school, unless he has proved himself as an investigator. If the ready availability of money for conducting research is added to this fact, one can see how great the pressures are on ambitious young physicians.

Thoughtlessness and carelessness are inexcusable in the abuse of children and other human subjects. The nature of human research itself may have been evil before the field of bioethics developed. And even though these research projects often ended in useful information or treatments, one cannot ignore morality in their actions. Thus, individuals willing to put others in harms way lack empathy or have an ego placing research as more important than the rights of individuals.

Most of the experiments described above occurred over 50 years ago, and while there are still occasional reports of abuse in human experiments, more regulations are in place to try to curtail those studies before they start. For example, experiments on prisoners are highly restrictive and even banned in some states. Ethics professor Arthur Caplan wrote [32], "If you want to see a place where ethics has gotten somewhere, human experimentation is a good place to look. I'm not going to tell you that everybody comes to the right answer and that everything is rosy, but it is much better."

The consequence of the evil experiments carried out in the name of science years ago, and the rogue scientists who continue such experiments today, can be deadly, as we saw during the COVID-19 pandemic. People do not trust science, and people do not trust vaccines. When science is performed correctly, the data are the facts. However, scientific interpretation of the data can vary leading to intellectual debates at conferences, during review processes, and even over the telephone. Scientists, like all people, must realize they can be wrong, and most early theories of science have been replaced as more knowledge is uncovered. But advances will continue as long as scientists stay true to their art.

Chapter 13

Human Research in the United States Associated with Wars

The first war the United States of America participated in was, of course, the American Revolution, (1775-1783) where the country fought for its freedom from Great Britain's reign. Ever since we gained our independence, it seems the U.S. spends more time in war than in peace. The Veteran's Association lists 13 wars fought by the United States [1].

After the American Revolution, we fought Great Britain again in the War of 1812 (1812-1815), this time fighting for trade and expansion. The Indian Wars were fought from approximately 1817-1898, when European colonies and the United States fought the indigenous people of America for the land. We fought Mexico from 1846-1848 to take ownership of California, Arizona, Nevada, Utah and New Mexico. We fought the Civil War from 1861-1865. The Spanish-American War was fought from 1898-1902 to end Spain incursion on the "New World." World War I (1917-1918) and World War II (1941-1945) were fought in Europe, but over 500,000 American servicemembers died in those conflicts. We then entered the era of "policing" in the Korean War (1950-1953) and the Vietnam War (1964-1975). More recently, we have fought wars in the Middle East, including Desert Shield/Desert Storm (1990-1991), War in Iraq, and War in Afghanistan, the latter two of which fall under the Global War on Terrorism (October 2001-September 2021).

That is a lot of wars, death, maiming, destruction, etc. for a relatively young country. And what for: land, power, riches, oil, democracy …

Many of the horrors of war are known, but not all. Besides the spoils to the winners, some good comes from war in the rapid developments in medicine and technology. Getting to the moon would likely have taken much longer if we did not have the technological advances of World War II. These advances do not make war good. In fact, many advances made during war time come at a great cost. This chapter will discuss some of the atrocities that advanced our knowledge during war time, and the next two chapters will describe the horrific actions of biomedical researchers in Germany and Japan during World War II.

Yellow Fever Experiments During the Spanish-American War

Christopher Columbus discovered Cuba October 27, 1492, and named the island Juana. Spain conquered the indigenous groups of the island and slowly colonized Cuba over three centuries. Not until the United States declared independence from England and Cuba became a major sugar producing island did Spain show great interest in Cuba. Spain had slaves brought to Cuba and helped build a prosperous agricultural region.

In 1895, Cubans rebelled against Spanish rule in their fight for independence. The United States sent its battleship, the USS Maine, to the Havana harbor to protect U.S. citizens in Cuba. In February 1898, an explosion occurred on the USS Maine and it sank; the United States blamed the Spaniards. (In 1976, U.S. naval investigators concluded that ammunitions aboard the USS Maine most likely caught fire and exploded by accident, not by enemy attack.) The phrase "Remember the Maine" became a rally cry to ignite the United States to help Cuba gain its independence.

In April 1898, United States and Spain declared war. The United States quickly defeated the Spaniards under the rule of U.S. Navy Commodore George Dewey and U.S. Army General William Shafter with his "Rough Riders." The war ended December 10, 1898, with the Treaty of Paris [2,3].

During the war, 385 Americans died from battle, with another 1,662 wounded. During that same time period, 2061 Americans died of diseases such as typhoid and yellow fever. With American troops occupying Cuba, concern grew over infections with tropical diseases. Thus, in 1900, the United States sent Major Walter Reed, a U.S. Army physician, and a team of medical researchers to Cuba to study the pathogenesis of yellow fever [2,3].

Cuban physician Carlos Finlay hypothesized that mosquitoes were the vectors for transmitting yellow fever; however, the U.S. team was skeptical. Although Major Reed headed the team, he left Cuba before human experiments began. Dr. Finlay harvested mosquitoes for use in the experiments. The mosquitoes would feed on infected patients and then be used to inoculate experimental subjects. Volunteers received payment in today's equivalent of between $1000 and $20,000, and for the first time in modern era, volunteers had to provide written consent [3].

Between August 6 and 16, 1900, U.S. Dr. Lazear inoculated five U.S. soldiers using bites from infected mosquitoes. On August 16, Dr. Lazear inoculated himself concerned how others would perceive the use of human subjects. After a total of nine inoculations of volunteers and that of Dr. Lazear, no subject developed yellow fever. On August 27, U.S. Dr. Carroll inoculated himself and developed yellow fever five days later and never fully recovered. A second case of yellow fever developed in an U.S. army volunteer who was bit by the same mosquito used to inoculate Dr. Carroll. After the second case, a mosquito bit Dr. Lazear and he got yellow fever and died one week later. In October, Dr. Reed presented to the American Public Health Association preliminary data suggesting that yellow fever was transmitted by mosquitoes [2,3].

On November 1, 1900, Dr. Reed returned to Cuba to perform more experiments on human subjects. Ten additional cases of yellow fever were induced through mosquito inoculations of two U.S soldiers and eight civilian volunteers. Reed and his team found that mosquitoes only picked up the yellow fever virus if they fed on a person during the first three days of infection. In addition, the virus required more days of incubation in the mosquito before it could be passed to another human through a bite [2,3].

Further human experiments tested whether yellow fever could be transmitted through injection of infected blood or serum into healthy individuals; four developed yellow fever. Dr. Reed conducted another experiment to determine if contact with material contaminated by an infected person could transmit yellow fever, but no transmission was detected [3]. Reed also studied if filtering blood of bacteria could protect against yellow fever, but it did not indicating that yellow fever was caused by something small enough to get through the filter - a virus. This was the first demonstration of a viral-induced disease in humans.

During all these experiments, the only death caused by mosquito inoculation was that of Dr. Lazear, and that may have been an accidental bite. Most cases of the induced yellow fever were mild. However, experimental subjects knew they could catch yellow fever and that people sometimes died from the disease.

One experimental subject described his willingness to volunteer [2]:

> Volunteering to Dr. Carroll for experimental yellow fever was, I can assure you, a cold-blooded business proposition. There were no heroics in it as far as I was concerned...I suspected that I would probably get it spontaneously anyhow, so I decided I'd rather have it under favorable circumstances.

In February 1901, a new phase of experimental trials were designed to use mosquito inoculations to induce immunity from yellow fever at Las Animas Hospital, the primary treatment center for yellow fever in Cuba. Twenty-three volunteers were inoculated 42 times with infected mosquitoes. Following the first 28 inoculations, only one mild case of yellow fever occurred. However, the next 14 inoculations resulted in 7 cases of yellow fever, 3 of them fatal. Due to fatalities, the experiments were terminated [3].

Finding the mechanism of yellow fever transmission led to a decline in the disease, not through the treatment of patients but by the eradication of mosquitoes. Dr. Walter Reed was bestowed great honors, including the naming of the hospital that serves the President of the United States after him. But at what cost did these honors come? People died during these experiments. The numbers were small, but these experiments were carried out on less than 100 participants. Subjects were well compensated, but there was no compensation for death [2]. Several volunteers were U.S. soldiers, and we will see shortly why soldiers are a vulnerable population.

American Mustard Gas, Nitrogen Mustard and Lewisite Experiments

For centuries, countries hesitated using chemical weapons during warfare because of their horrific effects, and chemical weapons were banned prior to World War I. However, both sides used chemical weapons such as chlorine, phosphene, and mustard gas, during World War I. In 1919, the Allies included in the Versailles Treaty a reaffirmation of the ban on using chemical warfare and forbade Germany from manufacturing or importing them [4].

At the Geneva Conference for the Supervision of the International Traffic in Arms, several countries signed the Protocol for the Prohibition of the Use in War of Asphyxiating, Poisonous or Other Gases, and of Bacteriological Methods of Warfare (also called the 1925 Geneva Protocol) on June 17, 1925, and it took effect February 8, 1928 [4,5].

Later chapters describe how Germany and Japan used chemical and biological agents during World War II on their enemies, including prisoners of war. The United States and other Ally countries (Great Britain, Canada, Australia and the Soviet Union) also tested chemical agents, but they used their own soldiers. One of these experiments tested the effects of mustard gas, nitrogen mustard, and lewisite on different races of United States soldiers.

Prior to World War I, physicians used mustard gas (sulfur mustard, known by the military as H, HD, and HT) and nitrogen mustard (known by the military as HN-1, HN-2, and HN-3) in treating psoriasis and to remove warts, respectively. Today, they serve solely as chemical weapons. The distribution and effects of nitrogen mustard are similar to those of sulfur mustard. Sulfur mustard can be distributed as a gas (mustard gas), in water, or as a liquid. Sulfur mustard can smell like garlic, onions, or mustard, but it can also be odorless and thereby go undetected by those exposed to it. In addition, adverse effects take 2 – 24 hours to begin and vary in type and severity based on the mechanism and level of exposure. Sulfur mustard acts as an irritant on several parts of the body. Upon initial exposure to sulfur mustard, itching and a redness of the skin occurs in a couple of hours. Later, the site forms blisters. The eyes become irritated and can swell with potential tearing of eye tissue. With severe exposure, temporary or permanent blindness can occur. People exposed often develop runny or bloody noses, shortness of breath, and a cough, and may develop chronic respiratory distress. The gastrointestinal system is also affected, with sulfur mustard causing abdominal pain, diarrhea, nausea and vomiting [6].

Sulfur mustard damages DNA, thereby affecting tissue with rapidly reproducing cells. Among these is the bone marrow, where red blood cells, most white blood cells, and platelets are produced. Thus, exposure to sulfur mustard could cause aplastic anemia (decreased oxygen carrying capacity of the blood due to reduced synthesis of red blood cells), weakness due to the anemia, increased bleeding due to the lack of platelets that are critical for blood clotting, and illness due to the lack of white blood cells that protect the body from pathogens [6].

There is no cure for sulfur mustard exposure, and treatments target the symptoms. However, death is rare [6].

Lewisite (referred to by the military as L) is another man-made irritant used in chemical warfare and with no medical purpose. Effects of lewisite resemble those of sulfur mustard, except lewisite also decreases blood pressure and has no effect on DNA and the bone marrow. Lewisite also contains arsenic and can produce symptoms of arsenic poisoning such as stomach ailments and low blood pressure [7].

During World War II, the U.S. Department of Defense utilized 60,000 of its own soldiers to determine if the effects of mustard gas, nitrogen mustard and lewisite varied by race, and how effective various protective devices were against mustard gas, also based on race. Military scientists tested the effects of mustard gas on African-American, Japanese-American and Puerto Ricans compared to those on white Americans, with the hypothesis that darker skinned races would be somewhat protected from the effects of mustard gas. If the hypothesis was true, the darker skinned races could serve more safely on the front lines. The hypothesis was false. As for Japanese-Americans, the military wanted to determine how the enemy would respond to mustard gas, which also was the same as white Americans. The U.S. government declassified documents relating to these tests in 1993 [8].

According to a 1993 Congressional Hearing on the Department of Defense Mustard Gas Testing, the Naval Research Laboratory reported [9]:

> From 1942 to 1945, The Naval Research Laboratory (NRL) conducted a program to test and evaluate protective clothing and applications (ointments, powders, etc.) for use against war gases. The tests involved the participation of naval enlisted volunteers from the Out-Going Unit (OGU) at the Naval Training Station at Bainbridge, MD. The volunteers were informed that they would be taking part in testing the effectiveness of protective clothing and ointments and, as a result, might come into contact with mustard-type gases.

Researchers carried out several experiments, the most frequent being *patch tests*. The purpose of patch tests was to determine the efficacy of protective or decontamination ointments, treatments for mustard agent and Lewisite burns, effects of multiple exposures on sensitivity, and the effects of physical exercise on the severity of chemical burns. Each of these experiments required liquid mustard agents applied directly to the skin of the subject [9,10,11,12].

Dr. Pechura, Study Director for the Institute of Medicine Committee to Survey the Health Effects of Mustard Gas and Lewisite, described the *chamber experiments* in her testimony to Congress [10]:

> Young men in Navy boot camps were offered extra leave and a change of scenery if they would agree to test summer uniforms. Once they had gotten to the test site, however, the men were dressed in various amounts of protective clothing, that is, impregnated with various chemicals to retard the penetration of gas. They were given gas masks.

At this point in her testimony, Dr. Pechura was interrupted by Representative Bilirakis to clarify that they were asked to test summer uniforms. Dr. Pechura clarified that this statement was correct [9,10]:

> Yes. Official documentation [Department of Defense documents], descriptions and summaries of these experiments oftentimes document the fact of what they were told at the time, and that was to test various kinds of uniforms. They were never told in the beginning about any gas.

Dr. Pechura then continued her testimony describing the chamber experiments [10]:

> …at the test site, they were given gas masks and put into gas chambers. They were actually locked into these gas chambers. The chambers themselves locked from the outside.
>
> Because there was interest in tropical conditions, the gas chambers were held at 90 degrees, 65 percent humidity. So they were hot and humid, and in some cases the concentration of sulfur mustard in these gas chambers would have been lethal without the gas masks.
>
> The men remained in the chamber for an hour, after which they remained in their protective clothing for various lengths of time. This scenario was repeated daily or every other day until the men's skin burned. That was the end point of the experiment, indicating that the protective clothing had failed.

Experimenters documented the location and extent of erythema, (reddening and burning) of the skin for each subject. The majority of soldiers exhibited widespread erythema, especially in areas of moist skin folds, the chest, shoulders, arms, and legs [9,10,11,12].

Some chamber tests were carried out in Panama, North Carolina, and Maryland under conditions designed to reproduce a battlefield. The subjects wore protective suits during drills for 1 day or up to 6 weeks, and then they entered the gas chamber as described above [10,11,12].

In addition to patch and chamber experiments, researchers performed *field tests* in the United States and Australia. Test subjects would be outside, with some wearing protective clothing and gas masks and others not. Mustard gas would be dropped on the subjects by planes at concentrations similar to those used in the chambers.

In the Congressional Hearing, Dr. Pechura described four aspects of concern about the experiments. 1) The soldiers were misled as to the purpose of the study and to what they were exposed. 2) Experimenters were told not to mistake laryngitis and conjunctivitis as symptoms of gas poisoning when in fact they were. 3) Those in charge were told to "dress down" those who did not want to continue in the experiments, and dressing down often included overt threats. 4) The participants were told to never reveal their participation in the experiments [9,10].

After the experiments were completed, participants left the test site and went on to other duties. Nobody monitored their health status, and their exposure to mustard gas was never added to their health record, preventing participants from later receiving health care from VA hospitals for conditions caused by the exposure, even though studies in 1933 showed adverse long-term effects of exposure to mustard gas [9,10,11,12].

Military personnel are one of several vulnerable groups for human experimentation, and these experiments show why. Soldiers follow orders of their commanders. Thus, can a soldier truly "volunteer?" Following are some excerpts from a National Public Radio presentation of the mustard gas research, including comments from experimental subjects [13].

Rollin Edwards, an African-American Army soldier during World War II, volunteered for the *chamber* experiment. He recalled, "they said we were being tested to see what effect these gases would have on black skins." Edwards also described what it was like in the gas chamber [13]:

> It felt like you were on fire. Guys started screaming and hollering and trying to break out.

> And then some of the guys fainted. And finally they opened the door and let us out, and the guys were just, they were in bad shape.

Edwards also participated in *field* experiments, where he described crawling around in the jungle after exposure of the soldiers and the jungle to mustard gas: "It took all the skin off your hands. Your hands just rotted." [13]

According to the NPR report, [13] Edwards, “never refused or questioned the experiments as they were occurring. Defiance was unthinkable, he says, especially for black soldiers.”

Lopez Negron, a Puerto Rican Army soldier, participated in the *field* experiment. He and other test subjects were in the jungle when the military sprayed them with mustard gas. He said that he and the other soldiers immediately became burned and ill. They had to move around in the jungle for several days after the exposure. Negron describes the experience to NPR [13]:

> We had uniforms on to protect ourselves, but the animals didn't. There were rabbits. They all died.
>
> I spent three weeks in the hospital with a bad fever. Almost all of us got sick.

For their troubles, the soldiers received a commendation. The commendation received by Louis Bessho, a Japanese-American soldier, read as follows [13]:

> These men participated beyond the call of duty by subjecting themselves to pain, discomfort, and possible permanent injury for the advancement of research in protection of our armed forces.

The military often used its own soldiers for experiments on potential weapons of mass destruction. Could the soldiers refuse to participate, or would that be disobeying orders?

Human Radiation Experiments

After dropping the atomic bomb on Japan twice, United States scientists studied the effects of radiation on humans in several experiments. Between 1945 and 1962, more than 210,000 civilians and military personnel were exposed to radiation without knowledge.

In 1994, then President Clinton acted by Executive Order 12891 to form the Presidential Advisory Committee on Human Radiation Experiments (ACHRE). Executive Order 12891 defined human radiation experiments as follows [14]:

> 1. Experiments on individuals involving intentional exposure to ionizing radiation. This category does not include common and routine clinical practices, such as established diagnosis and treatment methods involving incidental exposures to ionizing radiation.
>
> 2. Experiments involving intentional environmental releases of radiation that were designed to test human health effects to ionizing radiation, or were designed to test the extent of human exposure to ionizing radiation.

The executive order states the ACHRE shall [14]:

> … provide to the Human Radiation Interagency Working Group advice and recommendations on the ethical and scientific standards applicable to human radiation experiments carried out or sponsored by the United States Government.

The ACHRE examined radiation experiments on humans from 1944 – 1974. Not until 1974, the Department of Health, Education, and Welfare "adopted regulations governing the conduct of human research." (ACHRE) Almost four thousand radiation experiments were uncovered, but with limited documentation available. These experiments started with the Manhattan project.

The Manhattan Project

In August 1942, the United States government created the Manhattan Engineer District to develop an atomic weapon. The U.S. Army Corps of Engineers constructed three research facilities for this purpose: Oak Ridge, Tennessee; Los Alamos, New Mexico; and Hanford, Washington. The government recruited radiation researchers from across the country to study potential health risks to those working on the atomic bomb. The potential risk of radiation exposure became apparent in the 1930s when watch makers would apply radium paint to watch dials. The watch makers would lick the paint brushes in the process, thereby ingesting radioactive radium. These workers developed diseases of the jaw and bone cancer.

At the Los Alamos facility, where the atomic bomb was being made, Mastick, a chemist studying plutonium, accidentally ingested some plutonium when a vile of it exploded. This accident spearheaded the desire for biomedical research on the hazardous effects of the radioactive materials used to make the bomb (plutonium and uranium), and how to measure levels of radiation in the lungs, urine, and blood [15].

Between 1945 and 1947, scientists injected 18 patients with plutonium to determine how urine and feces levels could be used to estimate the level of plutonium in the body. The patients were chosen for the experiments because they had diseases that decreased their life expectancy. Wright Langam of the Los Alamos Laboratory Health described the selection of subjects [15]:

> As a rule, the subjects chosen were past forty-five years of age and suffering from chronic disorders such that survival for ten years was highly improbable.

The scientists believed the amount of plutonium injected was too small to cause considerable harm to the patients. However, two terminally ill patients were used to test higher doses of plutonium, with the idea that long-term effects would not occur during these patients' lifespan. Evidence of patient consent could only be found in one of sixteen patient records obtained by the ACHRE [15].

In 1997, the Department of Defense released the Report on Search for Human Radiation Experiment Records 1944 – 1994

Quaker Oats Radioactive Oatmeal Study of 1949

In 1949, Massachusetts Institute of Technology (MIT) teamed up with Quaker Oats for three studies utilizing radioactive substances [16]. The Atomic Energy Commission funded the experiments, which were completed at the Fernald State School, also known as The Massachusetts School for the Feeble Minded. Students at the school included mentally disabled children and children who had been abandoned by their parents. MIT recruited subjects by asking boys to join the Science Club and offering them gifts including trips to Boston Red Sox games, outings to the beach, Mickey Mouse watches, and free breakfasts. Approximately 74 boys, ages 10-17, *joined the club* [16].

The three experiments performed on the boys included eating breakfast consisting of Quaker Oats coated with radioactive iron tracers, drinking milk laced with radioactive calcium tracers, and direct injections of radioactive calcium. The first two experiments showed that oatmeal was no worse than farina (ingredient in competing Cream of Wheat) at inhibiting absorption of iron and calcium into the bloodstream. The third experiment showed that calcium in the blood goes to the bones [16].

The boys never knew they were part of an experiment until declassification of several Atomic Energy Commission documents in 1993. Fred Boyce said, “We didn’t know anything at the time. We just thought we were special.” [16] Charlie Dyer described his experience with the club [17]:

> They picked some of the oldest guys and asked us if we wanted to be in this club. We all got together and decided, why not? We'll get time off the grounds. It was like Christmas. Red Sox games, parties. I got a Mickey Mouse watch that I still have.

Charlie Dyer also described how he felt after finding out he was a guinea pig [17]:

> We thought something wasn't right, but we didn't know. They were using the kids who they were supposed to be helping.

In 1994, the Senate’s Committee on Labor and Human Resources held a hearing on the experiment. Senator Edward Kennedy, committee chair, addressed the committee [16]:

> Aren’t you appalled at the fact that the most vulnerable people in our society, which are young people, 7, 8 years old, that are in an institution, aren’t you appalled that they were the ones selected?

At the hearing, MIT claimed the level of radiation exposure was approximately equivalent to that of thirty consecutive chest x-rays and would have minimal if any effect on the children.

Thirty of the children used in the experiment later sued MIT and Quaker. In 1998, the case was settled out of court for 1.85 million dollars. President Clinton apologized to the former students in 1995 [16].

Guatemala Sexually Transmitted Disease Study, 1946-1948

Sexually transmitted diseases (STDs) have long plagued soldiers at war. In 1942, the Public Health Service, U.S. Army, and U.S. Navy endorsed a proposal by Dr. Carpenter to perform studies on the transmission of STDs and potential treatments in human subjects [18,19]:

> Because of the great prevalence of gonorrhea and its importance in the production of noneffective [sic] man-days both in the armed forces and civilian population, I believe that the human inoculation experiments proposed by Doctor Carpenter are justifiable if the human subjects are selected on a voluntary basis.

Following numerous legal debates, experiments on transmission of STDs started in September 1943 utilizing inmates in Terra Haute, Indiana prison, under the direction of Dr. Maloney, employed by the Public Health Service, and his assistant Dr. Culter. Subjects had to be 21 years old and receive detailed information of the experiment risks and benefits. They would be compensated $100.

The first subjects were used to determine a consistent technique to induce gonorrhea to enable future studies of prophylactic methods. Utilizing 243 test subjects and gonorrhea samples obtained from prostitutes and applied to the subjects' penises, experimenters concluded that gonorrhea could not be consistently induced in test subjects. Dr. Mahoney proposed termination of the experiments, but the NRC recommended continuing as the opportunity would unlikely occur again. Dr. Mahoney refused to continue the research, but Dr. Cutler would continue the study in Guatemala.

Dr. Mahoney had shown that penicillin quickly cleared syphilis and the U.S. Army adopted penicillin as the normal treatment for syphilis in 1944. However, Cutler wanted to know if syphilis could be prevented, and Dr. Juan Funes believed his home country of Guatemala would be a good place to find out.

U.S. and Guatemalan governments sponsored the research led by Cutler. More than 5000 subjects included U.S. soldiers, Guatemalan soldiers, prisoners, orphans, and mental asylum patients. Unlike the Terra Haute prison experiments, in Guatemala, researchers did not inform subjects of the experiment. A total of 1300 subjects, mostly Guatemalan soldiers, were exposed to syphilis through sexual intercourse with infected prostitutes or administration of bacteria obtained by scrapings of infected individuals. Approximately half of those infected received some form of treatment. Few children were infected with syphilis, but many were subjected to lumbar punctures. Following is the account of one experimental subject [18]:

> Berta was a female patient in the psychiatric hospital. Her age and the illness that brought her to the hospital are unknown. In February 1948, Berta was injected in her left arm with syphilis. A month later, she developed scabies (an itchy skin infection caused by a mite). Several weeks later,

> [lead investigator Dr. John] Cutler noted that she had also developed red bumps where he had injected her arm, lesions on her arms and legs, and her skin was beginning to waste away from her body. Berta was not treated for syphilis until three months after her injection. Soon after, on August 23, Dr. Cutler wrote that Berta appeared as if she was going to die, but he did not specify why. That same day he put gonorrheal pus from another male subject into both of Berta's eyes, as well as in her urethra and rectum. He also re-infected her with syphilis. Several days later, Berta's eyes were filled with pus from the gonorrhea, and she was bleeding from her urethra. On August 27, Berta died.

During the experiments, 83 subjects died, but Cutler attributed the deaths to tuberculosis, not the experiment.

The U.S. Presidential Commission for Bioethical Issues under President Barack Obama found the Guatemalan studies morally wrong and President Obama apologized in 1910 to then Guatemalan President Alvaro Colum [18,19]. However, no compensation nor treatment was provided to the subjects.

Studies of Dioxin and Agent Orange

The city of Philadelphia built Holmesburg Prison in 1896 and it remained in use until 1995. Originally, Holmesburg was a municipal prison designed for housing inmates short-term. The prison was built in a spoke and wheel style to provide maximum security, but the design was not successful. In the earlier years, criminal activity thrived at Holmesburg, with acts of rioting, rape, torture, corruption, and murder common. But other atrocities also happened.

Between 1951 and 1974, Holmesburg Prison gained notoriety for human experimentation on prisoners. Prison officials and physicians collaborated with the University of Pennsylvania, pharmaceutical companies, and the U.S. army to set up research experiments in Holmesburg [20]. Most of the experiments were performed on black men. Participants were paid, and they could participate in more than one experiment.

Dr. Kligman, dermatologist on faculty at the University of Pennsylvania, first thought of using prisoners for experiments when he went to Holmesburg Prison to treat an outbreak of athlete's foot in 1951. His various experiments continued until 1974. During an interview in 1966, he stated, "All I saw before me were acres of skin. It was like a farmer seeing a fertile field for the first time." [21]

Kligman received funds to test the effects of numerous chemicals on skin. In 1966, he received $10,000 from DOW chemicals to test the chemical dioxin, a component of Agent Orange [22]. The experiment tested for the minimum amount of dioxin that would elicit a skin reaction. Kligman applied small doses (0.2 to 16 micrograms) to the forehead or back of subjects, and found no notable effects. Thus, he increased the dose 486 times to 7,500 micrograms—not a good experimental technique in dose response tests. Seventy men were tested and left with permanent painful skin lesions. DOW chemicals had not approved this higher dose. Further, Kligman kept poor records and follow-up with the men tested was impossible [22].

Kligman took advantage of a vulnerable population during times of research uncertainty. He states, [22]

> It was years before the authorities knew that I was conducting various studies on [incarcerated] volunteers. Things were simpler then. Informed consent was unheard of. No one asked me what I was doing. It was a wonderful time.

In other experiments, Pennsylvania State and the U.S. Army collaborated with Drs. Kligman and Copelan to test the minimum effective dose (MED) to disable the mental ability of 50% of a population. Kligman and Copelan used 320 prisoners and paid them $12 for a medical screening and $25 for each medical test in which they participated. Chemicals tested included dioxin, other chemicals in Agent Orange, LSD, and other psychoactive drugs. Kligman also added in some chemicals he hypothesized could be used in dermatology. Kligman claimed the experiments had no harmful effects. However, U.S. Army clinical research expert Lt. Col. M.G. Bottiglieri said this about Kligman's report: [22]

> pure gibberish … absolutely useless … an attempt to provide a facade of competence and ability.

Once again, Kligman's records were incomplete and no follow-up with subjects was possible [22].

Pennsylvania State University denied harm to any subject, but offered free medical treatment to them. Few accepted. However, several former test subjects testified before the Pennsylvania State Legislation.

Joseph Smith participated in several experiments at Holmesburg prison and describes his experience [23]:

> During my stay at Holmesburg I participated in 5 different experiments. Another inmate in the prison told me about these experiments. I'd give the pharmacy guys – a group of inmates who

worked in the pharmacy – packs of cigarettes so they would sign me up on a list to do experiments.

One experiment I participated in was the patch test. I did this experiment three different times – once on my arm, once on my back, and once on my leg. The only thing that they told me before the test was that they were going to put some gauze on it with my initials. They said that that way they would know if I didn't keep the gauze on me, and then they wouldn't pay me for the experiment. Then, they applied patches to my skin that resulted in a lot of pain. At night I would lie in my bed with my arm in the air just shaking it so that it would ease the pain. Some of the guys would take the bandages off and put them on the wall – and then it would eat the paint on the wall.

Another experiment I participated in was one in which I took injections once every three hours for a period of three months. The people administrating the test never told me anything about what was going on. I didn't know what substance they were injecting or what they were trying to find out. Usually they injected me in the buttocks because that was the easiest, but they also injected me in other places as well. These injections made my buttocks swell up like balloons, which was very painful. Sometimes, though, because of all the injections my buttocks would be numb. I kept going back for shots because I received approximately two dollars for each shot.

A third experiment I participated in was the Army LSD test. For this experiment, I was take out into the Army trailers. I wasn't told that the substance they gave us was LSD, but I assumed that that was what it was because of the effects it had on me. …

Another experiment in which I was a subject was one that I call the "sight" test. Four dots were put on my shoulders and then my eyesight would go almost completely blind. Just as this happened they would give me a shot and it would bring back my eyesight.

The last experiment I recall participating in was a toothpaste test. Before the experiment they told us that they were trying to figure out which toothpaste worked the best. I had to brush my teeth with the toothpaste they gave me. The toothpaste tasted like mouthwash and I had to dilute it a little so it wouldn't tingle in my mouth quite as badly.

During all of these experiments I never saw a doctor. Inmates who wore nurses' jackets were the ones who administered the tests. … I signed a consent form before each experiment, but I never understood what it meant. … At the time I just felt lucky to be doing the experiment because it meant I could earn some money. …

After I got out of prison my mother finally took me to Embreeville State Hospital. At the hospital, [Name], a psychologist, examined me. … The doctor said that I had mild schizophrenic schizoid characteristics …

> I have been out of Holmesburg Prison for over 30 years. I think about these experiments and what they did to me all the time.

On January 11, 2023, The College of Physicians of Philadelphia issued a public apology [24]:

> From approximately 1951-1974, a significant number of inmates at the former Holmesburg Prison were test subjects in experiments that inflicted great and lasting pain upon them, pain that was both physical and psychological. These experiments were led in large part by the late Dr. Albert Kligman, a Fellow of the College of Physicians and a recipient of the College's 2003 Distinguished Achievement Award.
>
> The College of Physicians offers its deepest sympathies for those who suffered, including their families, and it apologizes for its silence in not expressing these sentiments sooner. The College has rescinded Dr. Kligman's Distinguished Achievement Award.
>
> Though this apology is long overdue, it is no less heartfelt for the delay. As former President Bill Clinton said in his apology for the U.S. government's role in the infamous Tuskegee experiments, the College recognizes that "[w]e can end the silence. We can stop turning our heads away. We can look you in the eye and finally say (we are) sorry...and (we are) sorry that this apology has been so long in coming."
>
> We also can commit the College to the work that our members do best; that is, to the task of healing.

> The College pledges to work with the Philadelphia Inmate Justice Coalition for the purpose of leading a broader public discussion about these issues to raise public awareness about the Holmesburg Experiments to ensure that no such practices recur in the future. …

If only we can trust that physicians and scientists have learned their lesson.

Chapter 14

Human Experimentation During World War II: Nazi Germany

Prior to World War I, Germany excelled in science and medicine. From the late 19th century until the end of World War II, German was the prominent language of western science. From 1901-1915, Germans received 4 Nobel Prizes in Physiology or Medicine, 6 in Chemistry, and 5 in Physics; more in science than any other country during the same time frame.

In the 19th century, Germany developed the model of universities serving as both educational and research institutions based on the concept that research provides the highest level of education. Other countries now use the same model. During this same timeframe, German education shifted toward more sciences and applied the scientific method to medical research. Germany also led the world in pharmaceutical developments.

Germany's leadership in science lasted through World War II, but in medicine, Germany fell to disgrace. We learned in Chapter 9 how eugenics came to Germany, and how German physicians contributed to genocide in the campaign to eliminate "*lebensunwertes Leben*," (life unworthy of life). But German physicians and scientists did much more than that. On May 12, 2012, the German Medical Association apologized for the atrocities physicians carried out under the Third Reich.

German Physicians Turn Socialist

The Germany National Medical Association was founded in 1873, with many Jewish doctors as members and leaders. However, the Weimar Republic decreased the monetary benefits of German physicians when, in 1883, Otto von Bismarck developed a socialized medical system. In 1888, Germany accepted its first women into medical school, and by 1933, approximately 6% (4,367) of physicians were female [1]. At that time, Germany also had approximately 6-7,000 Jewish doctors. With only 565,000 Jewish people in Germany in 1933, a little over 1 percent were physicians.

World War I further opened the door for female doctors, as male doctors served in the military. However, the doors for Jewish doctors were rapidly closing as antisemitism in Germany already existed and was expanding rapidly.

The Nazi party (National Socialist Party), founded in 1920, promoted both antisemitism and anti-feminism. They founded the National Socialist German Doctor's League in 1929. Prior to 1933, 3,000 German physicians (6%) joined the Nazi party. By 1942, more than half of German Physicians joined both the Nazi party and the National Socialist German Doctor's League, ultimately playing a critical role in Hitler's "Final Solution".

On January 29, 1933, Germany elected Adolf Hitler as Chancellor. Just two months later, the German Parliament passed the Law for Removing Distress for the People and the Reich, which permitted Hitler to make laws without support from the German governing boards; in essence, giving Hitler absolute authority in Germany. In April, 1933, Adolf Hitler enacted a law banning Jewish doctors. He followed with a ban on female doctors in June of that same year, although women doctors could work in midwifery. The National Socialist

German Doctor's League broke with the Hippocratic Oath and put in its place new medical ethics [2]:

> From the first day, we have made it clear that the major turnabout in the world view of our days, an essential portion of which is vanquishing the individual through experiencing the "people" [*Volk*] must be the guiding principle of the morality and ethics of the medical profession.

April 1, 1933 was boycott day, when Germans boycotted Jewish businesses. Nazi physicians joined in this boycott violently in Berlin when members of the National Socialist German Doctor's League broke into apartments of Jewish physicians and dragged the physicians outside where they beat them and put them on display at an exhibition area near the Berlin Lohrter Station. They continued to beat the Jewish physicians with sticks and to shoot at them. They kept the Jewish physicians for several days without food, and then turned them over to the Nazi SA (Sturmabteilung, or storm troopers) to be imprisoned.

The president of the Reich Health Office expressed the need for doctors to focus on the nation (at least the Aryans) over the individual [2]:

> The physician fights as a biological soldier for the health of the *Volk*. If the doctor fails in this struggle, then no form of government will be able to secure the future of this *Volk*.

German medical schools followed suit. Dr. Franz Hamburger coauthored a medical textbook. In a 1940 edition of the text, he added [2]:

> At all times should you be aware of the duties of the National Socialist physician, who keeps in mind not only the individual person, but the entire

> *Volkskorper* [people's body], in which the single person like the cell in the human organism is just a building block, just a cell of the people as a whole.

German Medicine Under Hitler

The removal of Jewish and female physicians from medical practice left Germany with inadequate health care. But that did not keep physicians from extending their responsibilities and support Hitler and the German military.

Hitler promoted health and physical fitness of all Aryan Germans, but wanted even more from the army – he sought to build an army of super humans. In 1938, Temmler Pharmaceuticals released its new drug, Pervitin, to increase alertness and energy, and to battle depression. Germany made Pervitin, an amphetamine, readily available to its people without the need of a prescription. Military physician Otto Ranke tested the ability of Pervitin to help 90 college students prepare for exams. He concluded from these studies that the drug would be useful for the German armed forces (Wehrmacht). In April 1940, 35 million Pervitin and Isophan (another type of stimulant) pills were sent to the front lines to assist soldiers invading France. Some claim that Pervitin was critical for the success of Germany's blitzkrieg warfare tactics, as the soldiers would often fight for days without sleep. When the Allies learned of Germany's use of Pervitin, they too turned to an amphetamine, albeit the much milder Benzedrine. Both Pervitin and Benzedrine kept soldiers awake as desired, but they also had serious side effects, most notably, addiction.

Pharmaceutical companies contributed to Nazi Germany's war in another, much more atrocious way. Tesch and Stabenow of Hamburg, and Degesh of Dachau produced a product called Zyklon-B, a poison gas that produced cyanide when exposed to air. Zyklon-B use in concentration camps is described below. Degesch was owned in part by I.G. Farben, which hired many concentration camp prisoners to work at their factories. The directors of these three companies were tried during the Nuremberg war trials. Bruno Tesch and the executive manager of Tesch Stabenow Karl Weinbacher were found guilty and sentenced to death. Both were hanged on May 16, 1946. Dr. Gerhard Peters, director of Degesch, was found guilty of accessory to homicide and was sentenced to five years in prison, but later acquitted. Thirteen members of I.G. Farben were found guilty of war crimes and sentenced to one to eight years in prison.

Hitler's personal physician, Theodor Morell, medicated Hitler with vitamins, herbs, extracts from animals, and various drugs to assist Hitler throughout World War II. Ohler suggests in his book, **Blitzed**, that Dr. Morell regularly injected Hitler with Pervitin, and during the latter years of the war, Eukadol (a very strong opioid) and cocaine [3]. Dr. Morell also provided "medications" to many of Hitler's advisors [3]. While Dr. Morell may have been a bad physician and contributed to Hitler's manic moments, other German physicians did much worse in the concentration camps.

Concentration Camps and the Nuremburg Laws

Hitler had the first concentration camp built in Dachau in 1933, shortly after he took power. Before Hitler killed himself in 1945, more than 44,000 camps were built in Germany and it's conquered territories. The United States Holocaust Museum estimates that Germans executed over fourteen million people during this time, six million of whom were Jewish [4].

Dachau was initially built to hold German political prisoners. Soon after, the camp housed gypsies, Jehovah Witnesses, and homosexuals [5]. Not until 1935, after the Nazi congress established the Nuremberg Laws, were Jews imprisoned in concentration camps.

In 1935, the Nazi party held its congress in Nuremberg where Hitler and other Nazi leaders formalized laws allowing the persecution of Jewish people, known as The Nuremberg Laws [6]. There were three main sections of the Nuremberg Laws: (1) Protection of German Blood and Honor, (2) Protection of Hereditary Health, and (3) The Reich Citizenship Law [7]. The first law forbade marriage or sexual relations between Jewish people and German citizens, and also denied Jewish people the right to display the swastika or other Nazi symbols. The second law formalized the eugenics and euthanasia programs already occurring in Germany. The third law banned Jewish people from being German citizens.

In response to the Nuremberg Laws, the Jewish community created their own leadership, the National Representation of the Jews in Germany [7]. The leaders established several principles for the Jewish people including providing education to the young; planning for increased emigration from Germany; taking care of the sick, needy, and elderly; trying to protect the means of income; and eventually form a Jewish Palestine. Recognizing the difficulty of carrying out these tasks, the leaders wrote [7]:

> In full awareness of the magnitude of the responsibilities involved and the difficulties of the task, the Reichsvertretung (National Representation) calls on the Jewish men and women, and on all Jewish youth, to join together in unity, to maintain high Jewish morale, to practice strictest self-discipline, and show a maximum willingness to make sacrifices.

According to the Virtual Jewish Library, the majority of Germans opposed the Nuremberg Laws as they might protect the Jews from persecution. The opinion in Berlin was described as follows [7]:

> The new laws — the citizenship law and the law for the protection of German blood and honor — that were passed by the Reichstag during the Party Rally of Freedom — have finally, after many years of struggle between Germandom and Jewry, created clear cut relations. Jewry is defined as a national minority and will be given the opportunity to create their own independent cultural and national life under the protection of the state. Any intervention in the life of the German nation will from now on be impossible and forbidden. The laws were received with satisfaction and enthusiasm.

History tells us that was not the case. Antisemitism grew in Germany. A tipping point for the persecution of Jews occurred on Krystallnacht (Crystal Night), November 9 and 10, 1938, when Nazi mobs destroyed synagogues and Jewish homes and businesses, as the Nazi party ordered the police and fire departments to stand down. Nearly 100 Jews died and thousands more were arrested and taken to concentration camps. Jewish emigration from Germany increased following Hitler's rise to power, leaving only about 200,000 Jews in Germany by September, 1939. But soon, persecution of Jews expanded beyond Germany.

After World War I, the Treaty of Versailles gave Poland the previously German provinces of West Prussia, Poznan, and Upper Silesia. It also forbade Germany from developing an army. In 1934, Hitler signed a nonaggression pact with Poland. In 1935, Germany defied the Treaty of Versailles by expanding its military capabilities. Germany then began expanding its territories by annexing Austria in 1938 and parts of Czechoslovakia in March 1939. Hitler next signed a nonaggression pact with the Soviet Union in August 1939.

World War II started with the Germany invasion of Poland on September 1, 1939. Based on a treaty of protection with Poland, England and France declared war on Germany on September 3. On September 17, the Soviet Union invaded Poland from the East, leaving Poland in desperation. Poland surrendered on October 6, with German-occupied Poland separated from Soviet Union-occupied Poland by the Bug River.

With the occupation of Poland, Nazi Germany controlled three million more Jews (15 times the number before the occupation). On October 8, 1939, the first ghetto was established in Piotrkow, Poland. Several hundred ghettos would be established over the next few years in Poland to concentrate the Jews in limited areas barring access to other people. The SS (Schutzstaffel, or protection squadron) established and controlled the ghettos from the outside, appointing a Jewish governing body (Judenrate) and police force to control the people inside under strict SS guidelines. As Germany continued its March east into other countries, including the Soviet Union, even more Jews came under their control. The number of Jews overwhelmed the ghettos, leading to the expansion of concentration camps and development of six extermination camps.

Hitler assigned Heinrich Himmler responsibility for the concentration camps, and Himmler appointed Theodor Eicke as the head administrator. As Eicke served as commander at the Dachau camp, it served as a model for others, including the guide for guards. Eicke named the guard units the Totenkopfverbände (Death Head Units), and they were trained to be harsh and cruel to the prisoners whom they considered sub-human. Thus, the prisoners were underfed, overcrowded, beaten, and tortured. Many concentration camps were work camps, where prisoners performed forced labor.

Extermination camps became part of the Final Solution to the Jewish Question, the elimination of all Jews, gypsies and other undesirable persons. Six extermination camps were established in Poland, and Jews were transported to them by train as Germany conquered more territory. The most notorious of the extermination camps was Auschwitz-Birkenau.

Physicians, both German SS and prisoners, played critical roles in concentration camps. At Auschwitz-birkenau the SS medical service, consisting of SS physicians, provided medical care to the SS workers at the camp and chose those prisoners who were to be exterminated. Some concentration camp physicians also performed experiments on the prisoners.

Choosing those to be exterminated consisted of two components: making selections as soon as prisoners exited the train and making selections inside the camp when it suited German leadership to do so. In Auschwitz, the selection procedure went as follows. As prisoners exited the train onto the platform, guards ordered them into two columns, men and boys 14 and older in one column and women and children in the other. The SS physician making selections that day would order individuals from a column to go either left or right. Those who went to the left would enter the concentration camp and be forced to work for the Germans. Those who went to the right would enter the "showers."

To keep prisoners calm, those selected to be exterminated were lead to a large facility with a sign indicating they were showers. The men, women and children had to remove all clothing and enter the shower region together. But the room was actually a gas chamber. A Nazi officer would drop Zyklon-B pellets through an opening in the roof, and the pellets would immediately turn into hydrocyanic acid, a deadly gas.

Zyklon-B was developed for exterminating rats and insects. However, the Nazis sought a better means of genocide than shooting people (which used too much ammunition) or gassing with carbon monoxide from the exhaust of trucks (it took too long). Thus, they tested the pesticide Zyklon-B, for its ability to rapidly and "cleanly" kill people. The first test took place in Auschwitz on September 3, 1941. Camp commandant at the time, Rudolf Hoess, had 600 Soviet prisoners of war and 250 Polish Jews who were in the Auschwitz hospital taken to one of the gas chambers. Hoess described the experiment as follows [8,9]:

> The first case of killing people with gas did not register much in my mind… I remember much better the killing of 900 Russians… in the old crematorium… the Russians were told to get undressed in the corridor, and then they completely quietly entered the morgue, because they were told that they had to go through disinfection. The morgue had just enough room for that many people. As soon as they all entered, the door was locked and through the holes in the ceiling the gas was poured in… Only several hours later the room was unlocked and ventilated. That was the first time I saw so many dead bodies killed by the gas… I felt ill at ease – I was horror-stricken… At that moment I did not reflect on the killing of the Soviet prisoners of war. I was given an order and I was obliged to carry it out.

After prisoners died in the gas chambers, Jewish concentration camp workers, called the Sonnderkommando, would remove the bodies and take them to the incinerators to be destroyed.

However, should a body have gold fillings, a prisoner dentist would remove the fillings for the SS officers (although sometimes SS guards would get to the gold first and pocket it).

Those prisoners who went to the left on the train platform might have escaped death, but for most, death was just delayed. The prisoners of concentration camps became forced laborers, sometimes working in factories such as I.G. Farber, and often simply given hard labor tasks to do for no particular reason (such as moving large rocks around camp). The prisoners had to do this on a diet of mostly bread and water, and little of that.

The SS doctors, meanwhile, had the responsibility of inspecting the camp hospitals and barracks. Often, they would perform selections of those unfit to work. This usually included all prisoners in the camp hospital, as well as prisoners in any barracks where an illness broke out. These selected prisoners faced the same fate as those who initially went to the right – death by Zyklon-B. Sometimes, a whole section of the camp (thousands of men, women, and children) would be chosen for extermination just to thin out the population.

Many SS doctors followed their orders and made their selections with much disdain and a large psychological toll. In his testimony at the Auschwitz trials in 1959 – 1963, Dr. Lucas stated the following [10]:

> Your honor! On the day on which against my will I was assigned to duty in the National Socialist concentration camp began a chapter in my life from which I will never recover. Forced to work on the ramp I naturally sought to save the lives of as many Jewish prisoners as possible. But today as then, I am torn by the question: "And what about the others?" I have fought my way toward the truth. I can only hope that you will arrive at a

> verdict that will help me to free myself from my bonds and begin a new life.

Several former Auschwitz prisoners testified of Dr. Lucas' compassion at the Auschwitz trials. Dr. Lucas was found guilty of joint murder.

In contrast, some physicians relished in their power over life and death. Dr. Herta Oberheuser at the Ravensbrück concentration camp writes in an affidavit of some additional actions of physicians at Ravensbrück [11].

> During my service at the concentration camp Ravensbrück I observed, that one of the physicians serving there, Dr. SONNTAG, severely mistreated prisoners who reported sick, by beating and kicking them.
>
> It was no rarity at Ravensbrück, that persons who were already approaching death, were killed by injections. I myself have given 5 or 6 such injections.

Dr. Oberheuser failed to mention that the injections she gave consisted of petrolin, a substance made from petroleum with properties similar to paraffin. The effects of petrolin injections were described by Dr. Rolf Rosenthal, also of Ravensbrück [12].

> Sometimes I have seen that Dr. OBERHAUSER administered petrolin-injections to the prisoners. She used a syringe for 10 cubic centimetres and administered in to the vein of the arm. The injection gave the impression of an acute apoplexy, the patients reared up and then suddenly collapsed. It lasted between 3 to 5 minutes from the time the injection was administered until they

> were dead. The patients were entirely conscious until the last moment.

Dr. Rosenthal mentions that he used morphine when patients were near death.

Nazi physicians (and some non-Nazi party German physicians), with the support of the Nazi regime, utilized concentration camp prisoners in experiments often designed to support military medicine, but sometimes to simply satisfy curiosity. Either way, the physicians violated the basic humanitarian principles of the Hippocratic oath, taken by those German physicians who earned their medical degree prior to 1933.

Almost 28,000 prisoners were used in experiments, with more than 4,300 dying from the experiment or being killed after the experiment for autopsy [13]. Heinrich Himmler described the need for some of the human experiments in a letter he wrote to Erhard Milch in 1942 [14]:

> These researches which deal with the behavior of the human organism at great heights as well as with manifestations caused by prolonged cooling of the human body in cold water, and similar problems which are of vital importance to the air force in particular, can be performed by us with particular efficiency because I personally assumed the responsibility for supplying asocial individuals and criminals who deserve only to die from concentration camps for these experiments.

High Altitude Experiments

From March to August, 1942, Dr. Sigmund Rascher performed experiments on Jewish prisoners in the Dachau concentration camp. The experimental purpose was to determine how high altitudes alter body function and cause death, and to better understand the altitude sickness many air force pilots face. Prisoners suited up in German Luftwaffe (Airforce) gear were placed in low pressure chambers that mimicked the pressures at up to 68,000 feet altitude (commercial airlines today typically fly at 30,000 – 35,000 feet). The prisoners hung from parachute-like equipment throughout the experiment, as scientists monitored physiological changes. Eighty of the 200 experimental prisoners died during the experiment, and the others were executed after the experiment for autopsies. In some cases, autopsies were initiated before the prisoner was dead.

Dr. Rascher requested permission from Heinrich Himmler to perform these experiments. In his request, he acknowledges that subjects may die. Following is an except from Rascher's letter of request: [14]

> For the time being I have been assigned to the Luftgaukommando VII, Munich, for a medical course. During this course, where researches on high-altitude flights play a prominent part (determined by the somewhat higher ceiling of the English fighter planes), considerable regret was expressed at the fact that no tests with human material had yet been possible for us, as such experiments are very dangerous and nobody volunteers for them. I put, therefore, the serious question: can you make available two or three professional criminals for those experiments? The experiments are made at Fornanont Luftwaffe Testing Station for Altitude Research, Munich. The experiments, from which the subject can, of course, die, would take place with my cooperation. They are essential for researches on high-altitude flight and cannot be carried out, as

> has been tried, with monkeys, who offer entirely different test-conditions.

In a secret report detailing the outcomes of the experiment, Dr. Milch, who worked with Rascher, shows little concern for the fate of his VP's (Versuchspersonen – human experimental subjects) [15]:

> Experiments which I carried out in this connection proved that with pure oxygen no lowering of the measurable raw energy (ergometer) was noticeable up to 13.3 kilometers. The VP's merely became unwilling since pains of the body cavities grew too severe, due to the lowering of pressure between body and thin air. When pure oxygen was inhaled bends occurred in all 25 cases only at heights above 14.2 kilometers…
>
> For the following experiments Jewish professional criminals who had committed "Rassenschande" (race pollution) were used: The question of the formation of embolism was investigated in 10 cases. Some of the VP's died during a continued high altitude experiment…

Autopsy procedures and results were then described. Dr. Milch continued describing what happened to other experimental subjects [15]:

> To find out whether the severe psychical and physical effects, as mentioned under No. 3, are due to the formation of embolism, the following was done: After relative recuperation from such a parachute descending test had taken place, however before regaining of consciousness, some VP's were kept under water until they died…

Freezing Experiments

Dr. Rascher also performed experiments on Dachau prisoners to determine the best way to treat pilots who were shot down and ejected into cold waters of the ocean. Some experimental prisoners were dressed in flight suits while others were left naked as they were exposed to cold for several hours either in the form of ice water tubs or placed outside in freezing temperatures. The experimenters recorded vital sign changes and other physiological responses until body temperature dropped to 79.7°F. At this point, the experiments compared different means of increasing body temperature, including exposure to hot water, covering the body in a sleeping bag, and putting the naked body next to that of a naked female prisoner. In some cases, the experimental prisoner was even forced to have intercourse with a female prisoner.

Some prisoners died before the warming up process began. In fact, death seemed to be the end point of one experiment described by Wolfram Sievers, administrator for the Reich Research Council, in an affidavit prepared for a post-war trial of the doctors involved in human experiments. Sievers wrote the following [16]:

> I witnessed a freezing experiment performed on an inmate of the concentration camp. He was a criminal who had been sentenced to death. Professor Dr. Hirth and Dr. Rascher were present at the Experiment. The experiment itself was conducted by Dr. Rascher. Thermometers were put on the prisoner and he had to don a flyers suit. He was given an Evipan injection, whereupon he became unconscious and was immersed into ice-cooled water. In the course of this experiment the

prisoner died and immediately after his death an autopsy was performed by Prof. Dr. Hirth.

Gas Gangrene and Sulfanilamide Experiments

Dr. Karl Gebhardt, chair of the department of orthopedic surgery at the University of Berlin, oversaw experiments on the effects of treating wounds with the antibiotic sulfanilamide, from July 1942 to August 1943 at Ravensbrück concentration camp. Dr. Herta Oberheuser, a dermatologist, assisted. Soldiers' wounds often became gangrenous due to exposure to uncleanly conditions. Previously, German, American, and British scientists showed that sulfanilamide was not effective to treat gas gangrene. Dr. Gebhardt hypothesized that surgical intervention as opposed to sulfanilamide would be an effective treatment. According to the New York Post, Hitler's close friend, Reinhard Heydrich, died of gas gangrene and Hitler was angry Heydrich was never treated with sulfanilamide [17].

To mimic battlefield wounds, doctors cut the prisoners' legs and rubbed dirt, cloth, glass, wood, and metal shavings into the wound, as well as spreading various bacteria in the wound. Blood flow to the area of incision was blocked by tying off blood vessels. The ensuing infection was treated with sulfanilamide. Dr. Oberheuser described the experiment in an affidavit prepared for the Nuremberg trial [18]:

> The operations were performed by Dr. Fritz FISCHER. Sometimes I helped and assisted at these operations, and had the duty, to attend the patients after the operation. Different series of experiments were conducted; in some, glass- or wood splinters and cloth fibers were forcibly brought into the wound, in order to cause

> gangrene. For this purpose, a cut on the calf of an approximate length of 10 centimeters was usually made.

Initially, the doctors experimented on 20 male prisoners at Sachsenhausen concentration camp. However, the male prisoners resisted due to the extreme pain associated with the experiments. Thus, the doctors turned to female prisoners at Ravensbrück [17]. Many of the female prisoners had been operated on several times, up to eight. Doctors operated on Barbara Pietrzyk five times in 1942 resulting in paralysis of her left leg [19].

Martha Hall Kelly interviewed some survivors from the sulfanilamide experiments for her novel, **Lilac Girls**, and she describes the procedure based on those interviews [17]:

> They would take groups of 10 women, keep them for a while and then use a different group. Some of them died during the experiments, and several were executed right after. Some hadn't healed yet, and had to be carried to the shooting wall.

After 60 prisoners had wounds inflicted on them, SS Chief Medical Officer Dr. Ernst-Robert Grawitz went to Ravensbrück to observe the experiments and decided that the wounds did not appropriately simulate gunshot wounds and thus needed revising [17].

> The normal result of all bullet wounds was a shattering of tissue, which did not exist in the initial experiments. As a result of the injury, the normal flow of blood through the muscle is cut off. The muscle is nourished by the flow of blood from either end. When this circulation is interrupted, the affected area becomes a fertile field for the growth

> of bacteria; the normal reaction of the tissue against the bacteria is not possible without circulation.

To better simulate bullet wounds, doctors experimented on 24 more prisoners, tying off the blood supply to a muscle causing necrosis. Bacteria were then administered to the area to cause an infection. The severity of the infections caused 5 prisoners to die – the surviving prisoners were given sulfanilamide treatments.

In all sulfanilamide experiments, the infections were painful and left surviving prisoners with permanent damage. One of the unfortunate subjects, Broel-Plater, testified against Dr. Oberheuser and stated the following [20]:

> My leg pained me; I felt severe pain, and blood flowed from my leg. At night we were all alone without any care. I heard only the screaming of my fellow prisoners, and I heard also that they asked for water. There was nobody to give us any water or bed pans.

Another witness, Karolewska, provided even more detail in her testimony [21]:

> I was in my room and I made the remark to fellow prisoners that we were operated on in very bad conditions and left here in this room, and that we were not given even the possibility of recovery. This remark must have been heard by a German nurse who was sitting in the corridor because the door of our room leading to the corridor was open. The German nurse entered the room and told us to get up and dress. We answered that we could not follow her order because we had great pains in our legs and could not walk. Then the German nurse

> came with Dr. Oberheuser into our room. Dr. Oberheuser told us to dress and go to the operating room. We put on our dresses; and, being unable to walk, we had to hop on one leg going into the operating room. After one hop we had to rest. Dr. Oberheuser did not allow anybody to help us. When we arrived at the operating room quite exhausted, Dr. Oberheuser appeared and told us to go back because a change of dressing would not take place that day. I could not walk, but somebody, a prisoner whose name I do not remember helped me to come back to the room.

Because of their need to hop, the prisoners in these experiments referred to themselves as rabbits.

The doctors conducting the experiments found the second technique to induce gunshot-like wounds successful. Thus, the experiments continued, comparing various therapies for their effectiveness in fighting the infection. Therapies included locally applied sulfanilamide powder, local sulfanilamide powder plus sulfanilamide tablets, and the last group received the local sulfanilamide powder and tablets plus an injection. There was also a control group that received no treatment. None of the treatments were effective [17].

In a similar experiment, SS physicians injected purulence into female prisoners' legs. A few days later, the women were given sulfonamide injections. All subjects developed serious infections with abscesses that required draining. The sulfonamide was again ineffective [17].

The experiments were repeated using surgical debridement of the infected tissues. The debridement was not effective [17].

Thirteen subjects died due to the gas gangrene induced by the experiment, and another six were executed to prevent them from testifying against the SS doctors. An inmate doctor who worked in the concentration camp hospital during the experiments, Dr. Zofia Maczka, testified about the experiments and also suggested that the women who died could have been saved by amputation [17].

Bone, Muscle, and Nerve Regeneration Experiments

SS physicians at Ravensbrück performed experiments to observe the process of bone, muscle, and nerve regeneration, as well as the possibility of transplanting bone or a limb from one person to another. Dr. Gebhardt and his adjunct Dr. Ludwig Stumpfegger headed these experiments.

In terms of muscle regeneration, physicians performed multiple surgeries on each female subject removing muscle tissue from the legs, one piece per surgery. No regeneration occurred.

Nerve regeneration experiments involved cutting parts of nerves of the legs to see whether the nerves regenerated. Some of the subjects of these experiments were previously used in the gas gangrene sulfonamide experiments described above [17]. Again, no regeneration occurred.

Bone and limb transplant experiments are self-explanatory. However, prisoner Dr. Zdenka Nedvedova-Nejedla describes who provided the transplant material and who received it [17]:

> High amputations were performed; for example, even whole arms with shoulder blade or legs with iliaca were amputated. These operations were performed mostly on insane women who were immediately killed after the operation by a quick

> injection of evipan. All specimens gained in operations were carefully wrapped up in sterile gauze and immediately transported to the SS hospital nearby, where they were to be used in the attempt to heal the injured limbs of wounded German soldiers. After operations, no one except SS nurses was admitted to the persons operated on, whole nights they lay without any assistance and it was not permitted to administer sedatives even against the most intensive postoperational pains. From the persons operated on, 11 died or were killed.

Poisoned Bullet Experiments

Some human experiments were not to help treat injured soldiers, but to help soldiers kill the enemy. Such is the case of the poisoned bullet experiments. Nazi doctors tested if bullets laced in poison were more effective in killing the enemy. A report prepared by Dr. Joachim Mrugowsky and sent to Dr. Widmann describes these experiments [22]:

> On 11 September 1944 in the presence of SS-Stbf. Dr. Ving, Herr Dr. Widmann and the undersigned experiments with "Akonitinitrate" projectiles were carried out on 5 persons already sentenced to death. These projectiles were filled with poison of crystalline shape, caliber 7.55 mm. While in the prone position, they were each shot at in the upper part of their thigh. On two persons' upper thighs the bullet went straight through. Even later no poisonous effects could be recognized. [As far as the experiment goes], these guinea pigs had to be disregarded. The 3 persons sentenced revealed

> respective symptoms, which were of amazing accord. At first, no peculiarities were evident. 20 to 25 minutes later a motor disturbance and a small flow of saliva set in. Both [of these symptoms] then again let up. After 40 to 44 minutes saliva flowed more strongly. The poisoned swallowed often, later the flow of saliva was so strong, that it could not be overcome by (mere) swallowing it. Foamy saliva flowed from the mouth. After that a sensation of strangulation and vomiting set in.
>
> After 58 minutes the pulse of two persons could not be recorded any more.

Sea Water Experiments

From July to September 1944, doctors at Dachau concentration camp experimented on prisoners in pursuit of making seawater potable. At the time, Germans experimented on two methods for use by the airforce (Luftwaffe) during emergencies: the I.G. Farber method and the Berke method.

In the Berke method, 16.5 grams of Berkatit are dissolved in 250 cc sea water.

Dr. Eppinger assigned Dr. Beigleboeck to carry out the experiments. Dr. Beigleboeck was hesitant, and communicated to a colleague his distaste in the human experiments at concentration camps. His colleague, Dr. Annemarie Speiss, wrote a letter to Gustav Steinbauer in defense of Dr. Beigleboeck, including the following description of her conversation [23]:

> ...On the occasion of this conversation, Prof. Dr. BEIGLBOECK condemned the principle of the performance of scientific experiments in concentration camps most strongly and at the

> same time reproached his chief Prof. Dr. Eppinger, for selecting him for this work.

Dr. Ignaz Bauer, prisoner at Dachau, assisted with the experiments. His affidavit read differently than those we saw from German physicians involved with human experimentation. In describing the purpose of the experiments, Dr. Bauer wrote [24]:

> The purpose of these experiments was:
>
> 1. To find out how long it takes until a healthy human specimen without water and food, what at most is given some seawater, will fall into agony.
> 2. To study thoroughly the symptoms of dying of hunger and of thirst.
> 3. To test the effects of a serum, the composition of which is unknown to me, on the experimental subjects who were already dying.

This varies considerably from developing a technique for making potable water from seawater. But even the German doctors stated that they already knew the Berke seawater was no more potable than plain seawater.

In terms of how the experiments were conducted, Bauer described the procedures as follows [24]:

> … Several series of experiments were carried out forcibly, each one on about 15 gypsies. These experiments were carried out in such a way that for 4 to 6 days those people did not get anything to drink or eat except some saltwater (seawater).

Dr. Bauer also described the results of the experiment [24]:

> Symptoms of death by hunger and thirst which were also perceptible to the layman were:
>
> Visible deterioration of the patients, nervousness, agitation which, in some cases mounted to the point of madness. (The raving people were tied to the beds.) Some of the patients showed signs of apathy and unconsciousness. In many instances, weakness of the heart could be observed.
>
> Subjectively the patients suffered from gnawing hunger and, above all, from terrible thirst, which, through the drinking of saltwater, became even more painful. The thirst was so terrible that some patients did not hesitate to drink dirty water used for washing the floor. I saw one of these poor devils falling on his knees, begging in vain for water. Doctor Beigelboeck was pitiless. One youth, who succeeded in getting some water to drink, was bound to his bed for punishment (whether by Beigelbock or another, I unfortunately do not know; in any case, Beigelboeck, as chief of this division, was responsible.

Dr. Bauer continued to describe their emotional state – how they feared death from the knowledge they had of other subjects' fate when the experiment was over. However, Dr. Bauer was not privy to the fate of the prisoners.

Typhus Experiments

Typhus devastated the German army and a vaccine was needed. Thus, the Germans used concentration camp prisoners at several camps as subjects for experimentation. At Buchenwald, prisoners received typhus bacterium, some following vaccination and others without vaccine. Political prisoner, Herbert Buhlick, described the procedure in an Affidavit [25]:

> Following the experiment with lice, I with nine other prisoners as mediums had a 2 ccm [cubic centimeter or milliliter] intravenous injection of typhus without vaccination. I was injected by then Kapo, Artur Dietzsch.
>
> After a lapse of three months I was appointed handyman. I was the only survivor of the ten persons experimented on.

At the Natzweiler concentration camp in 1943 and 1944, Dr. Eugen Hagen experimented on 300 prisoners. In addition, some prisoners received injections of typhus to serve as incubation sites, that is, to maintain the virus for future uses [26].

Dr. Joaquim Mrugowsky, oversaw several other typhus vaccine experiments at the Buchenwald concentration camp in 1942 and 1943. He described the results in an affidavit for the Nuremberg trials [27]:

> In the case of sick persons during a typhus epidemic, who have not been vaccinated, the average duration of fever has been calculated to be 17 days. The metabolism and the nervous system were considerably affected. The mortality was around 30%.
>
> During the same epidemic, 4 groups of experimental subjects were vaccinated with one each of the abovementioned vaccines. The experimental subjects were mostly in their twenties and thirties. Care was taken when selecting them that they did not come from typhus districts and also to ensure an interval of 4 to 6 weeks between the protective vaccination and the outbreak of

> the clinical symptoms of the disease according to experience, this period is imperative to achieve immunity.

In the four experimental groups that received vaccine, 2 people died. Most survivors suffered fever for more than 10 days and low blood pressure.

Inmates at Buchenwald devised a scheme to sabotage the typhus vaccine. Once effective vaccines were developed, they were produced by prisoners under the guidance of Dr. Mrugowsky. The first batch of vaccine did not work. Thus, the SS brought biologist Ludwik Fleck, a prisoner at Auschwitz, to Buchenwald to assist with the production. Fleck immediately identified the problem, inmate scientists mistook white blood cells for typhus bacteria. The inmates agreed to keep this from Dr. Mrugowsky and they produced an effective vaccine that they provided to camp prisoners. They also continued to produce the ineffective vaccine, but that vaccine was sent to the German soldiers. This is one of several sabotage efforts concentration camp prisoners carried out on their oppressors [28].

Similar experiments were performed on spotted fever. Investigators found that neither intramuscular nor subcutaneous injections of infected serum (blood without cells or platelets) caused infections in the experimental prisoners, and thus they tested the ability to infect people through superficial and deep cuts in the arm. All experimental subjects became ill with spotted fever. The physician performing the experiments, Dr. Ding, accidentally injected himself with the serum and became ill as well. This model of infection was then used to test various vaccines to spotted fever, most of which were ineffective. Several subjects died from the experiments. One vaccine, developed by Beringwerke, was 75% effective against spotted fever [29,30].

Poison Gas Experiments

For fear the enemy might use poison gas attacks, Germans tested the effects of such gases on concentration camp prisoners, and tested potential treatments of the resulting wounds. From 1939 to 1944, Germans tested the effects of "Lost" (mustard gas) on prisoners at the Sachsenhausen and Natzweiler camps. To carry out the experiments at Natzweiler, Dr. Hirt needed cooperation from the administration of the camp, which did not come easily. The prisoners were treated as commodity, as Dr. Sievers wrote in a memo to Dr. Brandt [31]:

> According to Hirt's report the administration of the Camp was not very cooperative. The preparations in Natzweiler were going very slow.
>
> Furthermore, the administration of the Camp wanted Hirt's institution to pay for the prisoners they chose as objects of their experiments during the whole time the experiments continue…
>
> …Sievers expresses his indignation with the lukewarm attitude of the administration of the Concentration Camp Natzweiler.

Dr. Hirt completed the experiments at the two camps. Drs. Grawitz and Sonntag describe the results of the experiments in a letter to Himmler [32]:

> The report describes the experiments – the technical procedure used in the creation of wounds, the treatment, the process of the growth of wounds and of their healing – and states results. The report states that the persons experimented upon suffered considerably under the wounds

> caused by Lost, "the arms in most of the cases are badly swollen and pains are enormous." In order to have more possibilities for observation, the wounds were created on both arms of the human guinea pig. In four cases the wounds were infected for the purpose of the experiment. In the cases of experiments involving infected wounds, "a strong influence on the general state of health of the patients was observed. The results of the experiments were negative." The report states that the drug 'H' proved not to be a specific medicine against wounds caused by '01-Q' (Lost). It gives the impression as if the wounds treated with the drug 'H' (Holzmann) expanded more than the wounds treated with 'F 1001' using a dry method.

Anthropology Experiments

The Nazis believed that the Aryan race was superior to all others, and they set out to prove it, especially with regards to Jews. German doctors performed anthropological studies on the facial and skeletal features of Jews compared to Aryans. Dr. Sievers wrote the following in a report he sent to Himmler [33]:

> We have a nearly complete collection of skulls of all races and peoples at our disposal. Of the Jewish race, however, only very few specimens of skulls are available with the result that it is impossible to arrive at precise conclusions from examining them. The war in the East now presents us with the opportunity to overcome this deficiency. By procuring the skulls of the Jewish-Bolshevik Commissers, who represent the prototype of the repulsive, but oborotoristic subhuman, we have

> the chance now to obtain a palpable, scientific document.
>
> The best, or practical method for obtaining and collecting this skull material could be handled by directing the Wehrmacht to turn over alive all captured Jewish-Bolshevik Commissers to the Poldpolizoi. They in turn are to be given special directives to inform a certain office at regular intervals of the number and place of detention of those captured Jews and to give them special close attention and care until a special delegate arrives. This special delegate, who will be in charge of securing the "material", has the job of taking a series of previously established photographs, anthropological measurements, and in addition has to determine, so far as is possible, the background, date of birth, and other personal data of the prisoner. Following the subsequently induced death of the Jew, whose head should not be damaged, the delegate will separate the head from the body and will forward it to its proper point of destination in a thermetically sealed tin can, especially produced for this purpose and filled with conserving fluid.

Himmler approved the experiments, but utilizing prisoners at Auschwitz instead of the front. The study procured material from 115 people. With the end of the war coming and fear of retaliation, the bodies were sent to the Anatomy department of the Reich University of Strasbourg.

Experiments to Decrease the Population of Undesirables and Increase Aryan Population

As the Nazis tried to eliminate all Jews, they also desired to increase the population of Aryans and decrease the populations of undesirables. One of the most notorious Nazi doctors, Dr. Mengele, experimented on twins, including children, at Auschwitz concentration camp, with the desire to increase the number of twin births for Aryans. His experiments earned him the title Angel of Death (see Chapter 15).

Doctors at both Auschwitz and Ravensbrück concentration camps experimented on prisoners to determine an efficient means of mass sterilization of the "inferior races". Brandt described the purpose of the experiments in an affidavit prepared for the Nuremberg trials [10]:

> Himmler was extremely interested in the development of a cheap and rapid sterilization method which could be used against enemies of Germany, such as the Russians, Poles, and Jews. One hopes thereby not only to defeat the enemy but to exterminate him. The capacity for work of the sterilized persons could be exploited by Germany, while the propagation would be eliminated…It was further desired that a procedure be found which would result in sterilization that was not immediately noticeable.

Although surgical means of sterilization were known, they were inefficient. The Nazis tested medication, uterine irritation, and X-rays for their ability to sterilize subjects. In a 1942 letter from Brack to Himmler, Brack describes the horrid of Nazi extermination of Jews combined with slave labor: [34]

> Among ten millions of Jews, in Europe there are, I figure, at least two to three million of men and women who are fit enough to work. Considering the extraordinary difficulties the labor problem

> presents us with, I held the view that these two to three million should be specially selected and preserved. This can, however, only be done if at the same time they are rendered incapable to propagate. About a year ago I reported to you that agents of mine have completed the experiments necessary for this purpose. I would like to recall those facts once more. Sterilization, as normally performed on persons with hereditary diseases, is here out of the question because it takes too long and is too expensive. Castration by X-rays, however, is not only relatively cheap but can also be performed on many thousands in the shortest time. I think that at this time it is already irrelevant whether the people in question become aware of having been castrated after some weeks or months once they feel the effects.

German Universities and Research Centers Utilize Concentration Camp Specimens

With the eugenics movement and the T4 program to kill useless members of society, German physicians became a killing machine. But they also had a reputation as excellent physicians and physician scientists that they wanted to uphold by enhancing the Aryan race, especially university physicians. At their disposal were hundreds of thousands of prisoners in the concentration camps to use as human subjects as well as enemy who came their way. Why kill the prisoners for nothing? They could be used for human research, to train future medical doctors, and to study anatomy and anthropology. Corpses were often delivered to universities and other research institutes to study anatomy of the healthy and unhealthy human body. A few cases of anatomical research are described.

Hermann Voss earned his doctorate in 1919 at the University of Rostock, where he became an Assistant Professor of Anatomy. In 1938, Voss became the chief anatomist at Reich University of Posen in occupied Poland, and later moved up to the position of dean of the medical department [35].Voss expressed disdain for the Polish people in his diaries stating quite simply, "the Polish people must be eradicated." Voss used the bodies of executed concentration camp prisoners for several purposes, including the sale of prepared skeletons, skulls, death masks, and busts. Some of these items were sold to the Vienna Museum of Natural History and displayed in the Race Gallery [35].

Voss also studied the function of the spleen, which required large volumes of blood be taken from it. He received his sample from fighters in the Polish Resistance who got caught and went to the guillotine where the Posen Gestapo decapitated them. Voss wrote the following in his diary [36]:

> Yesterday I looked at the mortuary morgue in the basement and the furnace in the crematorium. The furnace was originally intended to dispose of the remnants of the dissecting course for the students. Now it seems to incinerate executed Poles. Almost daily the gray limousine arrives with the gray men, the SS-men from the Gestapo, carrying material for the furnace. As it was not working yesterday we could look inside the furnace, where we could see the ashes of four Poles. How little remains from a human being after all organic material has been burnt! Looking into such a furnace makes me feel very calm and comfortable. The Poles recently have become increasingly impudent and thus our furnace is very busy. How nice it would be to chase the whole population through such

> furnaces! Then the German people finally would get some rest in the East.
>
> ...Today I wrote to Prof. Schoen in Gottingen and asked him to remember me when the chair for anatomy at the university becomes vacant.

After the war, Voss became a most proficient anatomist at the University of Jena in East Germany.

Hermann Stieve, director of the Berlin Anatomical Institute from 1935-1945, used women in concentration camps for anatomical studies on the reproductive system. He gathered historical data from camp doctors, then after their execution, obtained the bodies for dissection. (Seidelman, Winkelman) He also studied the male reproductive system but was able to obtain anatomical specimens from criminal executions in Germany. Winkelman proposes Stieve fabricated data in his published works [37].

Universities also obtained corpses from the concentration camps to train future physicians. According to Seidelman, bodies were delivered to every university institute of anatomy in Germany, and possibly in Austria [35]. These bodies were used for years after the war ended. In 1984, Gotz Aly questioned the source of human brain anatomical slides at the Edinger Neurological Institute in Frankfurt am Main. Julius Hallevorden prepared the slides when he was director of the histo-pathology department at the Kaiser Wilhelm Institute for Brain Research during World War II. Following his research on the topic, Aly reported that the brain tissue came from Nazi euthanasia victims, including 33 children [38].

The Max Planck Institute initially disagreed with Aly's conclusion, but then more evidence of Nazi-derived university material showed up. Geneticist Benno Muller-Hill alleged Josef

Mengele provided anatomical specimens to his mentor Otmar von Verscheur, director of the Kaiser Wilhelm Institute for Anthropology [38]. Mengele and Verschuer sought to prove differences in Jewish verses Aryan anatomy.

The Max Planck Institute still hesitated accepting that concentration camp body parts could still be found at teaching and research institutes until 1988, when medical students at the University of Tubingen protested the use of camp remains for training. Protests erupted at the German embassy in Israel. In 1989, the removal and burial of any questionable remains started at institutes in Europe and continued for at least two decades [38]. A full account of whose remains were used, how many concentration camp remains were used, and whether all such remains have been removed from teaching and research institutes has not been possible.

The case of neuropathologist Julius Hallervorden, Kaiser Wilhelm Institute for Brain Research in Berlin-Buch, started the investigations. The brain anatomical slides corresponds to his request for hundreds of brains from Brandenburg-Gorden for neuropathological studies. When interrogated in 1945, he stated [36]:

> I heard that they were going to do that, and so I went up to them and told them "Look here now, boys, if you are going to kill all these people, at least take the brains but so that the material could be utilized." They asked me: "How many can you examine?" and so I told them an unlimited number—"the more the better." I gave them fixatives, jars and boxes and instructions for removing and fixing the brains and then they came bringing them like the delivery van from the furniture company… I accepted these brains of

> course. Where they came from and how they came to me, was really none of my business."

Teaching and research institutes also used concentration camp prisoners for experiments. In 1943, a medical student worked in a Berlin laboratory studying the function of the spleen. She used herself and other students as subjects when an Air Force Physician recommended that she stop experimenting on herself and utilize the concentration camp prisoners [36]. She listened.

Ferdinand Sauerbruch, a leading German surgeon at the time, sat on the Research Review Committee of the Reichsforschungsrat (the Reich Research Council) to approve research experiments, including the Mengele twin studies at Auschwitz. Sauerbruch was tried at the Doctor's Trial as an accessory to the crimes committed by the doctors because he partook in discussions of the experiments in concentration camps without any mention of the ethics behind such experiments. Likewise, Surgeon Erwin Gohrbrandt, director of Surgery III at the University Clinic of Berlin, participated in a 1942 conference on the freezing experiments conducted at the Dachau concentration camp [36].

Other University doctors participated directly in concentration camp experiments. Gastroenterologist Kurt Gutzeit, professor of medicine at the University of Breslau, used Jewish children at the Auschwitz camp to search for the carrier of hepatitis. His assistant, Dr. Dohmen, initially opposed performing experiments on humans, but ultimately infected 11 children with hepatitis-infected serum. Dr. Dohmen punctured the livers of these children to collect tissue samples for analysis. Dr. Gutzeit pushed for infecting many more children to achieve more definitive results, but before that could happen, Germany lost the war [36].

Heinrich Berning, associate professor at Hamburg University, used Soviet prisoners of war to test the progress of famine by starving the prisoners, noting changes as they slowly died. Following death, autopsies were performed [36].

The Doctors Trial

After Germany surrendered, World War II allies (USA, England, France, Soviet Union) joined forces and held an International Military Tribunal in Nuremberg (USA, France, UK, and USSR v Hermann Goering) to hold those in high command responsible for the atrocities of World War II. However, Americans were disgusted by the concentration camps, genocide, and human experimentation and decided to hold twelve additional Military Tribunals in Nuremberg, but under the name of the United States. One of those tribunals was USA v Karl Brandt *et. al.*, better known as the Doctor's Trial. Twenty doctors who experimented on concentration camp prisoners, and 3 administrators who supported those experiments, were tried for their crimes against humanity in an American military tribunal. In the opening statement, Prosecutor Telford Taylor stated the following [39]:

> The defendants in this case are charged with murders, tortures, and other atrocities committed in the name of medical science. The victims of these crimes are numbered in the hundreds of thousands. A handful only are still alive; a few of the survivors will appear in this courtroom. But most of these miserable victims were slaughtered outright or died during the course of the tortures to which they were subjected.

> …We have still other responsibilities here. The defendants in the dock are charged with murder, but this is no mere murder trial. … To kill, to maim, and to torture is criminal under all modern systems of law. These defendants did not kill in hot blood, nor for personal enrichment. Some of them may be sadists, who killed and tortured for sport, but they are not all perverts. They are not all ignorant men. Most of them are trained physicians and some of them are distinguished scientists. Yes these defendants, all of whom were fully able to comprehend the nature of their acts, and most of whom were exceptionally qualified to form a moral and professional judgment in this respect, are responsible for wholesale murder and unspeakably cruel tortures.

Prosecutors faced a dilemma in the lack of an international code for human experimentation. The prosecution utilized the Hippocratic oath as a universal oath for physicians and a German oath of 1933 protecting animals from cruel treatment including during experiments [40]. However, the prosecution failed to utilize the 1931 German Guidelines for Human Experimentation, which were technically in effect until 1945 [41], but ignored when Hitler took power in 1933. German physicians and scientists violated these fourteen guidelines their own country established:

> 1. In order that medical science may continue to advance, the initiation in appropriate cases of therapy involving new and as yet insufficiently tested means and procedures cannot be avoided. Similarly, scientific experimentation involving human subjects cannot be completely excluded as such, as this would hinder or even prevent

progress in the diagnosis, treatment, and prevention of diseases.

The freedom to be granted to the physician accordingly shall be weighed against his special duty to remain aware at all times of his major responsibility for the life and health of any person on whom he undertakes innovative therapy or performs an experiment.

2. For the purposes of these Guidelines, “innovative therapy” means interventions and treatment methods that involve humans and serve a therapeutic purpose, in other words that are carried out in a particular, individual case in order to diagnose, treat, or prevent a disease or suffering or to eliminate a physical defect, although their effects and consequences cannot be sufficiently evaluated on the basis of existing experience.

3. For the purposes of these Guidelines, “scientific experimentation” means interventions and treatment methods that involve humans and are undertaken for research purposes without serving a therapeutic purpose in an individual case, and whose effects and consequences cannot be sufficiently evaluated on the basis of existing experience.

4. Any innovative therapy must be justified and performed in accordance with the principles of medical ethics and the rules of medical practice and theory.

In all cases, the question of whether any adverse effects which may occur are proportionate to the anticipated benefits shall be examined and assessed. Innovative therapy may be carried out only it if has been tested in advance in animal trials (where these are possible).

5. Innovative therapy may be carried out only after the subject or his legal representative has unambiguously consented to the procedure in the light of relevant information provided in advance. Where consent is refused, innovative therapy may be initiated only if it constitutes an urgent procedure to preserve life or prevent serious damage to health and prior consent could not be obtained under the circumstances.

6. The question of whether to use innovative therapy must be examined with particular care where the subject is a child or a person under 18 years of age.

7. Exploitation of social hardship in order to undertake innovative therapy is incompatible with the principles of medical ethics.

8. Extreme caution shall be exercised in connection with innovative therapy involving live microorganisms, especially live pathogens. Such therapy shall be considered permissible only if the procedure can be assumed to be relatively safe and similar benefits are unlikely to be achieved under the circumstances by any other method.

9. In clinics, policlinics, hospitals, or other treatment and care establishments, innovative therapy may be carried out only by the physician in charge or by another physician acting in

accordance with his express instructions and subject to his complete responsibility.

10. A report shall be made in respect of any innovative therapy, indicating the purpose of the procedure, the justification for it, and the manner in which it is carried out. In particular, the report shall include a statement that the subject or, where appropriate, his legal representative has been provided in advance with relevant information and has given his consent.

Where therapy has been carried out without consent, under the conditions referred to in the second Paragraph of Section 5, the statement shall give full details of these conditions.

11. The results of any innovative therapy may be published only in a manner whereby the patient's dignity and the dictates of humanity are fully respected.

12. Sections 4-11 of these Guidelines shall be applicable, mutatis mutandis, to scientific experimentation (cf. Section 3).

The following additional requirements shall apply to such experimentation:

(a) experimentation shall be prohibited in all cases where consent has not been given;

(b) experimentation involving human subjects shall be avoided if it can be replaced by animal studies. Experimentation involving human subjects may be carried out only after all data that can be collected by means of those biological methods (laboratory testing and animal studies) that are available to medical science for purposes of clarification and confirmation of the validity of the experiment have been obtained. Under these circumstances, motiveless and unplanned experimentation involving human subjects shall obviously be prohibited;

(c) experimentation involving children or young persons under 18 years of age shall be prohibited if it in any way endangers the child or young person;

(d) experimentation involving dying subjects is incompatible with the principles of medical ethics and shall therefore be prohibited.

13. While physicians and, more particularly, those in charge of hospital establishments may thus be expected to be guided by a strong sense of responsibility towards their patients, they should at the same time not be denied the satisfying responsibility (Verantwortungsfreudigkeit) of seeking new ways to protect or treat patients or alleviate or remedy their suffering where they are convinced, in the light of their medical experience, that known methods are likely to fail.

14. Academic training courses should take every suitable opportunity to stress the physician's special duties when carrying out a new form of therapy or a scientific experiment as well as when publishing his results."

The tribunal, unaware of or disregarding the 1931 Guidelines for Human Experimentation, developed their own code of ethics for human experimentation called, appropriately, the Nuremberg Code, and adopted internationally. Ghooi believed the Nuremberg Code was likely based on the 1931 Guidelines as evidenced by considerable overlap. The tenets of the Nuremberg Code were given in part in Chapter 12.

Clearly, the experiments described in this chapter violate the 1931 Guidelines for Human Experimentation as well as the Nuremberg Code. But could the experimenters be successfully held responsible by a military court? Definitely not all, as many had escaped capture following the war. However, 23 people were indicted in the Doctor's Trial, and other physicians were indicted in other Nuremberg tribunals.

Sixteen of the defendants were found guilty of war crimes and crimes against humanity. Ten were also found guilty of membership in an organization declared criminal by the International Military Tribunal (the SS). Seven were sentenced to death by hanging: Karl Brandt, Karl Gebhardt, Rudolf Brandt, Joaquim Mrugowsky, Wolfram Sievers, Viktor Brack, and Waldemar Hoven. They were hanged on June 2, 1948. The others were sentenced from 10 years to life in prison, although no one actually served the entire sentence. The one female physician convicted, Herta Oberheuser, was sentenced to 20 years and served 10.

Not much justice can be found in trying just 20 physicians and convicting only 13 for the evil perpetrated on the prisoners of concentration camps. It is easy for us to look back and say that any physician participating in any experiments on concentration camp prisoners is guilty of some crime. Of course, the most common defense is "I was just following orders." Would you follow orders to hurt an innocent person?

Were German Doctors Evil?

What the Nazi doctors did was evil, and any physician willing to do the same is also evil. The environment at the time, struggles for Germany following World War I and eugenics being an international movement, supported the rise of Nazism and the promotion of racial hygiene. However, if physicians held a moral stand, they could have slowed the progress of the holocaust.

Bioethicists agree the human experimentation performed at concentration camps was evil. What they do not all agree on is whether or not the data should be used today in current research. Although many of the experiments were mere pseudoscience, some of the experiments were carried out meticulously and provide insights on the physiology of soldiers at war time (for example, the high altitude experiments). Although the methods used to obtain the data were evil, some data could contribute to saving lives. Should we use it? Some data has already been used, but with hesitation and under special circumstances. The main argument against using the data is to avoid giving any credence to the experiments. To do so could open doors for future scientists.

Chapter 15

Dr. Mengele, Angel of Death

You have already read the name Dr. Josef Mengele in Chapter 14, as he was a Nazi physician who performed experiments on others. But he seemed to go a step further than other physicians in that he seemed to enjoy the selection of those to be exterminated and thrived on the opportunity for experimenting on humans.

Josef Mengele was born to an affluent family March 16, 1911, in Gunzburg Germany. His father Karl Mengele owned an engineering company for farm equipment that was the leading employer in the region. Josef and his two younger brothers were raised in a strict Catholic upbringing. The family focused more on respect than affection. Being the oldest, Josef was expected to take on the family business, but that was not in his plans [1].

People who knew Josef as a child described him as ambitious and competitive, especially with his brother Karl. His family and friends called him Beppo. In April 1930, Josef passed his high-school exams with an average grade. His father told him that what counted was "what one achieved, not what one set out to achieve." [1,2]

Mengele's father was deep into the Socialist party when Mengele went to college in Munich. His father even hosted Hitler at the Karl Mengele factory to give a speech on farming in Germany. While in college, Mengele became exposed to the National Socialist German Workers Party, including their dogma that *some life was unworthy of life*.

Mengele earned a doctorate degree, *summa cum laude*, in Physical Anthropology from the University of Munich in 1935 under the guidance of Dr. Theodor Mollison, a strong proponent for distinguishing between the races. Mollison proposed an anthropological exhibit on racial science that would, "support the most effective way the endeavor of the government to awaken an understanding of the racial question in the entire people." Mengele's thesis studied racial differences in jaw structure, which concluded, "The jaws of the examined racial groups indicate in their front sections such distinct differences that they permit one to distinguish between the races." The science behind the conclusion is questionable [2].

In 1937, Mengele began working under Dr. Otmar von Verschuer, head of the Institute for Hereditary Biology and Racial Hygiene in Frankfurt. Von Verschuer continued as Mengele's mentor through Mengele's time at Auschwitz, when Verschuer worked as a professor of racial hygiene and director of the Kaiser Wilhelm Institute for Anthropology, Human Hereditary and Eugenics. [2,3]

Mengele earned his doctorate of medicine degree from the University of Frankfurt am Main in 1938, the same year he joined the Nazi party and enlisted in the SA (Sturmabteilung or Assault Division). During the early years of World War II, Mengele served as a medical officer with the Waffen-SS in Russia and France. During combat in 1942, Mengele was injured and could no longer serve on the frontline. He received the Iron Cross First Class for bravery. After recovery, he was transferred to the Race and Settlement Office in Berlin, which was responsible for ensuring the racial purity of SS recruits and their wives. In 1943, Verschuer requested Mengele be transferred to Auschwitz and Heinrich Himmler appointed him as chief doctor at Birkenau, the sister camp of Auschwitz. His supervisor at Auschwitz-Birkenau was Dr. Eduard Wirths [2].

At the Institute for Hereditary Biology and Racial Hygiene, Verschuer mentored Mengele in anthropological studies focused on proving the Aryan race superior to others and finding physical features that could be used to verify the race of an individual. Verschuer and Mengele continued to work together at the Kaiser Wilhelm Institute for Anthropology, Human Heredity, and Eugenics in Berlin. Even after Mengele went to Auschwitz, he continued collaborating with Verschuer, who obtained research funding from the German Research Council to fund Mengele's research in Auschwitz. In return, Mengele provided tissue specimens, blood samples, and body parts from his victims to Verschuer's research lab in Berlin [2].

Selections at Auschwitz

Auschwitz was developed in 1940 as one of six extermination camps focused on the elimination of all Jews. Located in Poland, it became the largest concentration camp and was divided into several sub-camps, the primary being Auschwitz I, II, and III. Auschwitz I was the oldest part of the camp and held 15,000 – 20,000 prisoners. Auschwitz II, also called Birkenau, held over 90,000 prisoners. Most of the mass exterminations took place at Birkenau. Auschwitz III consisted of more than 40 labor camps, where prisoners were forced to work in various industries.

Jewish and other undesirable persons were shipped to Auschwitz by train. The prisoners would be packed into cattle cars, more than 100 per car. Some would die before the train arrived at the Auschwitz platform.

Once at the platform, the camp doctor would separate the prisoners into two groups, those who would go straight to the gas chambers and those who would be spared as long as they were capable of performing labor. The total number of Jews killed at Auschwitz was between 1.1 – 2.5 million. Exact numbers could not be ascertained as Himmler ordered documents destroyed after major exterminations.

Mengele arrive in Auschwitz in May, 1943. At that time, Auschwitz had 140,000 prisoners. One of his responsibilities was to decide which of those prisoners would survive.

Richard Bock, who grew up in Gunzburg and knew the Mengele family, drove trucks from the platform to the crematoria. He testified at the Frankfurt Auschwitz trial on August 3, 1964, and described one of the selection processes: [3]

> There were the tracks, and the train stood there, and over there I saw Mengele. And there were a couple of officers; I didn't know all of them. … And I saw Mengele there and the rest of them. … Then they talked a bit at first, and I leaned against my truck, leaned against the fender and looked on.
>
> Mengele … said, "Doctors and pharmacists forward!" … And then the men came slowly, real gentlemen came forward. Their suits are wrinkled, but they were better dressed, some wore glasses. … They were elegant people, it seemed. And then Mengele made a gesture like that.

Bock made the gesture with his own thumb pointing to the left [3].

> Then they went by him, over to the left.
>
> A blond woman was the very first to come. I looked at her so closely, I would recognize her even today. I immediately thought that was the film actress. Lilian Harvey. And then Mengele said to her, "How old?" And she said, "Twenty-nine." Then Mengele asked, "Are you pregnant?"

> Then she said, "Yes." Then Mengele said, "In what month?" Then she said, "In the ninth." And then Mengele did so.

Bock again made a gesture with his thumb [3].

> And she walked over there, over the tracks, over and to the right. … Then came a young man, twenty-four years old. "Healthy, strong?" Mengele asked. Then he said, "Yes." Then Mengele did so, with the thumb—always with the thumb he did so—over and over. Then he went over to where the doctors were standing

Boch then testified that Mengele showed boredom or impatience of the selection and decided to conclude the selection as follows [3]:

> And then after … half of the train had walked by him, Mengele suddenly said, "Ach was. …" Everybody went over there … women came with children, then the women came pushing carriages, and everyone is over there.

Over there was the left side of the platform, awaiting transport to the gas chambers.

Prisoner Dr. Mauritius Berner described his experience as he arrived at the ramp on May 29, 1944 [3]:

> And the flow of people went forward, and I said to my wife—I was with my wife and three children, three little daughters: "Don't do anything. The main thing is that we five stay together." And: "We'll see how it goes."
>
> I hardly said this when another soldier came between us and said, "Men to the right and women

> to the left," and separated us. I didn't even have the time to embrace my wife. She yelled after me, "Come, kiss us!" Perhaps out of some kind of woman's intuition, she sensed the danger that threatened us.
>
> I ran back through the cordon to them, kissed my wife and my three children. And then I was shoved back to the other side, and we went on in parallel but separated. Between the two tracks, between the two trains, parallel but apart. Then I got pushed by the crowd and I lost sight of them, my family.

Dr. Berner was called out to join the group of physicians and pharmacists. He then testified what he saw as the selection continued [3]:

> All of a sudden, I saw my wife and three children walking further away from Mengele. It occurred to me: I would ask Dr. Capesius [German acquaintance of Dr. Berner who worked at Auschwitz] one favor. I went up to him and said: Herr Captain …, I have two twins, who require more care. I will work as you wish, only please allow me to remain together with my family. I did not know why we were there, where we were supposed to go. He asked me, "Twins?" "Yes." "Where are they?" I pointed, "There they are." "Call them back," he told me.
>
> Whereupon I called out the names of my wife and children. And they turned around, and I indicated they should come back.
>
> They came back, and Dr. Capesius himself took them by the hand, the two children, and led us to

> Dr. Mengele. And at his back he said to me, "So tell him." And I said, "Herr Captain," I did not know his rank, "Herr Captain, I have two twin children," and wanted to continue to speak, but he said to me, "Later I have no time." And with a dismissive gesture, sent me away.

Dr. Berner's family went back to the left group, and he never saw his family again [3].

Most concentration camp doctors found the platform selections a horrendous undertaking. Not Mengele. Mengele took pride in his work, and his devotion showed in his willingness to do more than what was required of him, including extra shifts at the platform.

Dr. Ella Lingens described SS doctors' responses to platform selections [1]:

> Some like Werner Rhode who hated his work, and Hans Konig who was deeply disgusted by the job, had to get drunk before they appeared on the ramp. Only two doctors performed the selections without any stimulants of any kind: Dr. Josef Mengele and Dr. Fritz Klein. Dr. Mengele was particularly cold and cynical.

Dr. Lingers also explained her take on Mengele's personal contempt for Jews [1]:

> He once told me there are only two gifted people in the world, German and Jews, and it's a question of who will be superior. So he decided that they had to be destroyed.

Survivors of Auschwitz often remembered Mengele on the platform and commented about his demeanor [1]:

> How we hated this charlatan. He profaned the very word "science." How we despised his detached, haughty air, his continual whistling, his absurd orders, his frigid cruelty.
>
> Day after day he was at his post, watching the pitiful crowd of men and women and children go struggling past, all in the last stages of exhaustion from the inhuman journey in cattle trucks. He would point with his cane at each person and direct them with one word: "right" or "left." … He seemed to enjoy his grisly task.

Inmate doctors and others with health knowledge were often given assignments in the camp hospitals. While such a responsibility could be considered noble, the inmate doctors had little medicine or other supplies to treat the prisoners. Further, they received other duties not quite so honorable, such as assisting with experiments on the prisoners as described below. The inmate doctors were treated better than other prisoners, but only for about three months after which they were sent to the gas chambers and replaced by incoming inmate doctors.

Dr. Miklos Nyiszli, a Jewish physician sent to Auschwitz, was chosen by Mengele to work for him as well as care for prisoners. His special privilege included housing where he could observe the unloading of the trains. Dr. Nyiszli provides his interpretation of the unloading of prisoners [4]:

> Any person who had entered the gates of the KZ [concentration camp] was a candidate for death. He whose destiny had directed him into the left-hand column was transformed by the gas chambers into a corpse within an hour after his arrival. Less fortunate was he whom adversity had singled out for the right-hand column. He was still

> a candidate for death, but with this difference, that for three or four months, or as long as he could endure, he had to submit to all the horrors the KZ had to offer, till he dropped from utter exhaustion. He bled from a thousand wounds. His belly was contorted with hunger, his eyes were haggard, and he moaned like one demented. He dragged his body across the fields of snow till he could go no farther. Trained dogs snapped at his wretched, fleshless frame, and when even the ice forsook his desiccated body, then the hour of deliverance, the hour of redeeming death was close at hand. Who then—of our parents, brothers, children—was more fortunate, he who went to the left or he who went to the right?

Mengele's sadism exposed itself during selections, as described in a witness deposition to the U.S. Army [1]:

> Several times we noticed the hypocritical manner in which the grim doctor Mengele treated women and children alighting from the train. "Madam, take care, your child will catch cold. … Madam, you are ill and tired after a long journey; give your child to this lady and you will find it later in the children's nursery." On these days he was in a good mood, treating in a friendly manner the people whom he sent to death and who were very often reduced to smoke five or six hours after their arrival.

Mengele's son Rolf, visit Mengele when he was in Brazil, and Rolf later described Mengele's view on the selections [1]:

> He told me he did not "invent" Auschwitz and that he was not personally responsible for the incidents

> there. Auschwitz already existed. He wanted to help but that was very limited. He couldn't help everyone. For instance, on the platform, he asked me what was he to do when the half-dead and infected people arrived? It was beyond one's imagination to describe the circumstances there. He said his job was to clarify only "able to work" and "unable to work." He tried to grade the people as "able to work" as often as possible. He thinks he saved the lives of several thousand people in that way. He didn't order the exterminations and he was not responsible.
>
> He said the twins owe their lives to him. He said he never harmed anybody personally, and he got very excited at this point. He asked me if I—his son—believe the lies in the newspaper.

Prisoner Healthcare

Mengele took his many responsibilities at Auschwitz-Birkenau seriously, including managing the health of the inmates. Within a few days of arrival at Auschwitz, Mengele sent 507 Gypsies and 528 Gypsy women suspected of typhus to the gas chamber. Several months later, he repeated this action with Jewish women, as described by prisoner Dr. Ella Lingens statement [1,2]:

> In fact, typhus was only stamped out when *Hauptsturmfuhrer* Dr. Mengele, the ruthless cynic, became camp doctor. He collected the 1,500 worst cases among the Jewish patients in a hospital hut and sent them to the gas chamber. Thus he obtained an empty hut, which was disinfected and supplied with new paillasses

> [straw mattresses] and clean blankets. Then the patients of the nearest hut were de-loused, examined and taken naked into the vermin-free hut, which was put out of bounds. The same with the next hut, and so on, until everything was clean. Given the circumstances, this was the correct way of fighting and overcoming the epidemic. But that the camp authorities did not think it necessary to build a new hut for the purpose, that the cleaning-up began with the murder of 1,500 Jewish women—that was part of the horror of a situation in which everything was perverted from its meaning, in which evil was good and good evil.

After the war, West Germany set many indictments against Mengele. The following indictment related to Mengele's ordering of executions of sick inmates [1]:

> Josef Mengele is accused of having actively and decisively taken part in the selection of prisoners' sick blocks, of such prisoners who through hunger, deprivation, exhaustion, sickness, disease, abuse or other reasons were unfit for work in the camp and whose speedy recovery was not envisaged, and also of those who had contagious or singularly unsightly illnesses, such as a skin outbreak.
>
> Those selected were killed either through injections or firing squads or by painful suffocation to death through prussic acid in the gas chambers in order to make room in the camp for the "fit" prisoners, selected by him or other SS doctors in the aforementioned fashion. The injections that killed were made with phenol,

petrol, Evipal, chloroform or air into the circulation, especially into the heart chamber, either with his own hands or he ordered the SS sanitary worker to do it while he watched. He is alleged also to have supervised, in cases of camp and hospital block selections, when SS sanitary workers threw granules of prussic acid formula Zyklon into the inlet pipes of the rooms with people condemned to die hemmed together, or he threw it himself.

Human Experimentation: Twins

Mengele developed his own research institute at Auschwitz. He recruited doctors and pharmacists during ramp selections, and staffed his research with other prisoners with specialized training. Mengele developed his laboratory in a barrack at the Gypsy camp. The purpose of the laboratory varied from procurement of body parts for Dr. Vershuer to infecting inmates with disease and trying to find a cure.

Mengele is best known for his research with twins. When trains would arrive to the platform, twins would be separated from the group and taken to Barracks 14 of Camp F and provided better, but not good, treatment than other prisoners (at least until the experiments began). Medical files located in Nazi archives verified that Mengele performed research on over 1,500 twins, with only 200 survivors. Several of the studies he performed on twins are briefly described [1,2,4].

Mengele's interest in twins follows that of Nazi Germany and of his mentor, Dr. Verschuer, who delivered a lecture to the Royal Society of London titled "Twin Research from the Time of Galton to the Present-Day." Nazi Germany's interest in a purified Aryan race made twin research popular among German scientists. Hermann Siemens described the use of twin research in studying hereditary traits [3]:

> Identical twins, which proceeded from the cells of a single fertilized egg, were held to be genetically identical—accordingly, any differences between identical twins were conceived of as exclusively environmental. Fraternal twins, on the other hand, which proceeded from different egg cells fertilized at the same time or in short succession, shared just over half of their genomes—like other siblings, so that differences between fraternal twins were understood to be the result of the coaction of a genetic *and* environmental factor. … The twin method represented a "nearly universally utilizable instrument for the practical investigation of the problem of heredity vs. environment, "as every characteristic that could be measured or counted could be tested "for rates of concordance/discordance."

Verschuer called Siemen's twin method of research the "superior means of genetic research in humans." Nazi scientists used the twin method to determine the heritability of various systemic diseases, mental diseases, even criminality and likelihood of divorce. Funding of twin research was plentiful, and even foreign donors, such as the Rockefeller Foundation, provided funding to the Kaiser Wilhelm Institute of Anthropology. The only shortage in twin research was twins. This was not an issue at Auschwitz.

Mengele enlisted the assistance of an adult male twin, Zvi Spiegel, to look after the twin boys housed in barrack 14 of Camp VIIf. Spiegel was given the title *Zwillingsvater*, or twins' father. He described how he gathered information on the children as they entered their barrack [3]:

> The moment a pair of twins arrived in the Barrack, they were asked to complete a detailed questionnaire from the Kaiser Wilhelm Institute in Berlin. One of my duties as twins' father was to help them fill it out, especially the little ones, who couldn't read or write.
>
> These forms contained dozens of detailed questions related to a child's background, health, and physical characteristics, … the age, weight, and height of the children, their eye color and the color of their hair. They were promptly mailed to Berlin when they were completed.
>
> After the form was filled out, I would take the twins to Mengele, who asked them additional questions. Because many of the children spoke only Hungarian—and Mengele spoke only German—I would serve as the translator….
>
> Much of the information Mengele was seeking had to do with demographics—where the family was from, what the parents had done for a living.

The least innocuous experiments on twins involved measuring the physical characteristics of the pairs. These measurements would help distinguish identical from fraternal twins using the Polysymptomatic Similarity Diagnosis procedure. For this task, Mengele enlisted the assistance of inmate Dr. Puznya. For over six months, she made meticulous measurements on 250 pairs of twins, after which the twins would be available to Mengele for his experiments. She described her interactions with Mengele [1]:

> He wasn't very talkative. He just used to look at the charts that had been written up. He never told me what he was applying this work to, but I had an idea. He wanted every detail, especially any difference in twins, and their eyes.

Mengele would follow up Puznya's measurements with some of his own. One eleven-year-old twin at Auschwitz explained to an interviewer in 1995 what was the most uncomfortable part of the examination by Mengele. He responded [3]:

> [Mengele] would measure everything. … He did all sorts of measurements … everything from the head and the eyes. I know one of the most unpleasant things I can recall … was when he took impressions of our teeth, because I recall he used plaster Paris … and, of course, they filled our mouths and it was extremely unpleasant. I mean I was gagging for several hours until that thing dried and was taken out.

Mengele used twins in two types of infectious disease studies. In one study, one twin served as a control (no disease) while the sibling was given a disease. Comparisons were made between the condition of the healthy and the sick twin, until one twin died (by the disease or euthanasia). The healthy twin was immediately euthanized so that simultaneous autopsies could be performed. In the other disease studies, both twins were infected and observed for any differences in their responses to the disease.

Mengele seemed to have an obsession in the responses of one twin to distress induced on the other. The distresses included removal of genitals or other organs without anesthesia, amputations without anesthesia, and creating serious wounds that he then infected.

Mengele had an opportunity to collaborate with Verschuer's team at the Kaiser Wilhelm Institute in a study on eye color. The study actually began at the institute under the direction of Dr. Karen Magnussen, who had a strong interest in the phenogenetics of eye pigmentation, especially with heterochromia (different colored eyes). Verschuer brought her to the institute to complete her study, "On the Investigation of the Heritability of the Development of Eye Color as the Basis for Examinations of Race and Descent." Her studies at the Institute took a major fall when the subjects she wanted to study (Otto Mechlau's gypsy family, with lots of twins and high incidence of heterochromia) were arrested and transported to Auschwitz [3]. But Magnussen met Mengele once, when Mengele visited the institute prior to moving on to Auschwitz. What could have put a halt to her experiments now turned out to be a favorable substitution.

Mengele was a meticulous man at times, and he performed the experiments on the Mechlau family twins to his best ability. Mengele prepared a genealogical chart of the Mechlau family for Magnussen to use. He precisely recorded the physical attributes of each person.

During the course of these experiments, members of the Mechlau family developed pulmonary tuberculosis. Dr. Magnussen requested that "should" any of the family members die, would Mengele send her the autopsy reports and the "eye material."

Mengele sent eyes to Dr. Magnussen. Most were obtained after the original owner of the eyes had been executed.

In another eye experiment, Mengele injected one eye of 36 children with various chemicals to try to change the eye color. The injections often caused painful infections and sometimes blindness. The eye experiments Mengele performed mimicked those Magnussen performed on rabbits.

For twins with heterochromia, Mengele would inject a chemical into one eye to try to change the color of the eye so they would match. What chemicals Mengele actually injected was not documented, but believed to include hormones or other pharmacologically active substances that were hypothesized to play a role in color changes with age, such as adrenaline, physostigmine, atropine, and doryl. Several of these chemicals would have caused pain and potentially life threatening symptoms depending on the dose [3].

Prisoner Dr. Rudolf Diem, was ordered to give injections into children's eyes [3]:

> [Mengele] ordered the inmate-physicians, including myself, to put drops into the eyes of [people with different color irises]. I know that these drops contained adrenaline, something he [Mengele] spoke of himself. He reckoned that the use of these drops would cause a change in eye color.

Jewish prisoner Dr. Vexler Jancu described one of his experiences with prisoner eyes [1]:

> In June 1943, I went to the Gypsy camp in Birkenau. I saw a wooden table. On it were samples of eyes. They each had a number and a letter. The eyes were very pale yellow to bright blue, green, and violet.

Witness Vera Kriegel described the wall of eyes she saw in Mengele's laboratory [1]:

> They were pinned up like butterflies. I thought I was dead and already living in hell.

The eyes, like many other tissue collected by Mengele, were sent to Kaiser Wilhelm Institute of Anthropology. Institute technician Hans Nachtsheim wrote to a colleague in 1961 describing his experience [3]:

> It was the greatest shock that I had experienced during the entire Nazi period when one day Mengele sent the eyes of a Gypsy family that had been interned in Auschwitz. The family had heterochromia of the iris and a female colleague at the institute, who worked on heterochromia, had earlier shown interest in these eyes.

Mengele also performed numerous twin experiments of which a reasonable scientific explanation cannot be given [1,2,3]:

> Mengele gave one twin a disease and then transfused his blood to his sibling.
>
> Mengele swapped the blood supply of different pairs of twins.
>
> Mengele performed sterilizations on some twins and forced insemination on others.
>
> Mengele sewed twins together to make “Siamese twins.”
>
> Mengele forced twin sisters to have sexual intercourse with twin brothers.
>
> Mengele injected twins with blood from another set of twins.

Dr. Horst Fischer, another Auschwitz doctor, described Mengele's enthusiasm for the opportunities given him at Auschwitz [3]:

> … Mengele saying that not to utilize the possibilities Auschwitz offered would be "a sin", a "crime" and "totally irresponsible" towards science.
>
> … a unique opportunity that would never again be offered.

Noma Outbreak Provides Experimental Opportunity

Noma is a rapidly progressive infection of the face that occurs in unsanitary conditions and with starvation. The infection starts with inflammation of the gums and development of ulcers, and under normal conditions, can be cured with early treatments. But untreated, the person becomes ravenous and will eat just about anything. As the disease progresses, the infection becomes gangrenous and is often fatal.

Prisoner Dr. Lucie Adelsberger worked in the Gypsy camp at Auschwitz and wrote [5],

> In 1943 there was a widespread outbreak of Noma, confined almost entirely to children. They showed deep and extensive ulceration of the mucous membranes of the mouth and cheek, with slimy necrotic holes inside the mouth, upper lips missing, and holes in the cheeks: this did not deter them from eating anything on which they could lay their hands.

Mengele saw the opportunity to study the disease, and recruited help from Dr. Berthold Epstein, a world famous pediatrician and prisoner at Auschwitz. Mengele had the infected children moved to a separate barrack and assigned Epstein to research possible treatments. Dr. Adelsberger described Epstein's work [5]:

> At that time the disease was inevitably fatal: later, Prof. B. Epstein, working in the camp, clarified the problem of the condition's pathology; and by introducing treatment with a combination of sulphonamides and nicotinic acid provided an effective cure. At the same time, permission was received to give these children special high calorie diets, with meat and added vitamins. It was then possible, with nicotinic acid and injections of "Globucid" (suphanilamido-ethyl thiadiazol), to arrest a severe gangrene's stomatitis, even when oedema of the cheek indicated an impending perforation: in a few days a line of demarcation could be seen between the necrotic area and the adjacent healthy tissues, and the necrotic slough then separated.

Remarkably, some good came out of Auschwitz – an effective treatment for Noma. But the good did not last, for even if the children were recovering from Noma, they were eliminated so that the heads could be removed and sectioned histologically for further study.

Other Experiments by Mengele at Auschwitz

A horrendous experiment conducted not only by Mengele, but also in the Monowitz Labor Camp, was to determine how much electrical current a person could endure. Witness Ernest Michel describes what he observed relating to these experiments [1]:

> I saw Mengele standing there in his uniform, surrounded by three or four others. There was electrical machinery the likes of which I had never seen. As we brought in each girl, an officer would strap her down. We left quickly because we didn't want to be around Mengele very long. After a while the screaming inside stopped. When we took them out, two of the eight were dead, five were in a coma, one was strapped to the cot. Mengele was standing there, discussing it very casually. The only word I could hear was experiment.

Mengele exposed a group of prisoner Polish nuns to extreme X-ray to determine what damage would be caused.

Mengele collected blood from hundreds of inmates of various ethnicities and provided them to Verschuer to match induced proteins to certain races to assist in Aryan purification.

Mengele conducted bone marrow transplant experiments, which resulted in some amputations.

Mengele conducted sterilization and castration experiments on male prisoners.

As sadistic as these experiments were, Mengele supposedly also had the breasts cut off of female prisoners, deprived a baby of food by putting tape over the mother's breast to see how long it would survive, and would stand on pregnant women's stomachs to expel the fetus.

Mengele did not perform his experiments alone—he had help. He selected doctors from the platform and put them to work doing horrendous experiments. Many of these inmate doctors survived and spoke out about their ordeals.

Dr. Miklos Nyiszli, Inmate and Assistant to Josef Mengele

Jewish physician Miklos Nyiszli was transported in train cattle cars from Hungary to Auschwitz along with thousands of other Jews. On the ramp at Auschwitz, a call went out for forensic physicians. Nyiszli stepped forward and his life would be changed forever as he was chosen to work as a camp physician as well as Mengele's assistant. He wrote **Auschwitz: A Doctor's Eyewitness Account**, which documents his time at Auschwitz and many of the atrocities of Mengele [4].

Nyiszli was assigned his own autopsy/pathology lab, very well equipped. He described his first autopsy of twins as follows [4]:

> They contained files on two little twin brothers. The commando crew, made up entirely of women, set the covered coffin down in front of me. I lifted the lid. Inside lay a set of two-year old twins. I ordered two of my men to take the corpses and place them on the dissecting table.
>
> The twins died at the same time ... It was they who had to—or whose tiny bodies had to—resolve the secret of the reproduction of the race. To advance one step in the search to unlock the secret of multiplying the race of superior beings destined to rule was a "noble goal." If only it were possible, in the future, to have each German mother bear as many twins as possible! The project, conceived by the demented theorists of the Third Reich, was utterly mad. And it was to Dr. Mengele, chief physician of the Auschwitz KZ, the notorious "criminal doctor," that these experiments had been entrusted.

> Among malefactors and criminals, the most dangerous type is the "criminal doctor," especially when he is armed with powers such as those granted to Dr. Mengele.

The following day, Sonderkommandos brought Nyiszli four more pairs of twins under ten years old to autopsy [4]. Nyiszli found that the children were euthanized by injection of chloroform into the heart, which caused the blood in the heart to coagulate resulting in rapid death. Nyiszli also noted that three of the four pairs of twins were heterochromatic, that is, they had different colored eyes. One eye was brown and the other blue. Further, all eight children had tuberculosis and hereditary syphilis (also called DuBois tumors).

Following the autopsies, organs of research interest were preserved while the rest of the bodies were burned. Some of the organs were simply examined by Mengele, but others were shipped to the Anthropological Institute at Berlin-Dahlem for further study.

Later that day, Nyiszli made another horrifying discovery. He heard a noise in a room he used as a temporary morgue. There were 70 women there whom he thought were dead, only to discover a few were still alive. They were all shot in the back of the neck with a small caliber bullet from 1-2 inches away. A Sonderkommando explained to Nyiszli that seventy women from C section were killed in this manner each day. Nyiszli's day ended watching a game of soccer (football) between SS guards and the Sonderkommando [4]. Welcome to Auschwitz.

More twins would come Dr. Nyiszli's way, but Mengele also showed interest in Jews with deformities. The Nazis claim that Jews were inferior, including anatomically. Nyiszli recounts his examination of a father and son who each had a deformity. The father had a hunched back, and his teenage son had a deformed right foot. When they arrived by train, the father and son were picked out by Mengele and sent to Nyiszli for a physical and medical examination. After his examination, the pair were taken outside and Nyiszli heard two shots. The bodies of the father and son were returned to him still warm for autopsies. The father had suffered from retarded rickets causing deformation of the spine, a relatively common occurrence. The son had decreased muscle mass in his left foot, a condition called hypomyelia, also not a rare disease. After hearing this report, Mengele responded [4]:

> These bodies must not be cremated. They must be prepared and their skeletons sent to the Anthropological Museum in Berlin.

Mengele then ordered Nyiszli to clean the bones the quickest means possible, which was boiling. Nyiszli completed the orders utilizing two iron casks set over a fire in the courtyard. Some starving Polish prisoners were working nearby and thought the casks contained food. After consuming a small amount, Nyiszli and some Sonderkommandos saw them eating and immediately stopped them [4].

Nyiszli continued with the clearing of the tissue off the bone. Mengele was pleased with his work and believed that the Anthropological Museum in Berlin could use the bones as proof Jewish people had more physical deformities. Thus, the bones were shipped.

Mengele was not the only Nazi doctor to benefit from Nyiszli's proficiency as a pathologist. Mengele called a meeting with Dr. Nyiszli, Dr. Thilo (head surgeon of the KZ), and Dr. Wolfe (director of the General Medical Service). Dr. Wolfe wanted to study the cause of dysentery through dissections of 150 bodies. Dr. Nyiszli was to perform these dissections with the aid of three physicians who had been in the prison barracks. Nyiszli was glad to have the help, and these three physicians would be treated better for their remaining time in Auschwitz [4].

Dr. Nyiszli and his wife and daughter survived the ordeal at Auschwitz when the Russians freed them from Nazi control in January, 1945.

One could suggest that Nyiszli was evil for saving his life by participating in the taking of others. This is one of the toughest cases in the book. If Nyiszli and other inmate physicians had not survived, many of Mengele's actions would have gone unrecorded. But what type of person could participate in Mengele's experiments to record the evil at Auschwitz? I would like to say I would have gladly stepped into the gas chamber as opposed to participate in such atrocities. But the last chapter of this book will touch more on who is capable of evil acts.

Cruelty for the Sake of Cruelty

Mengele also showed signs of cruelty unrelated to any supposed medical purposes. Mengele threw a new-born child onto a pile of corpses and another onto a stove. During the selection procedures, Mengele hit an elderly man over the head with an iron bar splitting the man's skull and killing him. Mengele had 300 children burned alive in front of prisoners, one of whom, Annani Silovich Pet'ko testified for Mengele's indictment [1]:

> After a while a large group [of SS officers] arrived on motorcycles, Mengele among them. They drove into the yard and got off their motorcycles. Upon arriving they circled the flames; it burned horizontally. We watched to see what would follow. After a while trucks arrived, dump trucks, with children inside. There were about ten of these trucks. After they had entered the yard an officer

> gave an order and the trucks backed up to the fire and they started throwing those children right into the fire, into the pit. The children started to scream; some of them managed to crawl out of the burning pit; an officer walked around it with sticks and pushed back those who managed to get out. Hoess and Mengele were present and were giving orders.
>
> The first group of children were from Dnepropertrovsk. I was told by the zone commanders that it was difficult to poison the children in the gas chambers, therefore they were burned in the pit. They were all under five years old. I heard that they had brought either an entire kindergarten or an orphanage from Dnepropertrovsk. Later I was told that some of these children that were brought and burned were actually taken from their mothers.

In another incident, inmate Dr. Gisella Perl described how Mengele treated a woman who escaped the gas chambers six times by jumping off the truck to the chambers [1]:

> "You are still here?" Dr. Mengele left the head of the column, and with a few easy strides caught up with her. He grabbed her by the neck and proceeded to beat her head to a bloody pulp. He hit her, slapped her, boxed her, always her head—screaming at her at the top of his voice, "You want to escape, don't you. You can't escape now. This is not a truck, you can't jump. You are going to burn like the others, you are going to croak, you dirty Jew," and he went on hitting the poor unprotected head. As I watched, I saw her two

> beautiful, intelligent eyes disappear under a layer of blood. Her ears weren't there any longer, maybe he had torn them off. And in a few seconds, her straight, pointed nose was a flat, broken, bleeding mass. I closed my eyes, unable to bear it any longer, and when I opened them up again, Dr. Mengele had stopped hitting her. But instead of a human head, Ibi's tall, thin body carried a round, blood-red object on its bony shoulders, an unrecognizable object, too horrible to look at; he pushed her back into line. Half an hour later, Dr. Mengele returned to the hospital. He took a piece of perfumed soap out of his bag and, whistling gaily with a smile of deep satisfaction on his face, he began to wash his hands.

Dr. Perl described another incident that occurred when Dr. Mengele made an unexpected visit to the women's hospital and caught Dr. Perl and her colleagues cooking stolen potatoes [1]:

> The silence lasted only a second; the storm, when it broke, was all the more terrible. He ran around like a wild beast, smashing everything in his way. He kicked over the stove, stomped on our potatoes, overturned the operating table, screaming, shouting incessantly: "Yes, this is how I imagined a Jewish hospital. You dirty whores, you unspeakable Jew swine." Suddenly I conceived a desperate plan. I got up from the floor, went to the shelf, and took down a jar containing a fetus and approached Dr. Mengele. "Herr Hauptsturmfuhrer may be interested in this specimen," I stammered. "Only rarely is it possible to bring it out in one piece." He stopped raving and grabbed the jar out of my hand. His

> face, which the moment before had looked like the face of a raving maniac, assumed a cruel, satisfied smile. "Good, beautiful, take it to crematory number 2 tomorrow. We are sending it to Berlin." And as if he had forgotten what went on he turned around and left the hospital.

After Auschwitz

Marwell, Posner, and Underwood all obtained detailed accounts of how Mengele escaped being held accountable by the Allies after World War II. Some of this information came from Mengele's autobiography, which is currently owned by his family, and other information came from indictments of Mengele filed by West Germany. While fascinating, Mengele's escape will be discussed only briefly with details that pertain to Mengele's experiments.

In January 1945, Russians were advancing toward Auschwitz. The SS destroyed documents and blew up the gas chambers and crematoria. Many inmates were evacuated on death marches to camps in the west. Mengele and other SS left Auschwitz on January 17, 1945, ten days before the Russians liberated the camp. But Mengele packed more than his clothes for the trip, as told by inmate anthropologist Martyna Puzyna [3]:

> I remember well that in January 1945, a few days before the evacuation, Mengele appeared in the room [where his files were stored] and like a madman packed up all the notes and prepared them for being shipped. He did not speak a word and then disappeared. I never saw him again.

Mengele, Horst Fischer, and Konig met up in the Gross Rosen concentration camp, where conditions were overcrowded with both inmates and SS, so they moved on to Berlin. According to Fischer [3]:

> We all agreed that we would be brought to account for the crimes committed at Auschwitz and hoped either to die at the front or, if we were taken prisoner, to escape being identified as a physician at a concentration camp.

However, the German Army had not given up hope, and Mengele was sent back to Gross Rosen camp. By February 13, the Russians liberated Gross Rosen camp, but Mengele had already skipped town. He joined a former colleague, Dr. Otto-Hans Kahler, at a German military field hospital in Saaz, where he discarded his SS uniform and dressed as a German Army officer. The field hospital continually moved west, ultimately escaping into unoccupied territory between Russian and American lines. Mengele befriended Dr. Fritz Ulmann, who would play a role in Mengele's escape.

Mengele described his capture, and then release, by the American army in his autobiography [3]:

> With a few vehicles … we formed a column and managed to … reach the Bavarian area. In the vicinity of the first big city, we were … brought to an American prison camp. We had reached our objective just as our supply of gasoline ran out. The Amis [German slang for "Americans"] dragged us then from one camp to another in which the rations got steadily smaller and our hopelessness greater. Finally, as the remnant of a camp that was being closed, we were released into the American occupation zone, as it had become known.

Mengele and others were released because the American Army had over three million prisoners of war and did not have the rations to care for them. The American Army only kept SS and war criminals, and by this time, Mengele had discarded his SS clothing. Most SS could be identified by a tattoo of their blood type, but because Mengele was in charge of tattooing the SS in his early war days, he avoided getting the tattoo. Mengele was also on the Central Registry of War Criminals and Security Suspects list developed in 1944, but the list had not been completely distributed until October, 1945. Thus, Mengele was released by the Americans in August, 1945, with a discharge certificate proving he had been screened by American authorities. This document allowed prisoners to go home and be eligible for rations and other necessities. Dr. Ulmann received two copies of his discharge certificate and gave one to Mengele, who after leaving the POW camp, got rid of the document in his name and used Ulmann's.

When Americans took Mengele prisoner, he did not have documents he gathered at Auschwitz. Marwell proposed that Mengele could have left the documents with a nurse friend, whom he visited after his release from the POW camp. After visiting this nurse, Mengele went to work for three years on a farm in Mangolding, in the foothills of the Bavarian Alps.

While working on the farm, Mengele's brother Karl visited him and shared news from home. Mengele's wife Irene also visited a couple of times, but the visits left him pessimistic of a future with her. In fact, she had a priest in Gunzburg conduct a memorial mass for Mengele, putting on a ruse that he was dead. During this time, the Nuremburg War Trials were ongoing, including the Doctors Trial. In January, 1948, American chief counsel for the war crimes General Telford Taylor reported, "Our records show Dr. Mengele [sic] is dead as of October 1946."

On August 1, 1948, Mengele left the farm and started his long voyage to Argentina by way of South Tyrol, Italy, the same path taken by several Nazi war criminals. On the way to Italy, he formally abandoned his interest in the family business for fear that his name could put the company at risk. However, his family continued to support him financially through his escape and until his death.

On his way to Italy, Mengele's family sent him a package that included money and a "box of scientific specimens" to take with him on his journeys. Mengele purchased necessary documents to emigrate under the name of Helmut Gregor. While he faced some snags along the way, Mengele and his box of specimens got a ride on the *North King* and arrived in Argentina on June 20, 1949. With the assistance of Nazi sympathizers, he was housed and moved as needed. Mengele met many other Nazis, including Ulrich Rudel, a Luftwaffe colonel who was Hitler's most decorated soldier, whom Mengele became close to and relied on frequently for safety. Rudel's driver was Sassen, a ghost-writer for a German-language publication *Der Weg*, who later would become an informant to Israel's Mossad.

In 1953, Mengele published an article (under the name G. Helmuth) on genetics in *Der Weg*, titled Hereditary as a Biological Process. Also at this time, Irene filed for a divorce admitting to the courts that he is alive, but providing false information on when she last saw him [3]:

> I last saw the defendant in November 1944. Since that time, I have had no personal connection to him. We have corresponded with each other concerning maintenance support. I no longer believe that the conjugal partnership between the defendant and me can be resumed. Our marriage is broken, the defendant did not return to me and

> in fact emigrated to South America in 1948, and no longer cares for me. Under no circumstances am I prepared to resume the marriage with my husband.

Mengele agreed to the divorce and had lawyers from Buenos Aires represent him in Germany.

In 1955, Mengele returned to Europe, including Germany, still using the name of Helmut Gregor. Mengele's brother Karl had died, and his father wanted Mengele to marry his former sister-in-law in a scheme to keep her share of the family business in "the family." Mengele met Martha and her son in Switzerland. After, Mengele went to Gunzburg to visit his family, including his son, Rolf. Mengele also visited with former colleagues, including Dr. Verschuer.

In 1956, Mengele reassumed his original identity as Josef Mengele and in 1958 he married Martha, both of which required formal documents from Germany for proof of existence and proof of eligibility for marriage (i.e. divorced or widowed).

In 1957, Mengele joined forces with Dr. Ernesto Timmerman and Heinz Truppel to expand a pharmaceutical company called Fadrofarm. Here, Mengele performed experiments on new treatments of tuberculosis. Elsa Haverich met Mengele twice, and describes her second encounter at Fadrofarm [3]:

> He came into the office one afternoon. I called him "Dr. Gregor" at that moment because to me he was "Dr. Gregor." Then he corrected me and told me that, "No it is Dr. Mengele." Joking a little bit, I asked him why he changed his name. He answered that it was for political reasons; when he got out of the war, he had to do it with another name.

The **Diary of Anne Frank**, a book of a Jewish girl hiding in Nazi Germany with her family, played a significant role in getting the attention of German officials that Josef Mengele was alive and well in Argentina. Anne Frank wrote her diary until her family was caught and sent to Auschwitz. In Auschwitz, Anne, her sister, and her mother died. Her father survived and in 1950 published her diary.

Some questioned the authenticity of the diary and others wanted to know what happened after Anne was captured. Her father wanted to have a follow up to the diary, but to avoid the "authenticity" issue, he asked Ernst Schnabel to write the rest of the story. The book came out in 1958 and got the attention of a reader who was stunned by a sentence in the book that maintained, "no one knew … where Dr. Mengele was, whether he died or whether he still lived somewhere." The reader wrote a note to Schnabel on July 7, 1958 [3]:

> Apparently, some people do know. … Otherwise the older Herr Mengele in Gunzburg would not have told his former housemaid … that his son, who was an SS doctor, practiced medicine in South America under a different name. And because he was so homesick, Mr. Mengele sent the widow of another son over there.

Schnabel sent the letter to the state prosecutor in Ulm, a city close to Gunzburg with the accompanying note [3]:

> I feel compelled to bring to your attention several issues that could—I am obviously not competent to judge—potentially lead to an inquiry by your office.
>
> The letter concerns itself with the concentration camp physician Dr. Mengele, who is mentioned

> on page 138 of my book. Mengele appears to have been one of the most dreadful people in Auschwitz.
>
> The anonymous tip about Dr. Mengele is a tenuous reason to get involved in the case. One would usually throw such a letter in the trash. … [However,] if her message helps to prevent a monster, like Dr. Mengele from Auschwitz, from continuing to work as a physician anywhere in the world, it would accomplish something very good for humanity, for us Germans, and for our deeply sullied German Name.

Thus, indirectly, **The Diary of Anne Frank** lead West German officials to focus attention back in Mengele's direction.

West German police were able to track Mengele to Buenos Aires. An arrest warrant for Josef Mengele was issued in Freiburg Germany on February 25, 1959, and a second warrant on June 5, 1959. However, there was no extradition treaty between Germany and Argentina. It took more than a year since a formal request for extradition was filed before the Argentina government agreed to arrest Mengele. By that time, he had fled to Paraguay and became a citizen under the name Jose Mengele. Paraguay prohibited the extradition of its citizens to other countries [3].

Mengele spent his final years travelling between Paraguay and Brazil. His biggest concern now was Mossad, Israel's national intelligence agency. On May 11, 1960, Mossad kidnapped Adolf Eichmann in Buenos Aires, Argentina, and transported him to Israel where he had to sit trial—no extradition necessary. Israel courts found Eichmann guilty of crimes against the Jewish people and crimes against humanity. He was sentenced to death, and hung on May 31, 1962, in Israel. In his final breaths, he declared his loyalty to Germany, Austria, and Argentina.

The Mossad came close to capturing Mengele, even tracking him to the place of his death, Sao Paulo, Brazil [3]. On February 7, 1979, Mengele suffered a stroke while swimming in a River in Bertioga, Brazil, and died. His remains were buried in Sao Paulo as Wolfgang Gerhard, until his body was exhumed in 1985. Dental and DNA analyses proved it was Mengele [2].

Although one of the most evil physicians ever, Mengele escaped legal punishment for his actions. However, the two Universities where he got his doctorate degrees, University of Munich and the University of Frankfurt, rescinded his degrees in 1960 and 1964, respectively [3].

Chapter 16

Japanese Experiments During World War II

For 250 years, Japan remained an isolated country headed by an emperor and run by a shogunate. In the mid to late 1800's, the west treated Asian countries as inferior, using their militaries to enforce unfair trade. In 1853, Commodore Matthew Perry of the U.S. Navy lead four ships into the Tokyo harbor demanding Japan to open up its borders and allow trade with the United States. With no navy to resist, Japan signed a trade agreement. Other countries soon followed the United States with trade agreements, and Japan was no longer isolated.

The Japanese felt betrayed by the shogunate and a new centralized government was formed, with the emperor, Meiji, as its symbolic leader. Lands became the property of the emperor. Western influence resulted in social reforms, including freedom of religion and compulsory education. The military grew and infrastructure in the country became better established. But toward the end of the 1800's, Japan started turning away from western influence and returning to principles of Confucian philosophy, more importantly, they reinstated the state religion Shinto, with the Emperor considered a direct descendant of the sun goddess, Amaterasu Omikami, who created Japan, making the emperor a deity.

Under the Qing Dynasty, China dominated the East for almost 300 years, until 1894, when disagreements over Korea lead to the first Sino-Japanese War between China and Japan. Japan won decisively and claimed much of China as its own. However,

Russia, France and Germany intervened and forced Japan to return to China all territories except Taiwan.

During wars before the 1900's, more Japanese soldiers died of disease than combat. Thus, Japan's military development increased the authority of the medical corps during warfare, including expanded research on preventing diseases. Because Japan is an island country, isolated from foreign diseases, Japanese people were particularly susceptible to disease brought in by overseas travelers (or soldiers). As Japan invaded China, and later other countries, the soldiers were exposed to new diseases to which the Japanese had developed no immunity. The medical corps promoted better hygiene for soldiers including safe water supplies, which decreased the likelihood of disease.

Japan's next war, the Russo-Japanese War, started in 1904 over disagreements with Russia over the rights to Korea and Manchuria. Japan won, regaining territory taken away after winning the Sino-Japanese War, and annexing Korea five years later in 1910. U.S. Army physician Louis Livingston Seaman, who traveled with Japanese troops during the Russo-Japanese War, wrote of Japan's humane war tactics in his book, **The Real Triumph of Japan: the Conquest of the Silent Foe** [1]:

> The history of warfare for centuries has proven that in prolonged campaigns the first, or open enemy, kills twenty per cent of the total mortality in the conflict, whilst the second, or silent enemy [disease], kills eighty...This dreadful and unnecessary sacrifice of life, especially among the Anglo-Saxon races, is the most ghastly proposition of modern war, and the Japanese have gone a long way toward conquering or eliminating it ...

> I unhesitatingly assert that the greatest conquests of Japan have been in the humanities of war, in the stopping of the needless sacrifice of life through preventable disease …

Seaman also wrote of the care provided by the Japanese medical corps to seventy-seven thousand Russian prisoners sick and injured. As we will see, this humanitarian treatment did not continue into World War II.

Russia, on the other hand, used biological weapons against Japan. Having been subjected to biological weapons once, the Japanese Imperial Army believed they needed to be more prepared for any future wars.

When Emperor Meiji died in 1912, Taisho became emperor. Politics changed as the Emperor lost power to the Parliament and democratic parties.

On July 28, 1914, World War I started with the assassination of Austrian Archduke Franz Ferdinand. Japan joined the Allies, but mostly stayed involved in eastern conflicts. In addition, Japan presented the secretive Twenty-One Demands to Chinese President Yuan Shikai, threatening war if the demands were not met. These demands would strengthen Japan's position in Asia, giving it authority over certain Chinese territories and industries, and most egregious, requiring China to install Japanese advisors who could take effective control of Chinese government, economy, and military. To avoid war with Japan, China agreed to all the demands except installing of advisors. China and Japan already showed hostilities toward one another, and the Twenty-One Demands strengthened those hostilities, but also strengthened Chinese nationalism and weakened President Yuan's authority. But Japan also grew in power.

Japan prevented China from joining the Allies during World War I, but China managed to contribute to the Ally efforts by providing laborers to the Ally countries while the soldiers were away fighting. At the Paris Peace Conference following the end of the war, China had only two representatives compared to Japan's five. When the Allies awarded Shandong province to Japan, China felt cheated and became the only country that refused to sign the Treaty of Versailles.

After World War I, Japan continued expanding its territories until 1922, when Japan, the United States, England, and France signed the Four-Power Pact. The pact provided military protection to Japan, while limiting their navy to 60% the size of that of the United States or England. Three years later, Japan and Russia also reached an agreement to limit their aggression. Japan's extension to Asia and the Pacific stopped, at least temporarily.

In 1927, civil war broke out in China between the Kuomintang (KMT)-led government of the Republic of China and the Communist Party of China. With China at war with itself, Japan took advantage and invaded Manchuria in 1931. The League of nations criticized Japan's invasion of China, causing Japan to leave the League in 1932. By 1937, Japan controlled much of China and the second Sino-Japanese war officially started. Japan supposedly initiated both wars to increase living space and resources for its people. In half a century, Japan's imperialism increased its size from 147,669 to over 1,000,000 square miles [2].

Shiro Ishii

During the time of Japan's expansion, Shiro Ishii became a prominent scientist in biological warfare. In 1920, Shiro Ishii graduated from Kyoto University medical school, well known for its expertise in bacteriology. Ishii enlisted in the army after graduation, and in 1922, he was commissioned to the First Army Hospital in Tokyo and became a Professor of Immunology. Ishii initially studied ways to prevent disease during battle, but he desired a change. He soon focused his research on diseases and chemicals, and how to weaponize them. However, Japan signed the Geneva Convention agreement of 1925, which prohibited the use of biological and chemical warfare. From Ishii's viewpoint, banned weapons would be more effective in war [3]. In 1930, Ishii started research on biological warfare using human subjects at the First Army Hospital in Tokyo. Two years later, he convinced the Japanese Imperial Army to support the development of biological and chemical weapons in a more secluded research environment.

Japan utilized new territory in China to build its biological warfare capabilities prior to the beginning of World War II. In 1932, Japanese scientists, including Dr. Ishii, moved into Harbin China (close to the Soviet Union border) to develop a research facility to study vaccines. South of Harbin, in an area called Beiyinhe, the Japanese built another research facility to grow and weaponize bacteria, although they told the Chinese it was a prison. The Japanese army forced Chinese laborers to build the facility, called Zhongma Fortress, in one year. A local Chinese resident described the process as follows [3]:

> We heard rumors of people having blood drawn in there, but we never went near the place. We were too afraid.

> When construction started, there were about forty houses in our village, and a lot of people were driven out. About one person from each home was taken to work on the construction. People were gathered from villages from all around here, maybe about a thousand people in all.
>
> The only thing we worked on were the surrounding wall and the earthen walls. The Chinese that worked on the buildings were brought in from somewhere, but we didn't know where. After everything was finished, those people were killed.

The fortress could hold a thousand prisoners at a time, and contained laboratories for human experimentation.

Japanese Military in China

After World War I, the Japanese people were heading toward a democracy, but the numerous wars and treaties had the country in disarray. In 1926, Emperor Hirohito came to power, quickly ousting political parties from government, including removal of three prime ministers. Future prime ministers who served under Hirohito came from the military and not political parties. Several military secret societies worshipped Hirohito and became ultra-nationalists [4]. The Japanese, like most races, considered their race the superior, and followed the rules of Hirohito with passion. Racism throughout the world lead to cruel treatment of "inferior" opponents during war, using the concept of sub-human to justify the actions.

Under the rule of Hirohito during the second Sino-Japanese war and World War II, Japanese soldiers and doctors carried out atrocities against the Chinese as horrendous as those carried out by the Germans against Jews. A most notable military action took place during the invasion of the then Chinese capital city, Nanking. Shortly after over-taking Shanghai in late 1937, the Japanese army, commanded by General Matsui Iwane, headed toward Nanking. Aware of this, the Chinese moved their national troops out of Nanking, leaving behind only auxiliary troops and civilians. The Chinese government forbade evacuation of their citizens. Western businesses in the city were protected by their status and location. In an attempt to protect Chinese civilians, these businesses built a safety zone that the Japanese agreed to honor [5].

On the way to Nanking, the Japanese army killed many Chinese military and civilians. Some claim that two Japanese officers had a contest to see who could first kill 100 Chinese men with swords on the way to Nanking. However, both officers exceeded 100, and it was not certain who reached 100 first, so they continued the contest by increasing the number to 150. One officer supposedly said the contest "was fun." Although the two officers were tried and convicted of war crimes, whether the contest happened is still a subject of dispute [6]. But one cannot dispute the horror that happened in Nanking over the 6 weeks starting on December 13, when Japan's army reached Nanking and, shortly after, entered the safety zone.

In Nanking, Japan's army killed 200,000 – 300,000 Chinese men, women and children. Over 20,000 women were raped. Surgeon Robert Wilson, working at a hospital in the safety zone, wrote the following in a letter to his family[5]:

> The slaughter of civilians is appalling. I could go on for pages telling of cases of rape and brutality almost beyond belief. Two bayoneted corpses are the only survivors of seven street cleaners who were sitting in their headquarters when Japanese soldiers came in without warning or reason and killed five of their number and wounded the two that found their way to the hospital. Let me recount some instances occurring in the last two days. Last night the house of one of the Chinese staff members of the university was broken into and two of the women, his relatives, were raped. Two girls, about 16, were raped to death in one of the refugee camps. In the University Middle School where there are 8,000 people the Japs came in ten times last night, over the wall, stole food, clothing, and raped until they were satisfied. They bayoneted one little boy of eight who [had] five bayonet wounds including one that penetrated his stomach, a portion of omentum was outside the abdomen. I think he will live.

When the Japanese invaded Nanking, their orders were to take no prisoners. However, thousands of Chinese soldiers surrendered, and thus were prisoners of war. Even though the Hague Convention forbade executing prisoners of war, Japan executed many.

In the **Nanking Massacre: Fact Versus Fiction**, Higashinakano Shudo claims civilians were not killed [7]. Further, he claims the Chinese were at fault for most of the events in Nanking for the following reasons: the Chinese soldiers were so poorly trained that they killed each other, the soldiers hid in the safety zone, many Chinese soldiers tried to leave Nanking and were killed by other Chinese soldiers, orders to kill all prisoners of war by bayonet were mistakenly distributed by one of the battalions, and prisoners of war were too numerous and rowdy to be kept safely and thus had to be executed. According to Shudo, the Japanese dropped leaflets with the following message into Nanking before the invasion [7]:

> The Japanese Army, one million strong, has already conquered Jiangnan. We have surrounded the city of Nanking...The Japanese Army shall show no mercy toward those who offer resistance, treating them with extreme severity, but shall harm neither innocent civilians nor Chinese military personnel who manifest no hostility. It is our earnest desire to preserve the East Asian culture. If your troops continue to fight, war in Nanking is inevitable. A culture that has endured for a millennium will be reduced to ashes, and a government that has lasted for a decade will vanish into thin air. This commander-in-chief issues a warning to your troops on behalf of the Japanese Army. Open the gates to Nanking in a peaceful manner, and take the aforementioned action.

What the Japanese leaders ordered and what the troops did may be two different things. Many stories besides those of Nanking tell the horrific actions of Japanese soldiers.

Yuasa Ken, a Japanese military doctor stationed in China, described the warfare in China as follows [3]:

> Soldiers went to the comfort women, and they raped them. [Comfort women included over

> 200,000 women who were kidnapped from Japanese-occupied countries and used as sex slaves.] Then, the next day, they would regain their strength to attack the Chinese…
>
> It is said that there were twenty million victims of the war in China. But only ten percent of these were killed in gunfire exchange. Most—non-resisting old people, women, and children—were captured and slaughtered. Prisoners of war could not be taken to the front or allowed to escape, so they were killed in the manner of the Rape of Nanjing.

Yes, war is hell! But there are internationally accepted rules of engagement set forth before World War I in the Hague Conventions of 1899 and 1907, and the first two Geneva Conventions in 1864 and 1906. Before the second Sino-Japanese War and World War II, rules were updated at the third Geneva Convention in 1929. The final Geneva Convention took place in 1949, after World War II [8].

Japan's military violated the rules of engagement and those of handling prisoners of war. But even worst is what the Japanese did in the name of science. Japan had two military medical schools prior to World War II, Tokyo Army Medical College and the Kwantung Army Medical College in Japanese-occupied northeast China (Mukden). Graduates of these medical schools and civilian medical schools in Japan, unlike their German counterparts, took no ethical oaths such as the Hippocratic Oath [4]. There was no pretense of empathy in Japanese medicine.

Unit 731

During World War I, both sides used chlorine gas as a weapon. While efficient at killing and disorienting the enemy, the distribution of the gas varied based on the weather, and sometimes affected the side releasing the gas. After the war, countries forbade the use of chlorine gas and other chemical and biological weapons. During World War II, Japan feared the Allies would use such weapons and wanted to be prepared to use them as well.

In 1936, Emperor Hirohito created a new unit of the army, the Boeki Kyusui Bu, or Anti-Epidemic Water Supply and Purification Bureau, and Dr. Ishii was put in charge. In 1938, Japan established Unit 731 in occupied Harbin, China, under the guise of a water sanitation plant, Epidemic Prevention and Water Purification Department of the Kwantung Army of Japan. Contrary to promoting public health, under the leadership of Ishii, hundreds of doctors and scientists from Japan's universities and military performed horrendous human experiments, including the development and testing of biological agents and methods of their delivery. Dr. Shiro Ishii's obsession with biological weapons made him and those who worked for him into monsters. He told those working under him [4]:

> Our god-given mission as doctors is to challenge all varieties of disease-causing microorganisms … to block all roads of intrusion into the human body; to annihilate all foreign matter resident in our bodies; and to devise the most expeditious treatment possible.

But he then negated the healing role of physicians, telling his workers [4]:

> ...the dual thrill of 1), a scientist to exert efforts...probing for the truth in natural science and research into, and discovery of, the unknown world and 2), as a military person, to successfully build a powerful military weapon against the enemy

Unit 731 consisted of more than 150 buildings in a 2 square mile zone and specialized in biological warfare. More than 3,000 Japanese were employed there and at satellite units. Between 3,000 and 10,000 people were experimented on and killed in Unit 731.

In addition to doctors and scientists, Japan sent teenagers to Unit 731 for educational purposes. These teenagers formed the Youth Corps and were to be trained in microbiology and biological warfare. At a 1994 Expo in Marioka City, one former member of the Youth Corps spoke of his experience [3]. At the age of 14, he entered the Army Medical College in Shinjuku, Tokyo. He was assigned to Unit 731 for his first year of the four-year Youth Corps program, which consisted of extensive studying and assisting the workers of Unit 731. Ishii worked closely with the Youth Corps. This member participated in human experiments, including the testing of poisonous gas. These boys were being trained to be future Ishii's.

Another former Youth Corps member, Tomioka Heihachiro, described his training in Manchuria [3]:

> The object of training was to teach young people to be soldiers in the shortest possible time, and in order to do so, the men in charge hit the new recruits. There was not one day when we were not

hit. Open-handed slapping does not hurt much, so they used fists. Sometimes, they would use the soles of *tabi* [Japanese-style footwear] so that the rubber treads would leave their marks on our faces…

There was one type of rifle called a Type 38. It had the emperor's chrysanthemum crest on the front end. We were taught that each rifle was lent to us by the emperor, and that we had to treat it with respect…

Every day, without fail, we would get hit. And that's where the spirit of absolute obedience is born. It's like training a dog. Humans and animals are the same. If you hit them, they learn to obey.

One day we were told that instead of using straw dummies for bayonet practice we would use people. We were going to practice on five people who had been brought in by the *kenpeitai*. We were told that they were members of the anti-Japanese resistance movement, but when they brought in the prisoners, they were seventeen or eighteen years old.

We were lined up in columns according to our unit, and the prisoners were tied in place. We were ordered to fix our bayonets. The boy at the front of the line was first. The commanding officer gave the orders: Forward! Back! Forward! Thrust!

This was the first time I'd killed anyone. My legs were shaking. When you thrust, it should be done

> fast. I was afraid though, and I closed my eyes, so I don't know where I struck the person. About twenty-five of us in turn, one after the other, struck the prisoner. By that time, his shirt looked like a beehive with flat holes instead of round ones. That's how we killed. There was a concept in our education that one does not become an adult until he had killed someone.
>
> Before that boy's breathing stopped, I heard him crying, 'Mama, mama,' and I realized that it's the same in China as in Japan.

Japanese scientists performed most experiments on Chinese civilians, whom they dehumanized by calling them "*maruta*" or logs. The Japanese *Kenpeitai* (military police) would capture civilians and bring them to Unit 731 rolled up in straw mats to look like a truck of logs. Prisoners of war were also used in experiments, as were political prisoners, criminals, and women and children. One of the researchers at Unit 731, Ueda Yatara, described the experimental subjects as follows [3]:

> To the guards, the people in here have already lost all rights. Their names have been exchanged for just a number written across the front of their shirts and the name *maruta.* They are referred to only as 'Maruta Number X.' They are counted not as one person or two persons but 'one log, two logs.' We are not concerned with where they are from, how they came here.

One *Kenpetei* member described his role in bringing political prisoners to Unit 731 after interrogating them through water boarding [3]:

The victim is placed naked on the board face up. The head is held immobile inside a wooden box frame fixed at the end of the board. The arms and legs are strapped down. The victim is not able to move no matter how great the suffering. If a hose is used to force water into the mouth, as has been the case previously, the victim can evade some of the water. A different system was devised, an easier one whose results are more complete.

First, water is poured on the chest to prevent a heart attack. A cloth is placed over the mouth and nose, and water is trickled over it slowly. The cloth becomes saturated, and no matter how much the victim tries to fight, water enters the body and in five or six minutes the belly swells. This is continued until the desired confession is obtained.

…Sometimes people would not confess under any amount of torture. Their bodies would be broken down but they would not give in. If we released them, they would accuse of us torturing them, so we sent them to the Ishii unit.

…There were occasional parties to commemorate someone's work. A committee would decide on carrying out executions while everyone was drinking sake. There was one officer who was especially fond of executions. He would have about five persons sit in front of a ditch, then behead them and kick the bodies into the ditch. These particular executions apparently had no scientific aims.

Workers at Unit 731 described receiving these bodies with bloated bellies for use in experiments.

More than 50 different lethal experiments were performed at Unit 731. In most of the experiments below, the final step was vivisection, often without anesthesia. Vivisection is similar to autopsy, but the person is alive for the purpose of viewing living organs. One Unit 731 worker describes how he prepared the body for vivisection after the person was infected with a bacterium [9]:

> I was ordered to wash that person's body with a deck brush before he or she was taken into the dissection room naked by a member of a special team. The first time, I trembled. One team member was listening to the heartbeat using a stethoscope. One was standing holding a knife. The moment a stethoscope was removed from an ear, the knife went into the body. I did not know, but according to doctors, this timing was very important because if the timing was wrong, we could get blood all over us, and then we could get infected.

Further, if a prisoner survived an experiment, he would be used for another, and this would continue until the prisoner died, as stated by Kawashima [10]:

> If a prisoner survived the inoculation of lethal bacteria, this did not save him from a repetition of the experiments, which were continued until death from infection supervened. The infected people were given medical treatment in order to test various methods of cure, they were fed normally,

> and after they had fully recovered, were used for the next experiment, but infected with another kind of germ. At any rate, no one ever left this death factory alive. ...

In the end, the bodies were destroyed by crematoria much like happened in the concentration camps of Nazi Germany. Much of what we know about the experiments in Unit 731 came from Soviet Union trials of Japanese prisoners of war.

Frostbite Experiments

Frostbite was induced by placing the subjects' wet hands and arms in a freezer that could reach -50°F or by having the subjects stay outside when the weather was extremely cold. Many of the subjects lost fingers or all the flesh of the fingers leaving only the bone. Some of the frozen limbs were thawed and the limbs studied as the flesh rotted. Physicians experimented on other frostbit limbs to compare effectiveness of various treatments. Kurakazu testified the following at his trial in the Soviet Union [10]:

> . . . On each floor there were several rooms used as laboratories, and in the middle were the cells where the experimentees, or 'logs,' as Sergeant Major Tasaka told me they were called in the detachment, were kept... I remember clearly that, in addition to Chinese, there were Russians among the prisoners. In one cell I saw Chinese women… All the people kept in the cells had chains on their legs... Three Chinese had no fingers, and in the case of others, the finger bones could be seen. ... Yoshimura told me that this was the result of

> freezing experiments which he had been performing. ...

Witness Furuichi testified [10]:

> ... A group of Russians, Manchurians, Chinese and Mongolians, with their legs in chains, were led out into the frost in parties of from two to sixteen and, on pain of being shot, were made to plunge their bare hands (one or both) into barrels of water, and then to keep their bare wet hands out in the frost for from ten minutes to two hours, depending on the temperature of the atmosphere. When freezing had set in, they were taken to the prison laboratory. In the majority of cases, these criminal experiments ended in gangrene, the amputation of the limbs, and the death of the people experimented on. The object of these experiments was to find means of preventing and treating frozen limbs in the projected military operations against the Soviet Union.

Seawater Experiments

To determine if seawater (average 3.5% salt) could substitute for saline (0.9% salt solution used in intravenous injections), physicians injected subjects with seawater.

Liver Experiments

The liver is a remarkable organ in that it can regenerate. To determine how long a person could live following loss of part of the liver, physicians surgically removed portions of the liver on experimental subjects.

Point of Death Experiments

Several experiments tested how long a person could survive harsh conditions including centrifuging at high speeds, exposure to high pressure chambers, water deprivation, and lethal doses of x-rays. Although scientists took some physiological measurements on these prisoners as they were dying, most data pertained to how the prisoner reacted to the oncoming death [9].

Explosion Experiments

Japanese scientists wanted to know how much damage occurred from shrapnel following an explosion based on how close the person was to the explosion. Subjects were tied to stakes and placed in a spiral around a grenade or other explosive device. Following the explosion, surgeons performed vivisections on the subjects without anesthesia (those closest to the explosion died prior to autopsy).

As a twist to the shrapnel, the explosive was filled with biological or chemical agents to determine the effectiveness of explosives as a means of spreading germs or chemicals. On trial in the Soviet Union, Karasawa testified [10]:

> ... I personally was present on two occasions at the Anta proving ground when the action of bacteria was tested on human beings under field conditions. The first time I was there towards the end of 1943. Some ten persons were brought to the proving ground, were tied to stakes which had been previously driven into the ground five metres apart, and a fragmentation bomb was exploded by electric current about 50 metres away from them.

> A number of the experimentees were injured by bomb splinters and simultaneously, as I afterwards learned, infected with anthrax, since the bomb was charged with these bacteria. …

Nishi Toshihide also was tried by the Soviet Union, and testified [10]:

> . . . In January 1945, in my presence, Lieutenant Colonel Ikari, Chief of the 2nd Division of Detachment 731, and Futaki, a research official of this division, performed an experiment at the detachment's proving ground near Anta Station in infecting ten Chinese war prisoners with gas gangrene. The ten Chinese prisoners were tied to stakes from 10 to 20 metres apart, and a bomb was then exploded by electricity. All ten were injured by shrapnel contaminated with gas gangrene germs, and within a week they all died in severe torment.

Surgical Training

Doctors and veterinarians used vivisection of prisoners to practice various surgical procedures. Several surgeries were performed on one prisoner and then the prisoner was killed by lethal injection. One of the surgeons, Dr. Ken Yuasa, states, “I was evil. I was a devil. We all were.” Dr. Yoshio Shinozuka describes his vivisection [11]: “The first time, my legs were shaking so badly I could hardly stand up. At the vivisection, I could not meet his eyes because of the hate he had in his glare at me.” He also stated, “We called the victims ‘logs’. We didn’t want to think of them as people. We didn’t want to admit that we were taking lives.”

Some of the surgical training utilized Prisoners of War and were independent of any special "Unit." After Yuasa Ken graduated medical school, he wanted to be a village doctor, to serve those who had no medical care nearby. But he became an Army doctor to serve his Emperor and worked at an army hospital in Shanxi Province in China. He describes his mandatory surgical training [3]:

> About half the army doctors die not knowing how to use a scalpel. Each method of treating sick and wounded soldiers at the front had a bearing on military strength. If men were sent behind the lines for treatment, it reduced the army's fighting strength considerably. One function of the army hospital was training doctors. Vivisection provided practice for performing operations at the front lines. I operated on living Chinese for whom I had no hatred whatsoever to gain surgical ability in order to win the war.

Another demonstration of surgical procedures occurred in the Philippines in 1942 [12]. A surgeon ordered some soldiers to bring a healthy Filipino male into a field where some students were gathered to observe the procedure. The surgeon spread a sheet on the field, placed a mask over the nose of the victim, and anesthetized him. He then surgically opened the man's abdomen, "removed his appendix and sewed him back up. Then, the lesson over, the surgeon pulled out a gun and shot and killed the patient."

In another example from 1942, a surgeon demonstrated surgical procedures at a municipal hospital in the Shanxi province of China. The "students" included seven doctors, a dentist, a pharmacist and an accountant. Harris describes the events that took place [4]:

> Two healthy Chinese men were brought into the operating room. One was given certain procedures that the gathering did not observe. The other was anesthetized by a female nurse who also cooed to the victim in Chinese, "sleep, sleep, sleep." One of the medical observers asked the surgeon who was about to administer… lumbar [sic] anesthesia if he wasn't going to disinfect the point of injection. "What are you talking about? We are going to kill him," he replied. Once asleep, the patient's healthy appendix was removed. Then, one of the doctors amputated one of the patient's arms. The Japanese doctor also practiced techniques on this patient for suturing intestines.

Japanese army doctor Yuasa Ken served at an army hospital in Shanxi province, where he participated in surgical practice and vivisection [3].

> I also saw vivisections. Once I saw about forty doctors gathered. There was a man bound and squatting. The guard asked the doctors, "Are you ready?" and the prisoner was laid out and, without anesthetic, two cuts were made down his belly. The victim made a few gasps—the dissection was a botch—and he died soon. I saw four people dissected that way.
>
> Once, at the Shanxi First Army Headquarters, there were some forty army doctors gathered from base and field hospitals. There was a lecture on

> military medicine, and afterward we were led to the prison cells. There were two Chinese in a cell. The jailer took out his pistol and fired two shots into each of their bellies. One of them was vivisected right there in the room. There was no anesthesia. While this was going on I heard four more shots fired. That meant two more people. Our object was to keep the person alive until the bullets were removed. Since we neither tried to administer ether nor stop the flow of blood, the men died soon.

A mother of one of the Chinese victims wrote Yuasa Ken a letter while he was in prison for war crimes [3]:

> I saw you people take him away. I was choked with emotion, and I ran after you with my bound feet. Later I learned that he was taken to the army hospital and cut up alive. I cried. I couldn't eat.

When a newspaper reporter interviewed Yuasa Ken after his release from prison, Yuasa recalls the reporter's comment that "in spite of what I had done, I am still active as a doctor. 'In Germany,' he told me, 'you would have been placed on trial [3].'"

Biological Warfare

In 1925, the Geneva Protocol banned the use of biological and chemical weapons in war. However, many countries continued development of potential weapons during World War II, including the United States and England, with both countries carrying out their experiments on animals. Japan, however, performed their biological and chemical weapons research on humans.

Shiro Ishii, physician and microbiologist, lead the biological warfare experiments at Unit 731. In 1932, Ishii became head of the Army's biological warfare division.

Yamada Otozoo served as the Commander-in-Chief of the Japanese Kwantung Army starting in 1944 until the surrender of Japan. Yamada supervised the activities at Unit 731 and Unit 100, in preparation for biological warfare, and with the authority to utilize biological weapons.

Workers at Unit 731, in addition to carrying out experiments, mass produced biological agents. Karasawa Tomio stated his group could produce 100 kilograms of plague, 200 kilograms of anthrax, 300 kilograms of typhoid, 300 kilograms of parathyroid "A", 330 kilograms of cholera, and 300 kilograms of dysentery. For one particular mission, his group prepared 70 kilograms of typhoid bacteria and 50 kilograms of cholera bacteria [10].

The biological weapons created at Unit 731 were tested both at the Unit and in the "field", the field being China. Ishii particularly wanted to weaponize bubonic plague, which killed 75% of those infected. Ishii's major research focus was an efficient delivery method. Several techniques were tested utilizing infected fleas and rats infested with infected fleas. While China was the primary target and "practice" site, Japan also planned on using biological warfare against the United States and the Soviet Union [10].

Chinese farmers were paid by Unit 731 to breed rats. After the war ended, the farmers denied knowing that the rats would be used to carry disease. Likewise, Chinese prisoners at Unit 731 were used to breed fleas by serving as the food for the fleas. The

Japanese would then infect the fleas with a disease, infest the rats with the fleas, and release the rats in China [10].

One of the diseases used on China was the plague. The Japanese utilized airplanes that sprayed plague-infected fleas directly over territories in China. Ishii went on several of these missions, including one in 1940, infecting the Nimpo area of China [10]. The mission was successful, as an epidemic of plague broke out in Nimpo.

After Japan launched a biological attack, scientists would go to the territory attacked and choose people to be vivisected in the fields to obtain data as quickly as possible after infection [9]. During an attack on Changteh in 1941, the spray method spread plague to not only 10,000 Chinese, but also to 1700 Japanese troops who died, leading to an end of this technique in 1942 [13].

Karasawa testified at his trial in the Soviet Union about another biological attack on China [10]:

> …As a preliminary to this expedition, 130 kilograms of paratyphoid and anthrax bacteria were prepared under my direction, again on the orders of Major Suzuki. According to my information, fleas were also used as epidemic carriers in this expedition.
>
> ... To carry out the expedition, General Ishii left with a party for Central China, where the Japanese troops were retreating at that time. The members of the expedition took advantage of the retreat to disseminate bacteria in the abandoned territory in order to cause outbreaks of epidemic among the advancing Chinese troops.

The Japanese army also utilized sabotage of water supplies to deliver bacteria to the enemy, as stated by Kawashima in his testimony [10]:

> In 1942, detachments 731 and 100 carried out special reconnoitering operations in the border areas of the Soviet Union in preparation for bacteriological warfare. For several years before that, on the orders of the Headquarters of the Japanese Kwantung Army, Detachment 100 had systematically sent bacteriological parties to the borders of the Soviet Union, where they had contaminated water sources, notably in the Tryokhrechye area, as a sabotage move.

Japan also delivered infected fleas with ceramic bombs that would be dropped by airplanes over their targets in China.

In their plans to spread plague and other diseases in the United States, Japan utilized these ceramic bombs and hot-air balloons (called Fugu balloons) [14]. Starting on November 3, 1944, over 9,300 hot-air balloons armed with traditional incendiary and anti-personnel bombs targeted the United States to start fires and directly kill Americans. Approximately one thousand reached North America, travelling as far east as Michigan, as far north as the Arctic, and as far south as Mexico. The first balloon found by Americans landed in the Pacific Ocean near San Pedro, California, on November 4, 1944, just twenty-four hours after its launch. The United States requested the press to refrain from giving information about the balloons to keep Japan from learning of their reaching the States, as Japan might use more and the potential for forest fires was great.

On May 5, 1945, children found a balloon that landed near Lakeview, Oregon. Five children and a woman were killed when the bomb went off as they were dragging the balloon. On May 22, the United States decided to report the balloons to the public and warn them of the dangers [14]. The greatest concern to the United States was if Japan loaded the balloons with biological weapons. Using Fugu balloons, Japan planned to launch its first plague attack on the United States in September, 1945. However, the Americans dropped an atomic bomb on Hiroshima on August 6, 1945, and another on Nagasaki three days later, forcing the Japanese to surrender before they could launch their plague infested balloons.

The Japanese also tried to spread anthrax in China. One method was developed by Japanese Colonel Oota, who had chocolates laced with anthrax and gave them to Chinese children.

As with German physicians tried in Nuremberg, some Japanese also used the excuse of just following orders, as Kikuchi Norimitsu says in his testimony [10]:

> I knew that the Japanese Kwantung Army was preparing to conduct bacteriological warfare against the Soviet Union, that preparations for this were also being made by Detachment 731 and Branch 643, which helped Detachment 731 in its preparations to conduct bacteriological warfare, and that the object of my work in the 1st Research Section of the branch . . . was also to prepare to conduct bacteriological warfare. Consequently, I myself took part in the preparations to conduct this warfare, and to this I plead guilty.

> I committed this crime not of my own free will, but by order of my superiors whom I obeyed. I could not refuse to obey their orders, for I would have been punished for refusal to obey orders.

Other Japanese Research Institutes During World War II

Although Unit 731 is the most infamous, Japan built approximately 25 biological and chemical warfare institutes in China during World War II.

Unit 516, in Qiqihar China, specialized in chemical weapons, producing 1200 tons annually. Japan used several chemical weapons against the Chinese including phosgene, hydrogen cyanide, bromobenzyl cyanide and chloroacetophenone, diphenyl-cyanoarsine and diphenylchlorarsine, arsenic trichloride, sulfur mustard and lewisite. Workers at Unit 516 produced seven million bombs with chemical weapons, four million of which are still missing. On August 4, 2003, 36 people were injured and one killed when construction workers in Qiqihar discovered five drums of mustard gas [15].

Unit 100 (Wakamatsu Unit) in Changchun was called the Anti-Epizootic Protection of Horses Unit. Scientists experimented on biological and chemical warfare, specializing in poisonous plants and snake venoms. They also studied plant diseases for crop destruction and animal diseases for livestock destruction. Various infectious agents and chemicals were tested for their effects on animals and humans, including bubonic plague, anthrax, smallpox, paratyphoid A and B, cholera, syphilis, botulism, dysentery, glanders, yellow fever, tularemia, scarlet fever, diphtheria, salmonella, infectious jaundice, epidemic cerebrospinal meningitis, and pesticides. Often, scientists performed vivisections outdoors.

Fukuzumi Mitsuyoshi, who served in Detachment 100 as a veterinary surgeon, testified [10]:

> Being an experimental unit, Detachment 100 had a research staff of bacteriologists, chemists, veterinaries and agronomists. All the work of this detachment was conducted in preparation for bacteriological sabotage and warfare against the Soviet Union. The personnel of the detachment and its branches carried on research ... in methods of employing bacteria and virulent poisons on a large scale for the mass extermination of animals and human beings.
>
> . . . In order to ascertain the effectiveness of these poisons, experiments were performed on animals and living people.

Researchers at Unit 1644 (Tama Unit), in Nanking, China, focused on livestock bacteria, but also used humans in research. One civilian researcher described the treatment of humans as follows [3]:

> Everyone connected with human experiments wore a special button on the side of his hat. The *maruta* were kept in cages on the top floor of a three-story building. The cages were referred to as *rotsu* [probably from the Japanese katakana character *ro*, which is square in shape], and the floor where they were kept was always just called the "third floor."

> ...Inside the door, the room was about ten by fifteen meters with cages all in a row. Most of the *maruta* in the cages were just lying down. In the same room were oil cans with mice that had been injected with plague germs, and with fleas feeding on the mice. These were not the usual types of fleas, but a transparent variety. Around the perimeter of the room was a thirty-centimeter-wide trough of running water. [The purpose of the trough, which is wider than the distance over which the fleas can leap, is apparently to keep the fleas from going outside the room.]

Scientists of Unit 1644 also tested chemical weapons: nitrile prussiate, acetone, cyanide hydric, arsenic, snake venom, tetrodotoxin from pufferfish, and trikabuto poison.

In addition to special military units, biological warfare laboratories existed on university campuses throughout Japan. Kitano Masaji, Professor of Microbiology at Mukden Army Medical College, tested biological weapons on human subjects and published the findings using the generic "monkey" (as opposed to a species of primate) to let his readers know the experiments were performed on humans. Kitano used several hundred people in his experiments, especially Chinese and Koreans [4].

At Kyushu Medical University, experiments were performed on 8 American POW's. According to Harris: [4]

> Operations on the men were performed on at least 4 separate days in May and June 1945. In one operation, a lung was removed from each of two prisoners. On a second occasion, doctors removed

> the stomach, heart, and liver from two other POWs. The third experiment led to the death of an airman whose brain was damaged in the course of surgery to examine the function of the trigeminal nerve. Three American fliers were used in the fourth and final test. The doctors operated on stomachs, gall bladders, livers, and hearts. All eight men used in the experiments died on the operating table, which, as Professor Roland notes, "was a poorly equipped dissecting room in the anatomy department."

American POW's were also used in experiments by Surgeon Captain Iwanami Hiroshi and Surgeon Lieutenant Nabetani on Dublon Island, Truk [4]. Two prisoners had tourniquets tied tightly around their arms and legs for 7-8 hours. When the tourniquets were removed, the prisoners died of shock. The surgeons dissected the bodies, and Dr. Iwanami kept the skulls as souvenirs.

Harris describes vivisections used for teaching purposes and medical students participating in human experimentation at several Japanese medical schools [4]:

> For example, the medical school of Kyushu Imperial University (located in the city of Fukuoka on Kyushu Island) was one of the most important medical training facilities in Japan. Its professors were considered to be among the finest scholars within the Empire. Senior professors took part in the vivisection experiments, while young interns either assisted their superiors in the experiments, or observed the activity. These experiments included: replacing blood with sea

> water; excising lungs, stomachs, livers, and other organs from POWs; interrupting blood flow from arteries of the heart to determine the time death would occur from such a procedure; and drilling holes in craniums, then inserting scalpels into the brain to determine what, if anything, medically useful could be discovered from the procedure.

In another example, researchers studied venereal disease in women forced into sexual service [4]. They first injected women with syphilis and then used vivisection to learn about the various stages in the development of the infection. However, syphilis did not develop. Next, they used infection through sexual contact, using prisoners as the sexual partner. They paired a healthy partner with one who had syphilis, and then monitored the progression of the disease for a few weeks. Doctors performed vivisections on the subjects at different stages of the disease to examine how different organs were affected.

Possibly most disturbing is how the medical school curriculum changed during the war, as described by Nakagawa Yonezo, who became a student at Kyoto Imperial University in 1945 [3]:

> When we first started studying at Kyoto, our instructor was an army doctor who told us that the practice of medicine is not for healing the sick and injured. Japan, he said, was fighting the world, and medicine itself must become a weapon.
>
> The instructor told us that animal tests alone were not sufficient for medical studies, and that human tests were also necessary. He said that such tests had actually been carried out, and showed us 16-mm movies that had been taken in Manchuria.

> One movie showed an experiment in which air was injected into the arms of living subjects to produce air embolisms. The films showed the victims in progressive stages of the condition, as they suffered to death.
>
> Another film the instructor showed us was a beheading with blood spurting from the body.

Japanese doctors also used prisoners of war in occupied territories to carry out experiments. For example, in Rabaul on the island of New Britain, Dr. Hirano sought a cure for malaria [4]. In one experiment, he injected nine malaria-free POWs with malaria-infected blood to determine whether the infected blood would protect the POWs from malaria. It did not. In another experiment, Hirano used blood from local villagers known to be immune to malaria, and injected this blood into POWs thought to be carriers of malaria. The purpose of the experiment, as he told one prisoner, was to "see what would happen." Two of the men died shortly after receiving the injections.

In another experiment, Drs. Hirano and Shigeo tested the minimum amount of food humans need to stay alive [4]. Assistants weighed 13 prisoners at the start of the experiment. The prisoners were fed what the doctors' thought would be a life sustaining diet consisting of 660 grams of peeled cassava root, one-quarter pint of soup, and one-half pint of water per meal for 30 days. By the end of the 30 days, two of the prisoners had died. The remaining 11 prisoners were weighed a second time, then placed back on the 30-day diet. Two more men died. The experiment ended as the hypothesis that the diet would sustain life was proven false.

Attempted Coverup

When Japan knew they would lose the war, Unit 731 physicians and scientists destroyed most documents, but kept some to bargain for freedom if captured. Cyanide pills were distributed to all personnel, including the Youth Corps, should they be captured. Youth Corps members were instructed to destroy the site, include the *maruta* [3]. It took them three days, leaving little evidence that Unit 731 existed. Ishii ordered all workers at Unit 731 to take it to the grave with them. One nurse at Unit 731 recalls Ishii telling her [16]:

> I'm sending you all back home. When you get there, if any one of you gives away the secret of Unit 731, I personally will find you, even if I have to part the roots of the grasses to do it.

Physician and bacteriologist Nishi Toshihide, head of Branch 673 of Unit 731, describes his role in the destruction as follows [16]:

> On August 11-12, 1945, owing to the advance of the Soviet Army, and with the object of concealing the fact that weapons for conducting bacteriological warfare had been manufactured in the Japanese Kwantung Army and that Branch 673 of Detachment 731 under my command had been involved in these criminal activities, on my orders all the branch's service premises and living quarters, equipment, materials and documents were destroyed by fire, and for the same purpose, on my orders, on August 14, 1945, poison in the shape of potassium cyanide was issued to the entire personnel (120 men) to be taken by them to

> commit suicide in the event of the danger arising of their being captured by the Soviet forces.

While effective at destroying documents and equipment, workers of Unit 731 left dead bodies. Soviet soldiers found 10,000 dead bodies near Hailar, and another 10,000 in Harbin. Experimental animals, however, were left alive in their cages [16].

On August 8, 1945, the Soviet Union declared war on Japan and shifted over a million troops from the European war (which ended when Germany surrendered unconditionally on May 7, 1945 in Reims, France, and again May 9 in Berlin, Germany) to Manchuria. This came two days after the United States dropped the atomic bomb on Hiroshima and one day before another atomic bomb was dropped on Nagasaki. Japan's emperor Hirohito announced the unconditional surrender of Japan on August 15, 1945, and signed the surrender documents on September 2, 1945. The Soviet Union claims Japan surrendered because of their assault on Manchuria [10], whereas the United States claims Japan surrendered because of the nuclear bombs.

Why would anybody choose to work at facilities such as Unit 731? Civilian employee Ogasawara Akira of the Unit describes his reasoning [3]:

> In 1937, when the Sino-Japanese War started, there was a pronouncement to the effect that the first reserve troops stationed in Manchuria would not be mobilized… Civilian employees would not be called up, and the salary was better than back in Japan. Employees could bring their families to live with them, and people working in university medical labs in Japan could dispense with the worry of being drafted into military medicine…

> I did not want to experiment on *maruta*. The major reason a lot of people joined was to protect health with hygiene, and the pay was good. At eighteen and nineteen years of age, we were getting higher salaries than the teachers who had educated us a long time ago, back in school.
>
> Manchuria from 1938 to 1940 was like heaven for Japanese. We heard that on the mainland even matches were government rationed, while we had plenty of everything… When I worked on the Ningbo biological attack, I was getting a salary of one hundred twenty yen a month. In 1940, the principal of an elementary school did not make that much.

United States Makes a Deal with the Devil

The United States knew of the Unit 731 atrocities. Although the United States held some German physicians accountable for the experiments performed in concentration camps at the Nuremberg Doctor's Trial, the United States had different plans for the Japanese physicians and scientists.

United States scientists were sent to Japan shortly after the end of the war to analyze the experiments of Unit 731 and determine whether or not data would be useful to the United States. The scientists also wanted to meet their Japanese "colleagues" [8]. By spring, 1948, General Douglas MacArthur worked out a deal with Ishii and other Japanese scientists to provide immunity in exchange for the data. One U.S. scientist mentioned what a bargain they got in obtaining the data [8]. A June 22, 1947 Top Secret document, declassified by the Nazi War Crimes and

Japanese Imperial Government Disclosure Acts, clearly shows that Japan used biological agents on the Chinese [17]:

> REURAD WAR 80671, 22ND JUNE 47, HELD ANOTHER CONFERENCE WITH [redacted] OF IPS WHO REPORTS THE FOLLOWING:
>
> 1. ON 27TH OCTOBER 1940 JAPANESE PLANES SCATTERED QUANTITIES OF WHEAT GRAIN OVER NINGPO. EPIDEMIC OF BUBONIC PLAGUE BROKE OUT 29TH OCTOBER 1940. KARAZAWAI AFFIDAVIT IN PARAGRAPH 3 BELOW CONFIRMS THIS AS ISHII DETACHMENT EXPERIMENT. 97 PLAGUE FATALITIES.
> 2. STRONG CIRCUMSTANTIAL EVIDENCE EXISTS OF USE OF BACTERIA WARFARE AT CHUHSIEN, KINGHWA AND CHANGTEH. AT CHUHSIEN JAPANESE PLANES SCATTERED RICE AND WHEAT GRAINS MIXED WITH FLEAS ON 4TH OCTOBER 1940. BUBONIC PLAGUE APPEARED IN SAME AREA ON 12TH NOVEMBER. PLAGUE NEVER OCCURRED IN CHUHSIEN BEFORE OCCURRENCE. FLEAS WERE NOT PROPERLY EXAMINED TO DETERMINE WHETHER PLAGUE INFECTED. AT KINGHWA, LOCATED BETWEEN NINGPO AND CHUPUIEN, 3 JAPANESE PLANES DROPPED A LARGE QUANTITY OF SMALL GRANULES ON 28TH NOVEMBER 1940. MICROSCOPIC EXAMINATION REVEALED PRESENCE OF NUMEROUS GRAM-NEGATIVE BACILLI POSSESSING THE MORPHOLOGICAL CHARACTERISTICS OF PETEL PESTIS. INOCULATION TESTS WERE PERFORMED ON GUINEA PIGS IN JANUARY 1941 WITH NEGATIVE RESULTS. NO EPIDEMIC OF PLAGUE OCCURRED

IN KINGWHA. AT CHANGTEH A JAPANESE PLANE DROPPED WHEAT AND RICE GRAINS, PIECES OF PAPER, COTTON WADDING AND UNIDENTIFIED PARTICLES ON 4TH NOVEMBER 1941. BETWEEN NOVEMBER 11TH AND 23RD, 5 CASES OF SUSPECTED PLAGUE OCCURRED AND 1 CASE DEFINITELY DIAGNOSED AS BUBONIC PLAGUE. NO EVIDENCE OF PLAGUE FOUND AMONG RATS AND CHANGTEH NEVER BEFORE AFFLICTED BY PLAGUE.

3. AT TIME OF FORWARDING OURAD C 53169 ONLY SOVIET SYNOPSIS OF AFFIDAVITS WERE AVAILABLE. SINCE RECEIPT URAD WAR 80671 FULL JAPANESE AFFADAVIT IT APPEARS AFFIANT WAS ENGAGED IN MANUFACTURE OF GERMS AT ISHII UNIT 1939 TO 1944. BETWEEN AUGUST AND DECEMBER 1940 ISHII, ACCOMPANIED BY 100 SUBORDINATES, LEFT FOR HANG-CHOW, CENTRAL CHINA, FOR AN EXPERIMENTAL TEST. AFFIANT MANUFACTURED 70 KILOGRAMS OF TYPHUS AND 5 KILOGRAMS OF CHOLERA BACILLI FOR THIS EXPERIMENT. 5 KILOGRAMS OF FLEAS INFECTED WITH PLAGUE WERE LIKEWISE PRODUCED. THE SUBJECT OF EXPERIMENT WAS SPRAYING OF BACTERIA BY AIRCRAFT IN THE ZONE OCCUPIED BY THE CHINESE ARMY. AFTER THE EXPERIMENT A PLAGUE EPIDEMIC OCCURRED AT NINGPO. AFFIANT WAS ADVISED OF THIS BY ENGINEER MINATO, MAJOR TAKAHASHI AND LIEUTENANT COLONEL IKARI WHO PARTICIPATED IN IT AND AFFIANT SAW CHINESE NEWSPAPERS DEPICTING INCIDENT, WHICH PAPERS WERE GATHERED BY PERSONS

SENT TO CHINA TO INSPECT RESULTS. IN WINTER 1939, ISHII TOLD AFFIANT THAT HE HAD EFFECTIVELY EXPERIMENTED WITH CHOLERA AND PLAGUE ON THE MOUNTAIN BANDITS OF MANCHURIA. ABOUT JULY, 1942, ISHII SENT TO CENTRAL CHINA AGAIN WITH ABOUT 70 SUBORDINATES CARRYING 130 KILOGRAMS OF TYPHOID AND PLAGUE BACILLI AND NECESSARY FLEAS INFECTED WITH PLAGUE. AT THAT TIME THE JAPANESE HAD EXECUTED RETREAT OPERATIONS AND ISHII INFECTED THE VICINITY OF CHUPSIEN AND YUSHAN WITH TYPHOID AND PLAGUE BACILLI. AT THE ACIA EXPERIMENTAL LABORATORY EXPERIMENTS WERE MADE 7 OR 8 TIMES DURING 1943 AND 1944 WITH PLAGUE AND ANTHRAX BACILLI ON HUMAN BEINGS. THE GANDARDES FURNISHED MANCHURIANS FOR THESE EXPERIMENTS WHO HAD BEEN SENTENCED TO DEATH. DIRECTORS OF THIS EXPERIMENT WERE ISHII AND IKARI. AFFIANT OBSERVED SOME OF THESE EXPERIMENTS AND WAS ADVISED REGARDING OTHERS BY IKARI, MAJOR HINDFUJI AND TAKAHASHI. MANY TONS OF MATERIALS FOR CULTIVATION OF BACILLI WERE ACCUMULATED IN 1944 FOR PREPARATION OF MASS PRODUCTION OF BACTERIA.

4. FROM KAWASHIMA AFFIDAVIT IT CLEARLY APPEARS THAT ISHII DETACHMENT HAD SECRET MISSION OF STUDYING USE OF BACILLI AS A WAR WEAPON. EXTENSIVE EXPERIMENTS WERE CONDUCTED BY DROPPING PEST-BACILLI INFECTED FLEAS FROM AIRPLANES

AND BY DROPPING BOMBS CONTAINING LIVE FLEAS. THESE EXPERIMENTS WERE CONDUCTED UPON ANIMALS AND 20 MANCHURIANS UNDER THE DEATH SENTENCES. KAWAHIMA AFFIDAVIT MAKES NO MENTION OF EXPERIMENTS AGAINST CHINESE TROOPS OR ON CHINESE CITIES. ACCORDING TO THIS AFFIDAVIT, MAJOR GENERAL KITANO SUCCEEDED ISHII AS COMMANDER OF THE UNIT IN AUGUST, 1942.

5. IPS OF THE OPINION THAT FOREGOING INFORMATION WARRANTS CONCLUSION THAT JAPANESE [ ] GROUP HEADED BY ISHII DID VIOLATE RULES OF LAND WARFARE, BUT THIS EXPRESSION OF OPINION IS NOT A RECOMMENDATION THAT GROUP BE CHARGED AND TRIED FOR SUCH. KARAZAWA AFFIDAVIT WOULD NECESSARILY NEED COLLABORATION AND TESTING FOR TRUSTWORTHINESS BY A THOROUGH INVESTIGATION BEFORE PROSECUTIVE ACTION IS DECIDED UPON.

The physicians and scientists who worked at Unit 731 (and other Units) re-entered civilian life, many in prestigious jobs at universities or research institutes. Three of the scientists founded the Green Cross, a Japanese blood bank. Some even found jobs with the National Institutes of Health in the United States. Ironically, Shiro Ishii could not find work, not even with those he worked for at Unit 731.

The Soviet Union charged twelve physicians and scientists from Unit 731 with war crimes, and trials were held in Khabarovsk, Russia, in December 1949. Eleven of the twelve defendants plead guilty to all charges against them, and the twelfth

defendant plead guilty to some of the charges. All twelve defendants were sentenced to prison, with time to be served varying from 2 – 25 years. Testimony from the Khabarovsk trial and testimonials from Unit 731 Exhibitions about 50 years later provides much of what we know today about Unit 731.
Japan denied allegations of Unit 731 until 1988.

One scientist who worked in Unit 731, Yoshio Shinozuka speaks of his regret 40 years later [18]:

> I am glad to be away from Unit 731. I was so stupid then. Being sent to a war crimes detention center helped me feel human again.

The Soviet Union, China, and North Korea accused the United States of utilizing what they learned from Ishii and Unit 731 in biological warfare during the Korean War. The United States denies these accusations.

Scientists carrying out these atrocities against the Chinese people and POWs were indeed evil, with Ishii seeming the Japanese equivalent of Mengele. But the U.S. government also showed its dark side in working with the evil physicians and scientists to obtain data and even employing many.

Chapter 17

Biomedical Research and Scientific Fraud

Over 50 years ago in the United States, scientists used vulnerable populations in their experiments (see chapter 12). While the government now protects these populations from scientific exploitation, scientists in colleges and universities continue to be pressured to publish peer reviewed journal articles and receive external grants to continue their research (and keep their jobs).

In the mid- to late-twentieth century, federal grant money for scientific research was relatively abundant compared to the last twenty years. The peak in funding occurred in the 1970's, when the United States spent more than 2% of its gross domestic product on research and development. By 2014, the amount was down to 0.78%. The money has dried up and I see colleagues who have been funded for years unable to get another grant, and others just starting out at universities struggling to get their first grant. Most research colleges and universities provide "start-up" funding to new faculty (ranging anywhere from $20,000 to $2,000,000 or more) to enable faculty to set up their labs and publish enough journal articles to make them competitive in grant applications.

In addition to federal grants, scientists can obtain research funds through non-profit institutions such as the American Cancer Society, American Heart Association, or the Bill and Melinda Gates Foundation. The research funds provided by these agencies are less than those by the federal government, but they are significant. For example, in 2020, the American Cancer Society (ACS) funded 660 research grants, a total of $391 million [1]. Each year, ACS receives about 1,000 new grant applications, and funds ten to fifteen percent. In contrast, the National Institutes of Health (NIH) provides $41.7 billion funding for over 300,000 research grants [2]. Thus, the U.S. government funds the majority of biomedical research conducted in research labs across the country.

With decreases in research funding comes a decrease in the number of faculty able to obtain grants, which corresponds to a more difficult road to tenure. At research-focused colleges and universities, tenure and promotions are granted primarily for publications and grants; teaching and service are required activities, but count little toward tenure or promotion.

I was lucky in that my first two grant applications to the National Institutes of Health were funded and I obtained tenure. However, I gave up on research for three reasons: 1) my teaching load doubled within ten years of being hired, decreasing the time I could devote to research, 2) I suffered mentally from the killing of rats in experiments, and 3) I preferred working with students through teaching and advising. By giving up research, I gave up my chance of promotion and retired as an Associate Professor (professorships in order are Assistant, Associate, and then Full). Being happier in my work was more important than promotion. But I could only make happiness a priority after I received tenure.

I can attest from my own experience and those of friends and colleagues that the "publish or perish" world of academia is a stressful one. For those who do not receive tenure, their lives are drastically changed as they often turn to teaching jobs in community colleges or high schools, usually associated with a salary cut. Some find jobs in industry, often with an increase in salary. If they are a health professional on faculty in a medical school or other health profession, they leave teaching and turn to clinical work where they make considerably more money. And sadly, some untenured faculty leave their field of specialty altogether.

Stress affects the way people make decisions, but the effects are not consistent. For example, stress negatively affects decision making when the person considers the stressor a threat. In contrast, a challenging stressor positively affects decision making. One person may consider the "publish or perish" concept a threat, whereas another might consider it a challenge. The one who considers it a challenge is more likely to succeed in that environment.

Unfortunately, some scientists feel a need to publish even if it compromises the quality of the science either because of the stress of tenure/promotion or for their ego. When poor quality or falsified science gets published, it sets the specific field back as scientists rely on the integrity of each other to utilize knowledge from previous studies to generate new hypotheses and design the next study to move science forward. A scientist who performed poor science deserves no trust. But he also brings a lack of trust to the field of science, and that hurts the scientific community and all of society. COVID-19 provides an example where people did not trust the science around the disease, the vaccines, and the treatments. Thus, the pandemic became worse than it had to, especially with the development of effective vaccines within a year. But how can one trust scientists developing vaccines after reading previous chapters?

Honorary Authorship

The most common form of scientific misconduct is the use of "honorary" authorship. Scientific articles often have more than one author because research is a collaborative effort. When I was an undergraduate student performing research at U.C. Davis, one of my mentors offered me authorship on a paper pertaining to research of which I completed many of the experiments. She explained to me that to be an author one has to contribute beyond the technical aspects of the experiment, including participation in data analysis and interpretation of the data taken in context with the current scientific literature. In addition, all authors have a responsibility of, at the minimum, reviewing the manuscript before it is submitted for publication. I agreed to participate at the level required and felt honored as a coauthor on a scientific paper as an undergraduate student.

Unfortunately, authorships are shared amongst colleagues to "beef up" their curriculum vitae. The International Committee of Medical Journal Editors (ICMJE) has a set of criteria for who deserves authorship associated with a given study. These criteria are [3]:

> Substantial contributions to the conception or design of the work; or the acquisition, analysis, or interpretation of data for the work; AND
> Drafting the work or revising it critically for important intellectual content; AND
> Final approval of the version to be published; AND
> Agreement to be accountable for all aspects of the work in ensuring that questions related to the accuracy or integrity of any part of the work are appropriately investigated and resolved.

These guidelines follow what I was told as an undergraduate student. But studies show scientists often disregard the guidelines.

Matawlie and colleagues performed a survey of co-authors in seven journals related to pain research [4]. Only 22% of the authors responded to the survey. Of the respondents, 13.3% reported working in a department where the senior member received authorship on all papers from that department [4]. The survey asked if there were any honorary authors on the article in question, and 13.5% responded yes, perhaps the ones who granted authorship to the head of their department. However, analysis of authorship showed that 40% of the authors did not meet the ICMJE guidelines [4].

Noruzi and colleagues performed a similar survey of authors in five cardiothoracic surgery journals [5]. Their results were more inciteful as 62.7% of the survey responders stated that at least one author had performed tasks below the level of deserving authorship, but only 25.3% of the same responders stated their article had at least one honorary author [5]. These findings indicate confusion on who deserves authorship, even though 77.1% of the responders knew of the ICMJE guidelines. Like the study of Matawlie, 34.5% of responders stated that a senior member of the department automatically received authorship on all papers [5]. A search of the literature shows similar studies in other scientific fields.

Why should honorary authorships matter? It goes back to publish or perish. Honorary authorships give the misconception of participation and knowledge of the experiment. Rewards in terms of tenure and promotion are based on these publications. Honorary authorship also decreases the integrity of the scientist(s).

Scientific Fraud

The National Institutes of Health (NIH) is the primary provider of federal funds for biomedical research. Scientists desiring research money write a detailed grant outlining the experiments they want to conduct, background information supporting the need for the research, proposed results expected from the research, and how those results contribute to our scientific knowledge, and possibly health care. Grants are peer reviewed by scientists appointed to committees by the funding organization.

The Office of Research Integrity (ORI) oversees research funded by federal agencies within the Public Health Services, including NIH. The ORI recognizes three forms of research misconduct [6]:

1) Fabrication is making up data or results and recording or reporting them.

2) Falsification is manipulating research materials, equipment, or processes, or changing or omitting data or results such that the research is not accurately represented in the research record.

3) Plagiarism is the appropriation of another person's ideas, processes, results, or words without giving appropriate credit.

If a research project is federally funded, and fabrication, falsification, or plagiarism occurs, that researcher has committed a crime and could be fined, lose eligibility for future federal funding, or be incarcerated. Only 2% of scientists self-reported research fraud on surveys, but another 33% self-reported participating in minor research misconduct. (Fanelli) When scientists were surveyed about the research practices of colleagues, 14% suspected colleagues of research fraud and 72% of minor research misconduct [7]. I wonder what these scientists consider *minor* research misconduct.

The following sections include a few examples of research fraud with major harmful effects. Some of these experiments are clinically based, meaning the experiments test the ability of a treatment to improve or cure a specific medical condition. Others study human behavior. Also included is basic biomedical research, which precedes research with direct applications to health care, as these studies form the scientific foundations on which clinical studies are based.

Clinical Scientific Fraud: Does the MMR (measles, mumps, rubella) Vaccine Cause Autism and/or Crohn's Disease?

Clinical research generally focuses on potential causes or treatments of a disease and often includes human subjects; thus, fraud in these experiments can directly lead to harmful effects to humans. Dr. Wakefield (MD) carried out possibly the most notable fraud in clinical research. Dr. Wakefield and 12 colleagues at the Royal Free Hospital in London published two studies in the *Lancet* (a prestigious scientific journal): a 1995 study claiming that measles vaccine causes Crohn's Disease, and a 1998 study claiming a correlation between the MMR vaccine and autism in children [8,9].

Vaccines

Smallpox existed in the world for over ten thousand years, showing up first in northern Africa, then the middle east, and then spreading to the western world about 1500 years ago. In the 1700's over 400,000 Europeans died of smallpox annually. When the Europeans settled the Americas, a large portion of Aztecs, Incas, and American Indians died as they had not developed any natural immunity to smallpox. As early as 430 BC, Eastern countries recognized that those who were infected with the smallpox virus and recovered could not be infected a second time [10].

Before the invention of vaccines, China utilized the process of variolation (early form of inoculation) with smallpox virus. A lancet or needle was used to take smallpox material from a pustule of an infected individual and then transfer it to a healthy individual through an induced cut to provide protection from the virus. Variolation worked and quickly spread west. However, the practice had several risks including severe infections, which are much less likely with vaccinations given intramuscularly (IM) today [11].

Vaccines are the safest and most effective protection against infectious diseases. Our immune systems include *specific* and *nonspecific* mechanisms of protection. The nonspecific mechanisms protect against any infectious agent, but is not as robust as the specific mechanism. During the specific immune response, certain blood cells and molecules in the body identify the presence of an antigen, which is an infectious agent or a specific portion of an infectious agent. For example, with COVID-19, the antigen was a portion of the spike protein. On the first exposure to an antigen, the nonspecific immune response reacts relatively slowly. However, that first exposure starts a protective mechanism whereby the immune system becomes more sensitive to that specific antigen through the production of *memory cells*, immune cells that specifically attack that one antigen. Thus, the specific immune response to a second exposure to that same antigen starts much quicker and is more robust, providing better protection against the infecting agent. The first exposure can occur naturally or through vaccination.

Vaccines are designed to activate the specific immune response, often for the first time. Thus, vaccines function as antigens that will activate the immune response, priming the immune system for a subsequent exposure to the antigen. Earliest vaccines utilized small amounts of the live infectious agent, but today, vaccines can be produced from dead infectious agents, portions of an infectious agent that will activate the immune response but not cause illness, or, as developed for the first time for the COVID-19 pandemic, nucleotides (such as RNA and DNA) that will code for synthesis of the antigen by our own cells. This new technique allows for the relatively rapid production of vaccine, and rapid alteration of the vaccine as new strains of the infectious agent develop.

In the 1960's, three vaccines were developed against three different diseases: Measles in 1963, mumps in 1967, and rubella in 1969. Measles is highly infectious with 90% of those coming into contact with the virus (if not vaccinated) catching the disease. Children are more susceptible than adults. The measles virus could survive in the air for over 2 hours after an infected patient left the room. Two out of every 1500 children with measles died of complications. Before the availability of vaccines, isolation of the patient decreased the spread. The first measles vaccine used live virus, but in a few years, a safer vaccine was developed. Mandatory vaccinations eliminated measles in the United States in 2000, but it has returned with frequent measles outbreaks as vaccine use has decreased [12]. As of October 2025, the CDC reported 1563 measles cases in the United States that year, including 1037 children [13]. Of those,

92% were unvaccinated. The World Health Organization estimates the measles vaccine saved over 60 million lives worldwide from 2000 to 2023 [14].

Dr. Maurice Hilleman developed the mumps vaccine after one of his daughters got infected. He collected a sample of virus from his infected daughter, made the virus weaker, and then inoculated his other daughter as well as other subjects [12]. The vaccine was effective. Mumps has never been fully eradicated in the United States, but vaccination decreased mumps infection rates to less than 1% of the pre-vaccination rate [15].

Rubella, or German measles, is most dangerous to pregnant women, causing miscarriages, still births, or various birth defects. Dr. Hilleman, who developed the mumps vaccine, also helped develop the rubella vaccine. Through vaccination, rubella was eradicated from the United States in 2004, but like measles, has returned [16].

Dr. Hilleman developed a combination measles, mumps, and rubella (MMR) vaccine in 1971, and Dr. Plotkin improved it in 1979. Today, the CDC recommends that children receive two MMR vaccinations, at 12-15 months and 4-6 years of age. Older children not yet vaccinated should receive two vaccinations at least 28 days apart, and adults not yet vaccinated should receive at least one MMR vaccination [17].

Numerous other vaccinations are available including chicken pox, flu, hepatitis A, hepatitis B, tetanus, etc. The most controversial vaccine has been the COVID-19 vaccine. The controversies regarding the COVID-19 vaccine stem from political, scientific, and social media information and misinformation that left the public confused and skeptical.

I believe in vaccinations, and I participated in the Pfizer study for the COVID-19 vaccine. I was eager to participate in a study as I believed a vaccine could stop the pandemic. The effectiveness and complications of the COVID-19 vaccine and the controversies surrounding them are beyond the scope of this book, and the story may not yet be complete as the long-term effects of the vaccine are unknown, and the virus continues to mutate producing new variants. To date, alterations in the COVID-19 vaccine have kept up with the new variants, and I am still alive after receiving each recommended booster, and I have yet to catch any variant of COVID-19. I am currently enrolled in a study testing an oral form of the COVID-19 vaccine.

Autism Spectrum Disorder

Autism is a neurodevelopmental disorder affecting a person's social behavior and ability to learn, with a prevalence in the United States of 1 in 59 children and increasing [18]. It is a spectrum disorder meaning symptoms and severity vary considerably between individuals. Pediatricians usually diagnose Autism Spectrum Disorder (ASD) because symptoms appear as early as 18 months old and continue developing with age. ASD has a high genetic predisposition, but no gender, ethnic, nor racial correlations. ASD cannot be cured, but treatments are available [19].

According to the American Psychiatric Association's Diagnostic and Statistical Manual, criteria for diagnosis of ASD include the following:

1) Persistent deficits in social communication and social interaction ranging from decreased or abnormal conversation reduced sharing of interests to reduced sharing of interests or

emotions to failure to initiate or respond to social interactions.

2) Persistent deficits in nonverbal communication.
3) Persistent deficits in making and keeping relationships with others.
4) Exhibit two of the following four types of restricted repetitive behaviors
    a. Stereotyped or repetitive motor movements, use of objects, or speech
    b. Insistence on sameness, inflexible adherence to routines, or ritualized patterns of verbal or nonverbal behavior
    c. Highly restricted, fixated interests that are abnormal in intensity or focus
    d. Hyper- or hyporeactivity to sensory input or unusual interest in sensory aspects of the environment

The severity of ASD depends on the level of social communication and interaction impairments and restricted repetitive patterns of behavior.

ASD often occurs with other health problems, including sleep disorders, seizures, attention deficit hyperactivity disorders (ADHD), anxiety, and mood disorders [18]. The increase in ASD prevalence is partially due to a greater awareness of the disease, updated guidelines for diagnosis, and better record keeping [18].

In the 1990's, Dr. Andrew Wakefield studied possible associations between ASD and gastrointestinal disorders such as Crohn's Disease and between ASD and the MMR vaccine.

The Wakefield Studies

Andrew Wakefield graduated from medical school in 1981, initially working toward becoming a professor of surgery. Wakefield took a two-year fellowship at Toronto General Hospital in Canada, working under surgeons hoping to be the first to successfully transplant the whole bowel. However, Wakefield found himself more interested in laboratory work, preferring research to patient care [20].

In 1987, Wakefield generated an hypothesis that inflammatory bowel diseases, such as Crohn's Disease, were caused by a virus interfering with blood flow, thereby causing inflammation. When he finished his fellowship in Toronto in 1988, Wakefield took a research position at the Royal Free Hospital in London. Seeking the cause of Crohn's Disease, Wakefield and colleagues first discovered that intestinal tissue from Crohn's patients showed signs of inflammation and tissue death in the intestinal blood vessels. Next, Wakefield sought to determine what caused the blood vessel damage, and he hypothesized a viral infection. A review of viruses suggested measles as a possible culprit because it causes ulcers and inflammation in the gastrointestinal tract [20].

Wakefield put together a team of physicians at the Royal Free Hospital to determine whether measles or the MMR vaccine could cause Crohn's Disease, other gastrointestinal inflammation or ASD. In 1995, the team published a study showing damage in the gastrointestinal blood vessels indicative of a decrease of blood flow. Further, subjects with Crohn's Disease had evidence of measles virus that persisted in the intestines well past the infection or vaccination. Although Wakefield claimed the persistent virus was from a vaccine, they could not prove this with their technique [8]. Other scientists using more sensitive techniques for detecting measles virus or vaccine could not replicate Wakefield's findings [21,22].

In 1998, Wakefield's team published a study showing a temporal relationship in children with ASD and gastrointestinal symptoms with the MMR vaccine [9]. Specifically, they hypothesized that the MMR causes inflammatory bowel disease, which causes ASD. This was an *hypothesis*, a question that needs further study to determine if true or not. Regardless, this study powered the anti-vaccine movement, as evident in a large drop in children receiving the MMR vaccine in England and, subsequently, the world.

The day the study was published in The Lancet, February 28, 1998, the Royal Free Hospital held a party for Dr. Wakefield and his colleagues celebrating his accomplishments.

Two weeks after Wakefield's publication, Chen and DeStefano from the CDC published a commentary in the Lancet critiquing Wakefield's studies [23]. They stated the bias in selecting subjects and the dependence of parental recall for onset of symptoms as two methodological problems with the study. In addition, Wakefield never showed a causality association between the MMR, autism, and/or Crohn's Disease [24]. To date, no studies have shown a causal relationship between vaccines and autism [25]. Most important, Wakefield's publication in 1998 was retracted by the Editor of the Lancet for unethical practices during the research. Regarding this last issue, in 2004, 10 of 12 coauthors of the article published a retraction regarding the interpretation of the data. They stated, "no causal link was established between MMR vaccine and autism as the data were insufficient. [26,27,28]"

John Deer showed several areas of fraud with the 1998 Wakefield study, which were substantiated by the UK General Medical Council, and, to a partial extent, by the Editor of the Lancet, Dr. Horton [20].

1) A facility-based ethics committee must approve research on the facility utilizing human subjects. Wakefield, apparently failed to obtain proper approval prior to his study on the 12 children for the 1998 publication. Dr. Horton accepted Wakefield's claim that the ethics committee approved the experimental protocol, but there was confusion as to whether the approved protocol was for the experiments undertaken. Wakefield also argued that the procedures were all diagnostic in nature, and the experimental portion of the study was the biopsy of tissue. Although one could argue the need for gastrointestinal biopsies to appropriately diagnose the gastrointestinal symptoms exhibited by the children, MRIs and EEGs were not diagnostic in these cases.
2) Wakefield claimed that general practitioners referred all 12 children used in the study. However, most parents of the children sought out Dr. Wakefield for his skepticism of the MMR vaccine. In these cases, Dr. Wakefield requested the parents to get a recommendation from the general practitioner thereby creating a biased sample.

3) Biopsy data analysis was altered following the initial analysis. Wakefield claimed this was due to the necessity of reanalyzing the data with better trained pathologists.
4) Wakefield was hired by an attorney to help find an association between the MMR vaccine and some ailment for lawsuit purposes, creating a conflict of interest.

I consider the conflict of interest most critical in assessing Wakefield's intentions. Scientists are required to disclose any conflicts of interest when they publish research articles. The guidelines for conflicts of interest have become more stringent since 1998; however, Wakefield denies a conflict of interest existed in 1998. Whether the Lancet required a disclosure of the conflict at the time of publication might be arguable, but being paid to prove a condition exists is a conflict! When an agreement is made by the scientists carrying out the research that has the potential to bias those scientists, there is a conflict of interest. In this instance, Wakefield was paid to find a link between the MMR vaccine and some ailment. Wakefield first showed a link between the MMR vaccine and gastrointestinal inflammatory diseases in 1995 [8]. His next hypothesis stated that measles (through infection or vaccination) caused the gastrointestinal inflammatory disease, which in turn caused autism. However, many studies before and after his 1995 study found no correlation between the MMR and gastrointestinal inflammatory disease, and the methodology has been questioned by other scientists [20]. One aspect of the Scientific Method is that results must be reproducible.

In 1999, Mark Pepys became the new head of medicine at Hampstead Royal Free Hospital. He had 25 requirements before he would accept the job including Wakefield's removal from the department. However, Wakefield had tenure and his dismissal would be problematic. Thus, Dr. Pepys made Wakefield an offer to either take a one year paid leave of absence to work with his biotech business, or repeat his study using a higher number of subjects referred only from general practitioners, including controls in his study, using blind analysis of data (analysis which requires researchers to be unaware of the source of data they are analyzing), using definitive genetic sequencing, and all techniques to be used had to be approved by the ethics committee. In addition, the study had to be repeated at two outside institutions to reproduce the results. Dr. Llewellyn Smith, provost and president of Hampstead, also met with Wakefield supporting this option. Llewellyn said, “We urge you not to publish prematurely observations that you may have or be making in this field. [20]”

As a scientist, I would have jumped on this opportunity – an experiment already designed and paid for, experience to carry out the experiment, and a chance to retain being a respectable scientist, not to mention keeping a job. Wakefield initially accepted the proposal to repeat the experiments, but never followed through and ultimately resigned his position and moved to Texas to carry on his work, mostly promoting the dangers of vaccines [20].

Deer reports another area of conflict that Pepys also alluded to – Wakefield was developing and working with biotech companies. With the biotech companies, Wakefield hoped to develop a single shot measles vaccine to be used in place of the MMR. If people believed his work on the hazards of the MMR, such a product would prove highly profitable [20].

In 2007, the UK General Medical Council Fitness to Practice Panel held a hearing on Dr. Wakefield’s research. When the hearing concluded in 2010, the General Medical Council found Dr. Wakefield guilty of serious professional misconduct and stripped him of his license to practice medicine [28]. Wakefield claimed the actions of the General Medical Council were political in nature, to protect the MMR vaccine.

During the hearing, Dr. Wakefield provided no documents for defense, nor did he appeal the decision. However, he did write a book, **Callous Disregard, Autism and Vaccines—the Truth Behind a Tragedy** [29], which read as a defense against the actions of the General Medical Council, the Editor of the Lancet (Dr. Horton), and the allegations of John Deer. Further, Wakefield filed several lawsuits against media and John Deer, all of which came to no fruition.

In his book, Wakefield presented himself as a martyr, willing to stand up against "Big Pharma" and politicians to protect the children. Wakefield also expressed concern that giving combined vaccines, such as MR or MMR is much more dangerous than giving individual vaccines separated in time [29]. As mentioned earlier, Wakefield continues to promote the danger of vaccines to public health, often being paid to speak on his unwarranted concerns [20].

As Wakefield confronts complaints of his 1998 research paper, he frequently states they did not *prove* the MMR causes autism, the researchers simply formed a hypothesis [29]. Thus, his promoting dangers of a vaccine based on his forming an hypothesis is misleading.

Dr. Joel Harrison (PhD and Master of Public Health) wrote an article analyzing the science in Wakefield's book point-by-point [28]. Following are the major findings of Harrison [28]

Wakefield made claims that the UK started administering booster shots with measles vaccine in it prior to any safety studies. He even claimed communicating directly with Swedish vaccine researcher Dr. Christensen about any possible safety studies and, according to Wakefield, she responded that only a minimal blood study for antibodies had been done. Harrison responds that Christensen has two studies published in the *British Medical Journal* (one of which was cited by Wakefield in his claim) showing the safety of the measles vaccine and booster, and Harrison communicated directly with Christensen verifying her finding of safety. In addition, Wakefield cited in his book a study showing safety of the vaccines by a third Swedish study.

In 1997, Wakefield claims he wrote a 250 page "comprehensive review of all safety studies performed on measles, MR, and MMR vaccines and re-vaccination policies," which contributed to his opinion that the UK adopted its policy on both the MMR and its booster without demonstrated safety. This report has never been made public, although Wakefield did publish an article in 2000 citing 8 studies regarding safety issues. A rebuttal to his publication referred to about 30 safety studies, which reported few adverse effects [30].

Wakefield accused the UK of approving the use of a specific strain of mumps vaccine (Urabe strain) in the MMR after Canada withdrew this strain from their MMR vaccine because of association with meningitis. Canada approved usage of their

MMR vaccine in 1986, but recalled certain batches of the vaccine in the Ontario province in October 1988. The UK started using the MMR vaccine on October 1, 1988, whereas Canada withdrew their approval in May 1990. In fact, Canada cited studies in the UK (as well as Canada and Japan) in their withdrawal of approval. The UK withdrew its Urabe-containing MMR on September 14, 1992. Harrison also cites several studies showing the lack of severity and low rate of occurrence of the Urabe-induced meningitis, and he describes the UK vaccine monitoring process whereby physicians or individuals may report any adverse responses to a vaccination. The UK took longer to withdraw its approval of the Urabe strain MMR than Canada because the danger of vaccine-induced mild meningitis were considerably less than the danger of mumps itself, and the safer vaccine was not yet available to the UK. Once a safer strain of mumps vaccine was available for the MMR, UK withdrew its approval of the Urabe strain.

Wakefied claims that measles-containing vaccines cause anaphylaxis, which could be deadly. He writes, "Without prompt treatment, this reaction, which occurred in 1 in 558 recipients of an MCV, might well have been fatal… Trying to persuade parents of the merits of an MR campaign on the basis of up to 50 possible measles deaths while ethically warning them of the possibility of up to 14, 337 anaphylaxis deaths from the MR vaccine would have doomed the campaign to failure." However, a Swedish study reported only seven

> cases of anaphylaxis out of 588, 300 immunizations over two years with no deaths, one Finnish study reported no permanent harms observed with the immunization of 430,000, and a second Swedish study reported no "full blown case of anaphylaxis" with immunization of 350,000.
>
> Wakefield cites a Danish study by Plesner, *et. al.*, showing that ataxia (irregular movement) occurs following MMR vaccinations in up to 1/1000 or 1/4000, but not with any of the individual vaccinations. However, Plesner reported only 24 cases of temporary and mild gait disturbances of 362,000 vaccinations. Plesner states in his report, "Health authorities might not take note of this adverse reaction to measles, mumps, and rubella vaccine because it seems to be a slight and temporary condition." In a subsequent study by Plesner, *et. al.*, general practitioners reported 41 movement impairments following 533,000 vaccinations from 1987 to 1996, a frequency of 8 per 100,000. Plesner *et. al.* also stated that parental questionnaires had descriptions of possible movement disturbances in 1 per 1000-4000 vaccines. Wakefield also cites a Swedish study by Taranger *et. al.* supporting that the MMR causes ataxia, but they reported only 5 cases after 700,000 vaccinations, with symptoms disappearing in 1-3 days.

To a lay person, Wakefield's book may read as strong evidence that the MMR vaccine causes more harm than good. However, Wakefield takes from the scientific literature only pieces that support his claim, with many of these pieces misrepresented. A good scientist does not do this. Harrison shows the flaws in Wakefield's scientific reasoning. But Brian Deer does more. He showed the flaws in Wakefield's so-called science.

In 2004, Journalist John Deer published his first of several reports in *The London Sunday Times* providing evidence of unethical and fraudulent practices by Dr. Wakefield [31].

Deer published a more thorough analysis in the *British Medical Journal* that proved Wakefield committed deliberate fraud [32]. Deer interviewed parents of some of the children in the study, and some of the indicators of fraud are provided below:

> Subject 11's father: "Please let me know if Andrew W has his doctor's license revoked," wrote Mr 11, who is convinced that many vaccines and environmental pollutants may be responsible for childhood brain disorders. "His misrepresentation of my son in his research paper is inexcusable. His motives for this I may never know."
>
> Subject 2's mother: "According to the paper, child 2 had his 'first behavioural symptom' two weeks, not six months, after MMR. This was derived from a Royal Free history (citing 'headbanging' and 'screaming' as the start) taken by Mark Berelowitz, a child psychiatrist and a coauthor of the paper. He saw Mrs 2 during the boy's admission, at age 8, after she had discussed her son's story with Wakefield."

> Subjects 6 and 7 were brothers: "The paper did not reveal that two of this trio were brothers, living 60 miles south of the hospital. Both had histories of fits and bowel problems recorded before their MMR vaccinations. The elder, child 6, aged 4 years at admission, had Asperger's syndrome, which is distinct from autism under DSM-IV, is not regressive, and was confirmed on discharge. His brother, child 7, was admitted at nearly 3 years of age without a diagnosis, and a post-discharge letter from senior paediatric registrar and Lancet coauthor David Casson summarised: "He is not thought to have features of autism."

In **Callous Disregard**, Wakefield denies these accusations by Deer.

Robert Horton, editor of The Lancet, published a statement with several allegations of scientific misconduct and fraud [33]:

> (1) That, contrary to a statement in the Lancet paper, ethics approval for the investigations conducted on the children reported in the study, some of them highly invasive (e.g., lumbar puncture), had not been given.
> (2) That the study reported in The Lancet was completed under the cover of ethics approval for an entirely different study of 25 children with "A new paediatric syndrome: enteritis and disintegrative disorder following measles/rubella vaccination".
> (3) That, contrary to the statement in the Lancet paper that children were "consecutively referred to the department of paediatric gastroenterology"

at the Royal Free Hospital and School of Medicine, children were invited to participate in the study by Dr Andrew Wakefield and Professor John Walker-Smith, thus biasing the selection of children in favour of families reporting an association between their child's illness and the MMR vaccine.

(4) That the children who were reported in the Lancet study were also part of a Legal Aid Board funded pilot project, led by Dr Wakefield—a pilot project with the aim of investigating the grounds for pursuing a multi-party legal action on behalf of parents of allegedly vaccine-damaged children, the existence of which was not disclosed to the editors of The Lancet.

(5) That the results eventually reported in the 1998 Lancet paper were passed to lawyers and used to justify the multiparty legal action prior to publication, a fact that was not disclosed to the editors of The Lancet.

(6) That Dr Wakefield received £ 55 000 from the Legal Aid Board to conduct this pilot project and that, since there was a substantial overlap of children in both the Legal Aid Board funded pilot project and the Lancet paper, this was a financial conflict of interest that should have been declared to the editors and was not.

Publicity of the study decreased MMR vaccinations in children. As stated by Rao and Andrade, "Despite the small sample size (n=12), the uncontrolled design, and the speculative nature of the conclusions, the paper received wide publicity, and MMR vaccination rates began to drop because parents were concerned about the risk of autism after vaccination." [34] In fact, Dr. Wakefield became a leader among anti-vaccine groups.

In 2002, Wakefield and colleagues published a second paper linking MMR vaccine to autism, but this paper was also flawed in design [35]. Wakefield studied two groups of children, 91 with autism and 70 without autism. Intestinal biopsies were performed and the tissue analyzed for the presence of measles virus. Startling, 75 of the 91 children with autism showed measles virus in their intestines, whereas only 5 of 70 children without autism showed measles in the intestine. This study seemed more damaging to the MMR vaccine than the 1998 study until details were revealed. First, the autism and non-autism groups were not matched for whether or not they had the MMR vaccine, even though that data were readily available to the investigators. If not matched, the data are meaningless. Second, the technique used to identify the measles virus in the intestines, although very sensitive, could not distinguish between a measles vaccine or natural measles infection [36].

In comparison, Brent Taylor and colleagues studied the relationship between children with autism and the MMR vaccine. Taylor examined the incident and age of onset of autism in two groups of children from the North Thames region: those who received the MMR vaccine and those who did not. They reported three major findings: 1) The percentage of children vaccinated was the same in children with autism as in children without autism. 2) There was no difference in the age of diagnosis of autism in vaccinated and unvaccinated children. 3) The onset of symptoms of autism did not occur within two, four, or six months of receiving the MMR vaccine [36].

Anti-vaxxers

People opposed to vaccinations, "anti-vaxxers," were around before Wakefield's experiments. However, his publications and public appearances greatly decreased the number of MMR vaccinations given to children under 24 months of age in Great Britain. The country promoted public health and vaccinations in the 1970's and 1980's, such that in 1996, before Wakefield's publication, 91.8 percent of children were vaccinated with the MMR; this number dropped to just 79.9 percent in 2004 [37]. MMR vaccinations in the United States remained stable at near 91 percent during the same time frame [38] and is still about 91 percent today [39].

The WHO recommends higher vaccination rates to prevent future outbreaks, and numerous countries have reached 99 percent and above rates [40]. Western countries are scarce on that list for two reasons: 1) some countries on the 99 percent rate mandate vaccinations and 2) some western people are skeptical of science and healthcare, as evidenced during the COVID-19 pandemic.

During his first term, President Donald Trump's "operation warp speed" led to development of a COVID-19 vaccine in about one year. However, skeptic people avoided the vaccines. Trust between people and their governments and health officials declined due in part to studies such as Wakefield's, and due to widespread misinformation. Even before COVID, anti-vaxxers distrust of science and government denied vaccination of their children to diseases that were almost eradicated from first world nations, such as measles and whooping cough [25]. As a consequence, these diseases are on the rise once again. In his second term, President Trump appointed Robert F. Kennedy Jr.

as United States Secretary of Health and Human Services, which is causing more distrust in healthcare.

Robert F. Kennedy, Jr.

Robert F. Kennedy, Jr. grew up in the prominent Kennedy family including his assassinated father (RFK, Sr.) and President John F. Kennedy. He lived a troubled youth, using cocaine and heroin from age 15 to 29. But he turned his life around and became a successful environmental attorney.

Kennedy won numerous major cases protecting the environment. He also helped develop non-profit environmental protection groups. He has made significant contributions to environmental protection. But his only real experience in healthcare is participation in the anti-vaxxer movement, and understanding some of the environmental factors affecting health. He has no formal education related to science or healthcare.

Robert Kennedy never too seriously pursued any political offices until the presidential election of 2024. On April 8, 2023, Kennedy filed his candidacy for the Democratic Presidential Nominee. However, on October 9, he entered the race as an Independent candidate. Running as a spoiler, Kennedy supposedly offered to withdraw from the race and endorse Kamala Harris, and later Trump, in exchange for a cabinet position. On August 23, 2024, Kennedy withdrew and endorsed Donald Trump for President.

On February 13, 2025, the Senate confirmed Robert Kennedy as the Secretary of Health and Human Services, but not with opposition. More than 80 organizations and 75 Nobel laureates recommended against Kennedy's confirmation. During the hearing, Senator Bill Cassidy, MD, of Louisiana stated the following concern [41]:

> As a physician who has been involved in immunization programs, I have seen the benefits of vaccinations. I know they save lives. I know they are a crucial part of keeping our Nation healthy.
>
> Now Bobby, I have learned you have got a tremendous following. … And there are many who trust you more than they trust their own physician. And so the question I need to have answered is what will you do with that trust?
>
> … I have constituents who partly credit you for their decision to not vaccinate their child. Now that is a real conversation. And I am hearing from them, and they want you confirmed.
>
> Now, you are going to tell us this week, … that you are pro-vaccine. So what will you tell the American mother? Will you tell her to vaccinate her child or to not, or to have a conversation with her doctor, but for many that will be permission to not vaccinate their child. We know that to be the case.
>
> Your past of undermining confidence in vaccines, with unfounded or misleading arguments, concerns me. … Can data and information change your opinion, or will you only look for data supporting a predetermined conclusion? … You

> have the responsibility to restore trust in our public health institutions in this position you have.

Parts of Robert Kennedy's response to Cassidy's (and Senator Bernie Sander's) comments follow [41]:

> Before I conclude, I want to make sure this Committee is clear about a few things. News reports and many in the hearing yesterday have claimed that I am anti-vaccine and anti-industry. Well, I am neither. I am pro-safety. I am pro good science. I worked for 4 years to raise awareness about mercury and other toxins in fish, and nobody called me anti-fish. All of my kids are vaccinated, and I believe vaccines have saved millions of lives and play a critical role in health care.
>
> In my advocacy I have often disturbed the status quo by asking uncomfortable questions, and I am not going to apologize for that. We have massive health problems in our Country that we must face honestly. And the first thing I have done every morning for the past 20 years is pray to God that he would put me in a position where I can end the chronic disease epidemic and protect our children. That is why I am so grateful to President Trump for the opportunity to sit here before you today and seek your support and your partnership in this endeavor.
>
> I will conclude with a promise to you, Mr. Chairman, to the Members of this Committee, to

> the President, and most of all to the tens of millions of parents across America, especially the moms who have propelled this issue to center stage. Should I be so privileged as to be confirmed, we will make sure our tax dollars support healthy foods. We will scrutinize the chemical additives to our food supply. We will remove the financial conflicts of interest from our agencies. We will create an honest, unbiased, science-driven HHS, accountable to the President, to the Congress, and to the American people. We will reverse the chronic disease epidemic and put the Nation back on the road to good health.

On June 9, 2025, Robert Kennedy fired all 17 members of the CDC Advisory Committee on Immunization Practice, claiming too many had pharmaceutical company ties [42]. The panel was replaced with vaccine skeptics, conspiracy theorists, and others all hand-picked by Kennedy. Needless to say, major changes in vaccination policies are occurring as this book is being written. Most importantly, Kennedy and his committee are pushing to remove the mandate for the MMR vaccine, and recommend separate vaccinations for measles, mumps and rubella. What's another four shots to babies? Further, policies on the COVID-19 vaccination are changing such that only elderly and those with underlying diseases are likely to have the vaccine recommended, and therefore, only they will have the vaccine covered by their insurance companies. Many healthcare providers are outraged, as they should be.

On June 10, the American Medical Association called for a Senate investigation of Robert Kennedy [43]. They also asked Kennedy to reverse his decision to remove all 17 members of the CDC Advisory Committee on Immunization Practice. I found a medical student delegate for the AMA to have provided keen observations in these excerpts from the AMA discussion [43]:

> We do not want to be on the wrong side of history," Goldman [President American College Physicians] said. "The country is burning down. Our infrastructure is burning down. Whether the outcome of this investigation is preordained should not determine our ability and desire to ... take a stand, fight for what we believe in, and ask the government to do their job."
>
> Priya Desai, a delegate who spoke on behalf of the medical student section, rejected the idea of a "poison pill" that might hurt the AMA's relationship with government.
>
> "The notion to not 'poison pill' the relationship we've had has proven to fail significantly," she said. "We did not speak up back in November with the nomination of RFK [Jr.] ... We did not speak up back in January ... when he was officially endorsed."
>
> "Congratulations, this poison pill we have swallowed. It is time for us to act now," she added.

On August 27, 2025, Robert Kennedy orchestrated the firing of CDC Director Susan Monarez. Several senior CDC officials resigned after his action. On October 11, 2025, during a government shutdown, Kennedy fired over 1000 Public Health Service scientists, doctors, and public health officials [44].

On October 6, 2025, HHS and the CDC announced changes to children vaccination schedules. The MMR vaccine had been combined with the varicella (V) vaccine for chickenpox, as the MMRV vaccine [45,46]. For children under 4, the MMR and V vaccines will be separate for the first dose, but the booster can be the combined MMRV. This is a minor change in that the separated MMR and V vaccine for the first dose was already recommended, but parents could opt to have the MMRV. More concern will be raised if the MMR vaccine also becomes separated into three vaccines: mumps, measles, and rubella.

More changes are likely during the next three years under Robert Kennedy Jr. Only time can tell what changes and whether the changes improve or hurt the health of the nation. If immunization rate decreases, as is likely due to a decrease in trust of vaccines simply with the appointment of Kennedy, then thousands of U.S. children are likely to die.

Clinical Scientific Fraud: Joachim Boldt Falsification of Data

Although these studies did not receive the public notoriety of Wakefield's, German anesthesiologist Joachim Boldt repeatedly performed clinical trials without appropriate approval. In 2011, sixteen journals questioned the ethics of Boldt's research in 89 of 102 publications [47]. According to the research watchdog Retraction Watch, by 2023, Boldt has the most retracted publications of any scientist, with 184 [48].

Dr. Joachim Boldt worked for the University Hospital of Giessen, Germany until 1996, then moved to the City Hospital of Ludwigshafen, Germany, where he served as chief anesthetist. Boldt was a prolific researcher, publishing more than 100 clinical research papers. Some changes in clinical protocols for surgery and intensive care patients were based on his research.

The research that ultimately exposed Dr. Boldt's misconduct studied methods of maintaining blood pressure with intravenous fluids for surgical and intensive care patients. Blood pressure must be maintained within a normal range as too low a blood pressure causes decreased blood flow to the brain and too high a blood pressure can result in damaged blood vessels or stroke. A person's blood pressure depends on basically two factors: how narrow the blood vessels are (narrower vessels means higher pressure) and the blood volume (more blood means higher pressure). Dr. Boldt's research focused on this second factor.

Water makes up approximately 60% of the human body. This water can occur in two places: inside cells (intracellular fluid) or outside cells (extracellular fluid). Blood contains both cells (red and white blood cells) and fluid (plasma), and thus has both types of body water. The extracellular fluid in blood contributes directly to blood pressure.

But extracellular fluid is not only located in blood. We have fluid scattered between our cells throughout our bodies. The extracellular fluid can move across the walls of blood vessels, thereby increasing or decreasing blood volume. If fluid enters the blood vessels, then blood volume and thus blood pressure increase. Dr. Boldt studied one of the mechanisms to enhance the movement of fluid into the blood to increase blood pressure under clinical conditions. Water moves by osmosis in the direction of higher concentration of solutes. Proteins in blood provide the primary osmotic force drawing water into the blood vessels, the most common blood protein being albumin.

When manufacturing an intravenous (IV) fluid, one must consider the osmotic force of any solutes within the fluid. For example, 0.9% saline solution has the same osmotic force as our extracellular fluid and thus administration of 0.9% saline IV will increase the volume of blood but not change its osmotic force. Likewise, 5% dextrose (dextrose = glucose) solutions have the same osmotic force as our extracellular fluid. Thus, 0.9% saline and 5% dextrose solutions are often given IV to patients.

In certain clinical cases, the IV fluid should exert a greater osmotic force to increase the movement of fluid into the blood vessels to maintain blood pressure. Saline and glucose are both substances utilized by our cells, and it is desirable to maintain normal concentrations. Thus, Dr. Boldt looked for substitute molecules to increase the osmotic force of intravenous fluids. The molecules he was most interested in were colloids, particles evenly suspended in the fluid.

Chemical manufacturing companies including B. Braun, Baxter, and Fresenius Kabi funded Dr. Boldt's research utilizing hydroxyethyl starch (HES) as a colloid for intravenous fluids [49]. He published numerous studies supporting the use of HES, and he gave paid speeches at medical conferences promoting HES [50]. In 2009, Boldt published a paper on the benefits of HES for cardiopulmonary bypass patients in the journal Anesthesia and Analgesia [51]. The editors of Anesthesia and Analgesia retracted this paper in 2010 because Boldt had not obtained prior Institutional Review Board (IRB) approval for his research, a clear violation of human rights [52]. This retraction started a domino effect. An investigation into Boldt's studies showed that not only did he rarely have IRB approval for his clinical studies, but he frequently falsified data [52].

Unfortunately, before Boldt was caught for research misconduct, his publications increased the use of HES in intravenous infusions. Meta analyses of his and other studies of HES showed that it was safe for use. However, on removal of his fraudulent studies, the same meta analyses found HES dangerous.

Boldt's fraudulent research created direct risk to patients. Thankfully, colleagues uncovered his fraud much quicker than the fraudulent work of Wakefield.

Chapter 18:

Psychological Experiments on the Potential for Evil

We can argue about what makes people evil, but some psychologists believe we are all capable of it.

After the end of World War II, Nazi doctors and soldiers who participated in the extermination of Jews or unethical experiments on Jewish and other prisoners avoided (or tried to avoid) prosecution in the Nuremberg War Trials by claiming they were "following orders." Allies, meanwhile, said they would never have participated in the execution of Jews, and following orders was not a valid excuse. Two psychologists headed studies on human behavior to see if people would follow orders to the point of harming others. Milgram tested whether ordinary people would harm others out of obedience to a higher order. Zimbardo tested how people responded to being given authority over others. Both experiments suggested *homo sapiens* are not as human as we would like to think. In addition, both experiments faced ethical backlash for the design of the studies.

The Milgram Obedience Experiments, 1960s

Stanley Milgram, a psychologist at Yale University, wrote [1]:

> Obedience is the psychological mechanism that links individual action to political purpose, the dispositional cement that binds men to systemic authority. Facts of recent history and observations in daily life suggest that for many persons obedience may be a deeply ingrained behavior

> tendency, indeed, a prepotent impulse overriding training in ethics, sympathy, and moral conduct.

Milgram tested whether a person would follow orders of an authority figure even if it meant going against their conscience and harmed others. He recruited male subjects through a newspaper ad looking for participants in a study on memory and learning. Forty subjects between the ages of 20 to 50 across different occupations were chosen to participate and paid $4.50 for doing so.

Milgram deceived the subjects regarding the purpose of the study by informing them the study would determine if punishment could improve learning. The experiment started with the supposed random assignment of subjects as a "teacher" or a "learner," with all subjects assigned as teacher. The learner was one of the experimenters. A third person participated in the experiment as the "authority figure" in charge of the experiment [2].

The experimenter had the subjects meet and draw their assignment from a hat, feigning the random assignment (both slips contained "teacher"). Immediately after their assignments, the experimenter took both the teacher and learner to an adjacent room where the learner was strapped into an "electric chair" and an electrode was attached to the learner's wrist. The experimenter told the teacher and learner that the electrode was connected to a shock generator in the other room. The learner then asked about the shocks, and the experimenter responded, "Although the shocks can be extremely painful, they cause no permanent tissue damage." [1]

The experimenter took the teacher to another room, had him sit in front of a machine called "Electric Shock Generator," and gave him instructions to apply electric shocks when the learner made mistakes. The experimenter stayed in the room providing guidance when needed [1].

The Electric Shock Generator showed the intensity of voltage indicated with each button (from 15 to 450 volts, in 15 volt increments), as well as the danger level of the voltage ("slight shock," "moderate shock," "strong shock," "very strong shock," "intense shock," "extreme intensity shock," "Danger: Severe Shock," and "XXX." Prior to the experiment, the teacher was subjected to a mild shock (45 volts) to demonstrate how the machine works and what the learner would feel [1].

The task required the teacher to read a series of paired words to the learner. Then the teacher read one of the first words of the pairs along with four words and the learner was to choose the word that made the correct pair. If the learner was incorrect, the teacher would apply an electric shock, increasing the intensity with each incorrect response. The teacher informed the learner the intensity of each shock just before applying it [1].

The learner's responses were normalized to be the same for each teacher. Many of the responses were incorrect as the purpose of the study was to see how far the teacher would go with applying the shocks (although no shocks actually occurred). If the teacher hesitated to give a shock, the experimenter would prod him with up to four different verbal commands given in order: 1) Please continue, 2) The experiment requires you to continue, 3) It is absolutely essential that you continue, and 4) You have no other choice but to continue. In addition, some teachers asked questions of the experimenter and special prods were used to answer them. If the teacher asked about possible permanent physical injury, the experimenter answered, "Although the shocks may be painful, there is no permanent tissue damage, so please go on." If the teacher said the learner did not want to continue, the experimenter answered, "Whether the learner likes it or not, you must go on until he has learned all the word pairs correctly. So please go on." [1]

As the experiment proceeded, the learner verbally responded to the electric shocks. At 75 volts, the learner grunted when the shock was applied. Verbal responses at higher intensities included comments of a bad heart and having heart pain. Next, the learner insisted on being released. At 285 volts, the learner started screaming. At 330 volts the learner made no more sounds, not even responding to the question. At this point, the teacher was instructed to treat a no response as an incorrect response and apply the next shock, up to 450 volts (the XXX) [1].

Although Milgram, his colleagues, and some psychology students expected most teachers to refuse to continue with the experiment, 26 (65%) proceeded to the 450 volts. All 40 teachers went to at least 300 volts. But no teacher proceeded without objection. The 14 who quit the experiment did so in an angry state, some giving back their $4.50. Those who continued to higher intensities showed stress during the process, including, "profuse sweating, trembling and stuttering" and "nervous laughter, which in some Ss [subjects or teachers] developed into uncontrollable seizures." These teachers showed signs of great relief when the experiment ended [1].

Milgram performed another 70 similar experiments under slightly altered conditions, and all showed the same results. Several psychologists repeated his experiments and showed similar results. Thus, are the majority of people bound to authority over their own morality?

What would you have done in the experiment? Most people say they would not have continued shocking the learner, but the data say differently. If you did shock a person with 450 volts, believing that you were harming him at the time, then how would you feel afterwards? Probably pretty bad. That was true of the subjects of these experiments. Milgram had to lie to the subjects for the experiments to work, but this does not fit the rule of "informed consent." Thus today, many ethicists find the Milgram experiments to have been unethical.

The Stanford Prison Experiment, 1971

Stanford University psychology Professor Philip Zimbardo studied the influence society had on individual human behavior. At the onset of his famous Stanford Prison Experiment, he stated [3]:

> Most people go about their daily life assuming that they have more control over their behavior than they actually do, … We are often unaware of the tremendous power which social situations exert upon us to shape, guide, and manipulate our behavior.

Zimbardo designed the experiment to "see what the psychological effects were of becoming a prisoner or prison guard." [3,4] To carry out the experiments, Zimbardo had the basement of the Stanford psychology department transformed into a make-shift jail in 1971, and recruited male students to participate in a two week study. The advertisement read:

> Male college students needed for psychological study of prison life. $15 per day for 1-2 weeks beginning Aug. 14. For further information &

> applications, come to Room 248, Jordan Hall, Stanford U.

Twenty-four college students were chosen for the study and randomly divided into two groups: prisoners and prison guards. Only nine of each group started the experiment, with six students available for backup if needed. The prison guards were given uniforms, dark sunglasses, a whistle and a billy club to help them "feel" the part [3,4].

Participants in the study had no privacy during the experiment as much of the "prison" was monitored by an intercom system and video equipment.

During the first day for the prisoners, they were picked up by Palo Alto police, patted down, handcuffed, and taken to the precinct where they were fingerprinted. Participants in the study were not informed the Palo Alto Police Department would be involved or that they would be arrested in their neighborhoods. They were then blind-folded for the trip to "prison" in the basement of the Stanford Psychology Building [3,4].

At the prison, prisoners were turned over to the warden (David Jaffe, an undergraduate student researcher). Guards prepared the prisoners for jail, by searching them, having them stripped naked and delousing each prisoner with a spray. Prisoners received a stocking cap to wear on the head to represent being shaved. They wore a white smock with their prison ID number, and had no undergarments to humiliate them and make the participants feel more like real prisoners. A chain was strapped to one ankle to remind them they were oppressed prisoners. Rubber sandals served as footwear [3].

Three prisoners occupied each 7x10 foot cell. They were addressed by numbers only (no names). Guards worked in groups of three for eight-hour shifts, although at times, guards were kept over to maintain security. The guards received instructions, but were also given much freedom in how they would keep order in the prison. Zimbardo described the instructions to the guard in a slide presentation of the experiment [3]:

> The guards were given no special instructions or training on how to be guards. Instead they were free, within limits, to do whatever they thought was necessary to maintain law and order in the prison and to command the respect of the prisoners. The guards made up their own set of rules which they then carried into effect under the general supervision of Warden David Jaffe, also an undergraduate student. They were warned, however, of the potential seriousness of their mission and of the possible dangers in the situation they were about to enter, as, of course, are real guards who voluntarily take such a job. As with real prisoners, our prisoners expected some harassment and to have privacy and some of their other civil rights violated while they were in prison.

David Jaffe gave the guards a six-hour orientation the day before the experiment began. Included in this orientation were the following instructions [5]:

> I guess when you come into prison you have to sit and wait and wait and wait and so we'll have that, the waiting with the uniforms, stocking cap and all this business. The idea is to order them around,

> you don't request them to do anything, you tell them, and I think that all of you could probably put on some sort of a firm policemen like voice. I don't know but we'll probably try to role play that a little bit this afternoon. … If we don't do anything, then we're just playing a game, and nothing happens — we don't learn anything. So your inventiveness and whatever else — stick-to-it-ness as far as devising things, working out a schedule, and then rewards and punishments and all of this kind of thing are very important part of making the thing run. To a large extent we will either create this prison environment or not on the basis of what you do during the shift.

Thus, Jaffe put pressure on the guards to keep a harsh prison environment from the beginning. The guards did not simply evolve into their characters. Guards were instructed, however, to not exert any physical abuse on the prisoners nor to withhold any food or drink [5,6].

Some guards were concerned with these instructions and expressed those concerns during the orientation. Jaffe responded, "We have access to the Cowell Health Center Facilities … and should we have need for that sort of thing, we can get them over [t]here very fast." [3]

One guard, nicknamed John Wayne, became the leader of the guards encouraging them to be tough on the prisoners. The guards became abusive of their power very quickly, subjecting the prisoners to a 2:30 am wakeup call on day 2, complete with whistles and clanging their billy clubs. The prisoners revolted, removing their numbers from their smocks, taking off the stocking caps and refusing to leave their cells. The guards sprayed fire extinguishers at the prisoners, took away the prisoner's clothes and mattresses, and even put some prisoners in "solitary confinement" in a closet. It was all downhill from here [3,4].

On the third day, the guards played the prisoners against each other by awarding privileges to those who cooperated and punishment to the troublemakers. The guards had random call outs for pushups, and took away bathroom privileges providing a bucket in each cell instead. One prisoner, #8612, requested to leave the experiment, a right of any participant in human experimentation. Instead, Dr. Zimbardo encouraged #8612 to become an informant and receive special treatment. Shortly after, #8612 was released from the experiment [3,6].

Guards continued to become more forceful and prisoners felt helpless. Prisoner #810, left the experiment under duress on day four. As #810 was meeting with Zimbardo, the guards made the remaining prisoners shout out that "Prisoner 810 did a bad thing. [3,4]"

On day five, family members were allowed to visit the prisoners. To mimic a real prison, they had long wait times and brief visitations under constant observation by guards. Some parents showed concern for their student prisoners.

As the experiment entered day 6, it was out of control but none of the researchers recognized it. Several guards were abusing prisoners both mentally and physically. Zimbardo himself was caught up in the experiment as one of the participants (prison superintendent) such that he no longer observed the experiment objectively. It took graduate student Christina Maslach (later to marry Zimbardo), who was not directly involved in the experiment, to explain to Zimbardo that the experiment needed to stop immediately – that these young men were being harmed. Zimbardo stopped the study on the sixth day.

Why did things go wrong so fast? Zimbardo's conclusion from the Stanford Prison Experiment is: [4,8]

> Human behavior, for many people, is much more under the influence of social situational variables than we had ever thought of before.

According to Zimbardo, the experiment showed that people will readily conform to the social roles they are expected to play [3,4,9]. In fact, Zimbardo testified for U.S. Army Military Police who abused prisoners at the Abu Ghraib prison in Iraq, claiming that his study suggests that these MPs responded to their environment and should not be held responsible. However, a flawed experimental design in the Stanford Prison Experiment negates this conclusion.

One of the more critical arguments against the validity (and ethics) of the experiment comes from an interview by David Jaffe with one of the guards who did not want to participate in harsh treatment of the prisoners. Jaffe explains to the guard why he has to be tough on the prisoners and act more like a "real" prison guard. In the process, Jaffe lies about the purpose (and importance) of the experiment when he states that the study will show what happens to prisoners when guards behave in this tough way. Jaffe states the purpose was to mimic the stereotypical guard [3].

Douglas Korpi, or prisoner #8612, told Zimbardo, "Release me. I want to leave." At that moment, 8612 should have been allowed to leave the experiment. Prisoner 8612 and the guard interviewed by David Jaffe mention that guards are "overacting their roles" and that their actions are "arbitrary." In a post-experiment interview, 8612 says that he signed up for the experiment with the idea that he could use his "prison" time to study for the Graduate Record Exam [10]. He was disappointed to find out he could not have books in the prison. During his meeting with Zimbardo, he asks why he cannot have books, and Zimbardo responded that new prisoners have no privileges. They have to "earn privileges". 8612 leaves Zimbardo's office, but returns even more upset that he could not leave. 8612 comes back and says, "I am f***ed up and I have to go. I cannot take it one more day." He also asks for a lawyer and apologizes because he "knows it's an experiment." During the conversation, he continually mentions the "experiment" and that they have no right to "f*** with my head" or "to keep me here." Although 8612 was ultimately permitted to leave the experiment, it was not immediate and he relayed to the other prisoners that they cannot get out of the experiment [4].

In interviews two months after the experiment, one of the guards described his discomfort at playing his role [9]:

> What made the experience most depressing for me was the fact that we were being continuously called upon to act in a way that was contrary to what I feel inside. I don't feel like I'm the type of person that would be a guard just continually giving out s*** and forcing people to do things and pushing and lying, and just, it didn't seem like me to continually keep up and put on a face like that is just one of the really most oppressive things you can do.

Another guard expressed a very different response to the experiment, finding himself doing things he never thought he would be capable of without remorse until the experiment ended [9]:

> I had really thought that I was incapable of this kind of behavior. I, I was surprised, no I was dismayed to find out that I could, I could really be a uh, that I could uh act in a uh, manner so, so absolutely unaccustomed to anything I would really dream of doing. And while I was doing it, I um I didn't feel any regret, I didn't feel any um guilt, it was only after, afterwards when I began to reflect on what I had done that this began to, this behavior began to dawn on me and I realized that this was uh, uh this was part of me that I hadn't really noticed before.

The guard who was nicknamed John Wayne and probably exerted the harshest behavior gave his opinion of the experiment [9]:

> Once you put a uniform on and are given a role, I mean a job, saying your job is to keep these people in line, then you are certainly not the same person you were in street clothes and in a different role. You really become that person once you put on that khaki uniform, you put on the glasses, you put on, you take the night stick, you know, you know you act the part. That's your costume and, uh, you have to act accordingly when you put it on.

This particular guard, in a discussion with one of the prisoners, explained how he was conducting an experiment of his own to see how much verbal abuse a person would take before objecting. He also said he was surprised that nobody objected or stopped him.

Zimbardo was not an evil researcher wanting to harm test subjects. The design of the experiment was flawed in that he and Jaffe were both researchers and part of the experiment. Zimbardo acknowledges this error in his book, **The Lucifer Effect**, and he has apologized for allowing the experiment to continue as long as it did [4].

Recently, Zimbardo released raw data including video and audio recordings of the experiment [3,8]. Other researchers studied these data and have different conclusions than Zimbardo. These researchers claim that the *situation* was not the reason the guard participants became cruel to the prisoners, but the *authority* of Zimbardo and Jaffe's instructions encouraged the guards to be cruel [5,6,7]. Zimbardo sticks by his original conclusion [10].

Conclusion

The Lucifer Effect describes how good people do bad things in certain situations. Milgram would claim they do so because they follow authority, as Nazi physicians and military claimed they were doing. But who is ultimately responsible for one's actions?

Each person must accept his own moral responsibility. Most of us have done things in our past that we regret. We hurt people, either purposefully or unintentionally. It is imperative that we learn from our mistakes and the mistakes of others.

But some people do not desire morality. These are the evil people—the ones we do not want treating our patients, teaching our children, policing the streets or giving us guidance in life.

References

Chapter 2

1. Gosselin, K.R. and Stacom, D., The victims, Hartford Courant, December 16, 2012, pg. A4
2. Ryckaert, V., Hays, H.V., and Bowman, S., Family, students praise teacher as hero, The Indianapolis Star, May 26, 2018, pg. A6.
3. Reilly, P., Teacher locked up for sex with 13-year-old student taught fellow inmates, New York Post, November 2021, https://nypost.com/2021/11/02/arizona-teacher-brittany-zamora-locked-up-for-sex-with-13-year-old-student-taught-fellow-inmates/
4. Hellring, S., Ariz. 6th Grade Teacher Groomed 13-Year-Old Student for Abuse with Naked Selfies, X-Rated Texts, People, June 12, 2019
5. Richardson, D. Wayne, Holmes County Assistant Principal Sentenced for Sexual Battery of Student, Office of the District Attorney, 4th District, Aprl 2, 2021, https://msdeltada.com/mississippi-delta-district-attorney-4th-district/holmes-county-assistant-principal-sentenced-for-sexual-battery-of-student/
6. Hauser, Christine, Nouman Raja, Ex-Officer is Convicted of Manslaughter in Shooting of Black Man in Florida, NY Times, March 7, 2019
7. FBI Crime Data Explorer, https://cde.ucr.cjis.gov
8. Forgrave, R. and Rao, M., A momentous day in the fight for justice, Star Tribune, April 21, 2021, pg. 1
9. Anderson, Chris, Cleveland police officer pleads guilty to urinating on 12-year-old Euclid girl, Cleveland 19 News, Dec. 16, 2019.

https://www.cleveland19.com/2019/12/16/cleveland-police-officer-pleads-guilty-urinating-year-old-euclid-girl/

10. Murdock, Jason, Police Officer Allegedly Filmed Himself Urinating on 12-year-old Girl Waiting for School Bus: 'He Will Be Held Accountable', Newsweek, Aug. 22, 2019. https://www.newsweek.com/ohio-cleveland-police-solomon-nhiwatiwa-arrested-urinate-school-girl-euclid-criminal-charges-1455646
11. DeBenedetti, Chris, Sexual abuse at St. Bede's relived, Oakland Tribune, Nov. 16, 2002, pg. 2
12. BishopAccountability.org, Collated USCCB Data: On the Number of U.S. Priests Accused of Sexually Abusing Children and the Numbers of Persons Alleging Abuse. 1950–2018, https://www.bishop-accountability.org/AtAGlance/USCCB_Yearly_Data_on_Accused_Priests.htm
13. AAMC, https://students-residents.aamc.org/anatomy-applicant/premed-competencies-resources

Chapter 3

1. https://www.merriam-webster.com/dictionary/evil
2. Kneeland, Douglas, Coroner Details the Tate Killing, New York Times, Aug. 22, 1970
3. Gardella, Kay, Actress and 4 Slain in Ritual, Daily News, Aug. 10, 1969
4. O'Neill, Tom, Chaos: Charles Manson, the CIA, and the Secret History of the Sixties, Little, Brown & Company, 2019.
5. Guinn, Jeff, Manson: The Life and Times of Charles Manson, Simon & Schuster, 2014.
6. Guinn, Jeff, Charles Manson's Family Reveals They Were 'Not Surprised' After Learning He Was the Mastermind

Behind the 1969 Murders, https://www.biography.com/news/charles-manson-the-making-of-a-serial-killer, September 11, 2020.

7. Bugliosi, V. and Gentry, C., **Helter Skelter**, **The True Stories of the Manson Murders**, W.W. Norton and Company, 2001

Chapter 4

1. Thilley, Frank, **History of Philosophy**, Henry Holt and Company, New York, 1914.
2. Plato, **Protagorus**, 532 BCE
3. Peccorino, P.A., **An Introduction to Philosophy, An On-line Textbook**, https://www.qcc.cuny.edu/socialsciences/ppecorino/intro_text/Chapter%202%20GREEKS/Socrates_Legacy.htm
4. World Health Organization. Eliminating female genital mutilation. An interagency statement, 2008, http://apps.who.int/iris/bitstream/handle/10665/43839/9789241596442_eng.pdf;jsessionid=889BDF51FE759AD14EB076A228E85BDE?sequence=1
5. Stanford Encyclopedia of Philosophy, Deontology Ethics, Oct. 30, 2020. https://plato.stanford.edu/entries/ethics-deontological/
6. Himma, K.E., Natural Law, Internet Encyclopedia of Philosophy, https://iep.utm.edu/natlaw/
7. Aquinas, T., **Summa Theologiae**, 1495
8. Confuscious, **The Analects**, 220 AD
9. Hobbes, T., **Leviathan: On the Matter, Forme, and Power of a Commonwealth Ecclesiastical and Civil**, 1651

Chapter 5

1. PEW Research Center, https://pewresearch.org/topic/religion
2. Mathewes, C., Why Evil Exists, The Great Courses, https://www.thegreatcoursesplus.com/
3. Qamar, Azhar Hameed. "The concept of the 'Evil' and the 'Evil Eye' in the Islam and Islamic faith-healing traditions". Journal of Islamic Thought and Civilization 3, no. 2 (2013): 44–53.
4. Aquinas, Thomas, **Summa Theologiae**, 1495.
5. Humes, D., **Dialogues on Natural Religion**, 1779.
6. Stanford Encyclopedia of Philosoophy, Leibniz on the Problem of Evil, Feb. 27, 2018 https://plato.stanford.edu/entries/leibniz-evil?
7. Merryman, Leigh, Do You Know the Basics of Hinduism, 2018, IMB, https://www.imb.org/2018/08/10/the-basics-of-hinduism/
8. **Basics of Buddhism**, PBS, https://www.pbs.org/edens/thailand/buddhism.htm
9. Evil and Suffering, BBC Bitesize, https://www.bbc.co.uk/bitesize/guides/znxpr82/revision/Buddhistsbelievethat,theirgreedangerandignorance.
10. Basics of Buddhism, PBS, https://www.pbs.org/edens/thailand/buddhism.htm
11. Pierre, Joe, Psychosis, A Differential Diagnosis of Demonic Possession, Psychology Today, Dec. 5, 2023, https://www.psychologytoday.com/us/blog/psych-unseen/202305/a-differential-diagnosis-of-demonic-possession
12. Blatty, W.P., **The Exorcist**, Harper Torch, 1971.

Chapter 6

1. American Psychiatric Association, **Diagnostic and Statistical Manual of Mental Disorders, 5th Edition**, with revisions, 2013. https://doi.org/10.1176/appi.books.9780890425787
2. https://www.merriam-webster.com/dictionary/psychopathy
3. Ye, Shuer, Zhu, Bing, et. al., Connectome-based model predicts individual psychopathic traits in college students. Neuroscience Letters, 769 (2022) 136387.
4. Kiehl, K.A. and Hoffman, M.B., The criminal psychopath: History, neuroscience, treatment, and economics. Jurimetrics, 51 (2011) 355-397.
5. Hare, R.D., **Without Conscience: The Disturbing World of Psychopaths Among Us**, Guilford Press, 1999.
6. De Brito, S.A., Forth, A.E., Baskin-Sommers, A.R. *et al.* Psychopathy. *Nat Rev Dis Primers* **7,** 49 (2021). https://doi.org/10.1038/s41572-021-00282-1
7. Dutton, K., **The Wisdom of Psychopaths: What Saints, Spies and Serial Killers Can Teach Us About Success**, Doubleday, Canada, 2012.
8. Cooke, D.J. and Michie, C., Refining the constructs of psychopathy: Towards a hierarchical model, Psychological Assessment, 13 (2001) 171-188.
9. Raine, A. and Yang, Y., The Neuroanatomical Basis of Psychopathy: A Review of Brain Imaging Findings, in **Handbook of Psychopathy**, Chapter 14, Ed. Patrick, C.J., The Guilford Press, New York.
10. Yang, Y., Raine, A., Lencz, T., Lacasse, L., and Colletti, T., Volume reduction in prefrontal gray matter in unsuccessful criminal psychopaths, Biological Psychology, 57 (2005) 1109-1116.

11. Ressler, K.J., Amygdala Activity, Fear, and Anxiety: Modulation by Stress, Biological Psychiatry 67 (2010) 1117-1119.
12. Raine, A., Lencz, T., *et. al.*, Corpus callosum abnormalities in psychopathic antisocial individuals, Arch Gen Psychiatry. 60 (2003) 1134-1142.
13. Seghier, M.L., The angular gyrus: Multiple functions and multiple subdivisions, Neuroscientist. 2013 Feb; 19(1): 43–61.
14. Glenn, A.L., Han, Hyemin, *et. al.*, Associations between psychopathic traits and brain activity during instructed false responding, Psychiatry Research: Neuroimaging 266 (2017) 123-137.
15. National Institute of Mental Health, Schizophrenia, https://www.nimh.nih.gov/health/topics/schizophrenia
16. Erickson, B., Deinstitutionalization through optimism: The Community Mental Health Act of 1963, American Journal of Psychiatry (online), 2021, https://ajp.psychiatryonline.org/doi/10.1176/appi.ajp-rj.2021.160404
17. National Institute of Mental Health, Mission, https://www.nih.gov/about-nih/what-we-do/nih-almanac/national-institute-mental-health-nimh
18. Joint Commission on Mental Illness and Health, Action For Mental Health: Final Report of the Joint Commission on Mental Illness and Health 1961.
19. CDC, Fetal Alcohol Spectrum Disorders, https://www.cdc.gov/ncbddd/fasd/index.html
20. Physicians for Social Responsibility, Environmental threats to brain development, 2016, https://www.psr.org/wp-content/uploads/2018/05/environmental-threats-to-brain-developtment.pdf
21. https://www.merriam-webster.com/dictionary/stress

22. Lopatina, O.L., Panina, Y.A., *et. al.*, Early life stress and brain plasticity: from molecular alterations to aberrant memory and behavior, Reviews in the Neurosciences on-line, https://www.degruyter.com/document/doi/10.1515/revneuro-2020-0077/html?lang=de
23. Basto-Pereira, M., Gouveia-Pereira, M., *et. al.*, The global impact of adverse childhood experiences on criminal behavior: A cross-continental study, Child Abuse and Neglect, 124 (2022) https://www.sciencedirect.com/science/article/abs/pii/S0145213421005287
24. Akers, R.L. and Jennings, W.G., Social Learning Theory, in The Handbook of Criminological Theory, Ed: Piquero, A.R., Wiley Blackwell, 2016, West Sussex, Chapter 14, pp. 230-240.
25. Krohn, M. and Ward, J.T., Integrating Criminological Theories, in The Handbook of Criminological Theory, Ed: Piquero, A.R., Wiley Blackwell, 2016, West Sussex, Chapter 17, pp. 319-335.

Chapter 7

1. Portenoy, R.K., Foley, K.M., Chronic Use of Opioid Analgesics in Non-malignant Pain: Report of 38 Cases., Pain 25: 171-186, 1986.
2. Rosenthal, Elisabeth, Patients in Pain Find Relief, Not Addiction, in Narcotics. New York Times, March 28, 1993.

3. Portenoy, R.K., Opioid Therapy for Chronic Nonmalignant Pain: Clinicians' Perspective, The Journal of Law, Medicine, & Ethics, 24: 296-309. 1996

4. Zee, Art Van, The Promotion and Marketing of Oxycontin: Commercial Triumph, Public Health Tragedy, Am. J. Public Health, 99:221-227, 2009.
5. Oxycodone and Oxycontin, Med Letters Drugs Ther, 43:80-81, 2001.
6. Porter, J., and Jick, H., Addiction rare in patients treated with narcotics, N Engl J Med. 302(2):123, 1980.
7. Clark, Katherine and Rogers, Hal, Report: Exposing Dangerous Opioid Manufacturer Influence at the World Health Organization, 2019. https://katherineclark.house.gov/_cache/files/a/a/aaa7536a-6db3-4192-b943-364e7c599d10/818172D42793504DD9DFE64B77A77C0E.5.22.19-who-Purdue-report-final.pdf
8. WHO, Ensuring Balance, a national policies on controlled substances: Guidance for availability and accessibility of controlled medicine, 2011 (Revised 2019).
9. Opioid Overdose Crisis, NIH National Institute of Drug Abuse, 2019, https://www.drugabuse.gov/drugs-abuse/opioids/opioid-overdose-crisis
10. U.S. Senate Homeland Security and Government Committee, Minority Staff Report, Fueling an Epidemic, Report Two, Exposing the Financial Ties Between Opioid Manufacturers and Third Party Advocacy Groups.
11. Blankenship, Kyle, Insys founder Kapoor found guilty in landmark opioid bribery lawsuit, Fierce Pharma, May 2019, https://www.fiercepharma.com/pharma/insys-founder-kapoor-found-guilty-landmark-opioid-bribery-lawsuit
12. Stanley, T.H., The Fentanyl Story, Journal of Pain, 15 (2014) 1215-1226.
13. Jannetto The Fentanyl Epidemic and Evolution of Fentanyl Analogs in the United States and the European Union Paul J. Jannetto,1 Anders Helander,2 Uttam Garg,3 Gregory C.

Janis,4 Bruce Goldberger,5 and Hemamalini Ketha. Clinical Chemistry 65:2, 242–253 (2019)

14. Graph of Opioid Overdoses., CDC, https://www.cdc.gov/drugoverdose/images/epidemic/3WavesOfTheRiseInOpioidOverdoseDeaths.png
15. MacMillan, Carrie, Why is Fentanyl Driving Overdose Deaths, Yale Medicine, March 18, 2024, https://www.yalemedicine.org/news/fentanyl-driving-overdoses
16. Keefe, P.R., **Empire of Pain: The Secret History of the Sachler Dynasty**, Random House, 2021.

Chapter 8

1. Inspector General, https://www.hhs.gov/sites/default/files/state/dab/decisions/alj-decisions/2004/CR1154htm
2. Data Brief 329. Drug Overdose Deaths in the United States, 1999-2017, Center for Disease Control, https://www.cdc.gov/nchs/data/databriefs/db329_tables-508.pdf#1
3. Chapman, M., *et. al.*, Research on Illegal Prescription Drug Market Interventions, 2015, www.ojp.gov/pdffiles1/nijgrants/248905.pdf
4. https://www.dea.gov/drug-scheduling
5. https://www.gov/controlled-substances-act
6. https://www.pharmacytimes.com/publications/issue/2007/2007-02/2007-02-6309
7. Hard Place: Can Physicians Prescribe Opioids to Treat Pain Adequately While Avoiding Legal Sanction?, American Journal of Law and Medicine, 2016 42(1) 7-52.
8. Nylander, Chorus, Dr. James Graves' son asking for 'mercy' in historic pill mill case, WEAR News 3, July 16, 2019.

https://weartv.com/news/local/dr-james-graves-son-asking-for-mercy-in-historic-pill-mill-case

9. https://www.latimes.co/archives/la-xpm-2002-feb-20-mn-28941-story.html
10. https://azdailysun.com/doctor-on-trial-on-manslaughter-charges-in-oxycontin-deaths/article_76b04c93-cc06-5032-bc2a-512e980d4f4c.html
11. https://www.cpmission.com/main/graves.html
12. Doctor's conviction sends message, DEA says, Tampa Bay Times, Feb. 21, 2002, https://www.tampabay.com/archive/2002/02/21/doctor-s-conviction-sends-message-dea-says/
13. WKRG, Pain management doctors arrested in Mobile on drug fraud charges, https://www.wkrg.com/news/2-pain-management-doctors-arrested-in-mobile-on-drug-fraud-charges/
14. US Attorneys Office, Two Mobile pain doctors arrested on drug and fraud charges, FBI, May 20, 2015. https://www.fbi.gov/contact-us/field-offices/mobile/news/press-releases/two-mobile-pain-doctors-arrested-on-drug-and-fraud-charges
15. Raymond, Nate, Alabama pill mill doctor tied to Insys gets 21 years in prison, Reuters, May 2017. https://www.reuters.com/article/us-alabama-drugs/alabama-pill-mill-doctor-tied-to-insys-gets-21-years-in-prison-idUSKBN18M2GK
16. Dr. John Patrick Couch sentenced 240 months, DOJ Press Release, May 15, 2017. https://www.justice.gov/usao-sdal/pr/dr-john-patrick-couch-sentenced-240-months
17. Harmon, K., SCOTUS throws out Mobile 'pill mill' doctors' case, sends down to Court of Appeals, NBC15 News, July 1, 2022,

https://mynbc15.com/news/local/scotus-throws-out-mobile-pill-mill-doctors-case-sends-down-to-court-of-appeals

18. Department of Justice, Former Mobile Pain Management Doctors Xiulu Ruan And John Patrick Couch Resentenced, July 17, 2024, https://www.justice.gov/usao-sdal/pr/former-mobile-pain-management-doctors-xiulu-ruan-and-john-patrick-couch-resentenced
19. SCOTUS, XIULU RUAN v. UNITED STATES CERTIORARI TO THE UNITED STATES COURT OF APPEALS FOR THE ELEVENTH CIRCUIT, https://www.supremecourt.gov/opinions/21pdf/20-1410_1an2.pdf
20. Kresnak, Christopher Testimony https://www.govinfo.gov/content/pkg/FR-2012-08-01/pdf/2012-18748.pdf
21. Orient, Jane M.D., Letter to Wexler, Leonard, Judge, Feb. 5, 2005. https://www.aapsonline.org/painman/hurwitzletter2.htm
22. McNeal, Brian, Jury convicts physician of illegally prescribing pills that led to deaths of four people, Drug Enforcement Administration https://www.dea.gov/press-releases/2011/05/10/jury-convicts-physician-illegally-prescribing-pills-led-deaths-four
23. U.S. Attorney's Office, Chicago Physician Receives Four Life Sentences for Illegally Distributing Pills That Led to Deaths of Four People, Feb. 14, 2012. https://archives.fbi.gov/archives/cincinnati/press-releases/2012/chicago-physician-receives-four-life-sentences-for-illegally-distributing-pills-that-led-to-deaths-of-four-people

24. Donald, Bernice B., Circuit Judge, UNITED STATES v. VOLKMAN, United States Court of Appeals, Sixth Circuit. Aug. 14, 2015. https://caselaw.findlaw.com/us-6th-circuit/1710544.html
25. Letter from Paul Volkman, MD, to Doctors of Courage https://doctorsofcourage.org/letter-from-paul-volkman-md/
26. Doctors of Courage Website https://doctorsofcourage.org
27. Linda Cheek's Website https://lindacheekmd.com/
28. Drug Enforcement Agency, Linda Cheek, M.D., Decision and Order, Oct. 23, 2011 https://wwwf.ederalregister.gov/documents/2011/10/28/2011-28002/linda-sue-cheek-md-decision-and-order
29. Cheek, Susan, **Target PainDoc**, https://lindacheekmd.com/writing/
30. Cheek, Susan, The Real Cause of Addiction, https://lindacheekmd.com/writing/
31. Cheek, Susan, Using the Law Illegally Against Doctors, https://lindacheekmd.com/writing/
32. Cheek, Susan, End the War on Drugs, Doctors, and Pain Patients.Take Control of Your Own Health and Well-being. https://www.sevenpillarstotalhealth.com/
33. https://media.ca11.uscourts.gov/opinions/pub/files/201712653
34. Chapman, Ronald, The Chapman Law Group Health Care Attorneys Website, https://www.chapmanlawgroup.com/team/
35. Department of Justice Opioid Enforcement Effort https://www.justice.gov/civil/consumer-protection-branch/opioid
36. Press Releas: Appalachian Regional Prescription Opioid (ARPO) Strike Force Takedown Results in Charges Against

60 Individuals, Including 53 Medical Professionals, April 17, 2019, Updated Feb. 5, 2025.
https://www.justice.gov/opa/pr/appalachian-regional-prescription-opioid-arpo-strike-force-takedown-results-charges-against

37. Press Release: Second Appalachian Region Prescription Opioid Strikeforce Takedown Results in Charges Against 13 Individuals, Including 11 Physicians, Sept. 24, 2019, Updated Feb. 5, 2025.
https://www.justice.gov/opa/pr/second-appalachian-region-prescription-opioid-strikeforce-takedown-results-charges-against-13
38. Anson, Pat, Virginia doctor sentenced for illegally prescribing pain meds, 2013,
http://nationalpainreport.com/virginia-doctor-sentenced-illegally-prescribing-pain-meds-8822083.html
39. Suda KJ, *et. al.*, Comparison of Opioid Prescribing by Dentists in the United States and England. JAMA 2(2019):e194303.
https://jamanetwork.com/journals/jamanetworkopen/fullarticle/2734067
40. Bay City News Partners, San Leandro Physician's Assistant Headed To Prison, Dec. 24, 2018.
https://patch.com/california/sanleandro/san-leandro-physicians-assistant-headed-prison
41. Rich, Ben A. and Webster, Lynn R. A Review of Forensic Implications of Opioid Prescribing withExamples from Malpractice Cases Involving Opioid-Related Overdose. 2011; Pain Med, S59:S59.
42. Edwards, H. and Bennett, M. Access to Opioids for Patients with Advanced Disease. Current Pharmaceutical Design, 2019, 25 (30): 3203-3208.
43. Foxwell, A.M., Uritsky, T., and Meghani, S.H., Opioids and Cancer Pain Management in the United States: Public Policy

and Legal Challenge, Seminars in Oncology Nursing, 35(2019) 322-326.

44. Barnes, M., Giampa, J., and Caron M., Opioid Prescribing and Physician Autonomy: A Quality of Care Perspective. Opioid Prescribing and Pain Management, 15 (2019) 20-26.
45. READ: Full text of Christie's 2017 State of the State, NJ.com, Jan. 10, 2017. https://www.nj.com/politics/2017/01/read_full_text_of_christies_2017_state_of_the_stat.html
46. Krebs, Erin E., Setting expectations, following orders, safety, and standardization: clinicians' strategies to guide difficult conversations about opioid prescribing, J. Gen. Internal Medicine, 35 (2019) 635-636.
47. Rosenblum, Andrew, et al. Opioids and the Treatment of Chronic Pain: Controversies, Current Status, and Future Directions. Experimental & Clinical Psychopharmacology, 405(2008) 405.
48. Flores, M., *et al.,* Examining racial/ethnic differences in patterns of opioid prescriptions and misuse. APHA's 2019 Annual Meeting and Expo (Nov. 2-Nov. 6). APHA, 2019.
49. Groenewald, C. *et.al.*, Racial differences in opioid prescribing for children in the United States, Pain 159 (2018) 2050-2057.
50. Peteet, B.J. Psychosocial risks of prescription drug misuse among U.S. racial/ethnic minorities: A systemic review, Journal of Ethnicity in Substance Abuse, 80 (2019)

Chapter 9

1. McLaren, A. **Our Own Master Race: Eugenics in Canada 1885-1945**. 1990, Toronto: McClelland & Stewart.
2. Goddard, H. **Human Efficiency and Leels of Human Intelligence**, 1920.

3. Kevles, D. **In the Name of Eugenics: Genetics and the Uses of Human Heredity**. 1985, Berkeley: University of California Press.
4. Largent, M. **Breeding Contempt: The History of Coerced Sterilization. 2011,** New Brunswick: Rutgers University Press.
5. Early Immigration Policies https://www.uscis.gov/about-us/our-history/history-office-and-library/our-history/overview-of-ins-history/early-american-immigration-policies
6. https://www.thecrimson.com/article/2018/10/18/immigration-restriction-league/
7. https://history.house.gov/Historical-Highlights/1901-1950/The-Immigration-Act-of-1924/
8. Jordan, D.S., **The Blood of a Nation: A Study in the Decay of Race by the Survival of the Unfit**. 1901.
9. Largent, M., Charles Davenport's Prediction for Biology, APS Library Bulletin, 2002 https://www.amphilsoc.org/bulletin/2002/davenport.htm
10. Better Babies Contests, http://www.eugenicsarchives.org/eugenics/
11. Fitter Family Contests, http://www.eugenicsarchive.org/eugenics/static/themes/8.html
12. Morgan, T., and Bridge, C., **Contributions to the Genetics of Drosophilia Melangaster**, 1919, Carnegie Institution of Washington.
13. Gur-Arie, R., Eugenical Sterilization in the United States (1922), by Harry H. Laughlin. https://embryo.asu.edu/pages/eugenical-sterilization-united-states-1922-harry-h-laughlin
14. Czarr, "Three Generations of Imbeciles are Enough"-The Case of Buck v. Bell, 2017. https://education.blogs.archives.gov/2017/05/02/buck-v-bell/

15. Seelye, D., Eugenics on Trial: The Sterilization of Ann Cooper Hewitt. https://nationalhumanitiescenter.org/education-material/eugenics-on-trial-the-sterilization-if-ann-cooper-hewitt/
16. Ann Cooper Hewitt Sues Her Mother; Heiress says she was duped into sterilization so parent could get income, NY Times, Jan. 7, 1936.
17. Pearl, R. The First International Eugenics Congress. *Science, 36* (1912) 395-396.
18. Schuster, E., The First International Eugenics Congress, Eugenics Review 4 (1912) 223-256.
19. Stillwell, D., Eugenics Visualized: The Exhibit of the Third International Congress of Eugenics, Bulletin of the History of Medicine, 86 (2012) 206-236.
20. Ploetz, A., **The Excellence of Our Race and the Protecting of the Weak.** 1895.
21. Stahnisch, F.W., The Early Eugenics Movement and Emerging Professional Psychiatry: Conceptual Transfers and Personal Relationships between Germany and North America, 1880s to 1930s. https://www.utpjournals.press/doi.pdf/10.3138/cbmh.31.1.17
22. http://eugenicsarchive.ca/
23. Weiss, S.F., William Schallmayer and the Logic of German Eugenics, Isis 77 (1986) 33-46.
24. Schallmayer, W., **Hereditary and Selection in the Life Process of Nations**, 1902.
25. Fischer, Lenz, Baur, **Principles of Human Heredity and Racial Hygiene**, 1921.
26. Binding, K. and Hoche, A., **Permitting the Destruction of Life Unworthy of Life**, 1920.
27. Hitler, A., **Mein Kampf**, 1925.
28. Kater, M.H., **Doctors Under Hitler**, 2000.

29. Roeder, T., Kubillus, V., and Burwell, A., **Psychiatrists: the Men Behind Hitler**, 1995.
30. http://ghdi.ghi.dc.org/sub_document.cfm?document_id=1528
31. Holocaust Encyclopedia https://encyclopedia.ushmm.org/content/en/article/euthanasia-program
32. Davies, K., Nature vs Nurture Revisited, PBS NOVA https://www.pbs.org/wgbh/nova/genome/debate.html
33. Syndrome and Social Capital: Assessing the Costs of Selective Abortion, March 18, 2022. https://www.jec.senate.gov/public/_cache/files/ade656cc-206b-4624-a51b-10eeca1d1f28/down-syndrome-report.pdf
34. White, M., The End of the Beginning, The Ochsner Journal, 11 (2011) 309-316.
35. FDA AquAdvantage Salmon Fact Sheet, 2025. https://www.fda.gov/animal-veterinary/aquadvantage-salmon/aquadvantage-salmon-fact-sheet
36. Editorial Staff, Aqua Bounty sued for nearly $1.6 million over unfinished Ohio salmon facility, 2025. https://www.salmonbusiness.com/aquabounty-sued-for-nearly-1-6-million-over-unfinished-ohio-salmon-facility/
37. NPR, Human Clones: Where's Baby Eve?, Dec. 30, 2003. https://www.npr.org/2003/12/30/1575178/human-clones-wheres-baby-eve#:~:text=On%20Dec.,that%20she%20is%20a%20clone
38. The Guardian, US scientist clones himself from skin cells. Jan. 17, 2008. https://www.theguardian.com/science/2008/jan/18/genetics.medicalresearch

Chapter 10

1. Patients Rights Council, The Real Jack Kevorkian, https://www.patientsrightscouncil.org/site/the-real-jack-kevorkian/, 2013.
2. Kevorkian, J., A Brief History of Experimentation on Condemned and Executed Humans, Journal of the National Medical Association, 77 (1985) 215-226.
3. Associated Press, 'Suicide doctor' has long been at the center of controversy,' The Ithaca Journal, June 7, 1990.
4. Rubin, Neal, Royal Oak: The Death Machine, Detroit Free Press Magazine, March 18, 1990.
5. Jackson, Nicholas, Jack Kevorkians death van and the tech of assisted suicide, The Atlantic, 2011. https://www.theatlantic.com/technology/archive/2011/06/jack-kevorkians-death-van-and-the-tech-of-assisted-suicide/239897/
6. Garnett News Service, Pathologist's 'suicide machine divides legal experts, Battle Creek Enquirer, October 26, 1989.
7. DeSimone, Bonnie, For machine's user, death won over loss of dignity, Lansing State Journal, June 10, 1990.
8. Mauro, T., and Basheda, V., Court set to rule on case with similar issues, Lansing State Journal, June 10, 1990.
9. Williams, Mike, Judge dismisses murder charge against doctor, Detroit Free Press, December 14, 1990.
10. CBS News, Dr. Jack Kevorkian's "60 Minutes" Interview, https://www.cbsnews.com/news/dr-jack-kevorkians-60-minutes-interview/, June 4, 2011
11. Charatan F. Dr Kevorkian found guilty of second degree murder. *BMJ*. 1999;318(7189):962.
12. Associated Press, Kevorkian released from prison, The Denver Post, June 1, 2007.

https://www.denverpost.com/2007/06/01/kevorkian-released-from-prison/

13. Schneider, Keith, Dr. Jack Kevorkian Dies at 83; A Doctor Who Helped End Lives, N.Y. Times, June 4, 2011. https://www.nytimes.com/2011/06/04/us/04kevorkian.html
14. Death With Dignity, https://deathwithdignity.org/states/
15. European Countries that Allow Assisted Dying, Medical Press, March 11, 2024, https://medicalxpress.com/news/2024-03-european-countries-dying.html
16. Humphrey, Derak, Farewell to Hemlock: Kill by its name, February 21, 2005, https://www.assistedsuicide.org/farewell-to-hemlock.html
17. Compassion and Choices, Strategic Plan, Our Roadmap to Success, 2017. https://compassionandchoices.org/about-us/strategic-plan/#:~:text=Our%20Mission%3A%20Compassion%20%26%20Choices%20improves,end%2Dof%2Dlife%20journey.
18. Not Dead Yet, Disability Groups Opposed to Assisted Suicide Laws, https://notdeadyet.org/disability-groups-opposed-to-assisted-suicide-laws/
19. Issa, Omayra, 25 years after conviction, Robert Latimer still believes he was right to kill his daughter, CBC News, Nov. 17, 2019. https://www.cbc.ca/news/canada/saskatoon/robert-latimer-25-years-later-1.5360711
20. Gibbens, Sarah, Hurricane Katrina Explained, National Geographic, 01/16/2019. https://www.nationalgeographic.com/environment/article/hurricane-katrina
21. Bailey, Ryan, The Case of Dr. Anna Pou: Physician Liability in Emergency Situations, AMA Journal of Ethics, 2010;12(9):726-730.

22. Fink, S., **Five Days in Memorial: Life and Death in a Storm-Ravaged Hospital**, Jan. 2016, Crown.
23. Pruitt, Sarah, How Levee Failures Made Hurricane Katrina a Bigger Disaster, https://www.history.com/news/hurricane-katrina-levee-failures
24. NPR, Author Interviews. During Katrina, 'Memorial' Doctors Chose Who Lived, Who Died, 09/10/13, https://www.npr.org/2013/09/10/220687231/during-katrina-memorial-doctors-chose-who-lived-who-died
25. Okie, Susan, Perspective: Dr. Pou and the Hurricane—Implications for Patient Care During Disasters, New England Journal of Medicine, 358 (2008) 1-5.
26. Bailey, Ryan, HEALTH LAW: The Case of Dr. Anna Pou—Physician Liability in Emergency Situations, American Medical Association Journal of Ethics, 12(9) 726-730, 2010.
27. Khan, Carrie, New Orleans Doctor, Nurses Charged with Murder, NPR – All Things Considered, July 18, 2006. https://www.npr.org/templates/story/story.php?storyId=5566057
28. Newsweek Staff, A Doctor Says She Didn't Murder Her Patients, 9/2/2007
29. Saavedra, S.M., Hurricane Katrina and the Deadly Choices at Memorial Medical Center, Nurselabs, October 19, 2017, https://www.nurselabs.com/hurricane-katrina-deadly-choices-memorial-medical-center
30. Canfield, Sabrina, Coroner Clears New Orleans Doctor Again, Courthouse News Service, March 12, 2012. https://www.courthousenews.com/coroner-clears-new-orleans-doctor-again
31. AP, Accusation of murder outrages some doctors, Daily Advertiser, Lafayette, LA, July 23, 2006, pg. 14.

32. CNN, Doctor cleared in Katrina hospital deaths, July 24, 2007. https://www.cnn.com/2007/US/law/07/24/katrina.doctor/
33. O'Reilly KB. Grand jury clears Dr. Pou; medicine wants to protect future disaster responders. American Medical News. August 13, 2007. http://www.ama-assn.org/amednews/2007/08/13/prl20813.htm
34. Scelfo, Julie, Vindicated Katrina Doc Tells Her Story, Newsweek, August 24, 2007. https://www.newsweek.com/vindicated-katrina-doc-tells-her-story-99333
35. LSU Cancer Center, Dr. Anna Pou, https://www.medschool.lsuhsc.edu/cancer_center/facultybio_pou_a.aspx
36. NBC News, Associated Press, Nurse who killed 29 sentenced to 11 life terms, March 2, 2006. https://www.nbcnews.com/id/wbna11636992
37. Medical Bag, Charles Cullen, The Killer Nurse: A Deeper Evil, Aug. 8, 2013. https://www.medicalbag.com/home/features/despicable-doctors/charles-cullen-the-killer-nurse-a-deeper-evil/
38. Assad, M. and Kraus, S., A slide into madness. More than 400 pages of previously sealed domestic violence records obtained by The Morning Call reveal a turbulent year in which Charles Cullen's family life was crumbling as his career as a nurse developed, The Morning Call, May 28, 2004. https://www.mcall.com/news/mc-xpm-2004-05-28-3537781-story.html

Chapter 11

1. FBI Behavioral Analysis Unit, Ed: Morton, Serial Murder: Multi-Disciplinary Perspectives for Investigators, https://www.fbi.gov/stats-services/publications/serial-murder#two
2. Aamodt, M. G. (2016, September 4). Serial killer statistics. Retrieved May 20, 2021 from http://maamodt.asp.radford.edu/serial killer information center/project description.htm
3. Kinnell H G. Serial homicide by doctors: Shipman in perspective, BMJ 2000; 321 :1594
4. Frieden, Joyce, Healthcare Serial Killings Down in U.S. but Up Globally, Medpage Today, October 28, 2014, https://www.medpagetoday.com/psychiatry/generalpsychiatry/48281
5. Yorker BC, Kizer KW, Lampe P, Forrest AR, Lannan JM, Russell DA. Serial murder by healthcare professionals. J Forensic Sci. 2006 Nov;51(6):1362-71. ,
6. Frank C. Health care serial murder: What can we learn from the Wettlaufer story? Can Fam Physician. 2020;66(10):719-722.
7. History.com Editors, Murder Castle, Jan. 23, 2020, https://www.history.com/topics/crime/murder-castle
8. Little, Becky, Did Serial Killer H.H. Holmes Really Build a 'Murder Castle'?, History, May 8, 2023, https://www.history.com/news/murder-castle-h-h-holmes-chicago
9. Mudgett, H.W., **Self Portrait of a Serial Killer, H.H. Holmes in His Own Words**, Burk and McFetridge Co., Philadelphia, 1895.
10. St. Paul Globe, Telltale Skulls, July 28, 1895.
11. Associated Press, Stranger than Fiction, The Herald (Los Angeles), Nov. 21, 1894.

12. The San Francisco Call, Buried in a Cellar, July 16, 1895.
13. The Evening Times (Washington DC), Holmes His Own Lawyer, October 28, 1895.
14. History.com Editors, Serial Killer H.H. Holmes is hanged in Philadelphia, May 5, 2020, https://www.history.com/this-day-in-history/a-serial-killer-is-hanged
15. Yang, Allie, Utah doctor's life of lies unravels after some of his children are convinced he murdered their mother, ABC News, June 13, 2019, https://abcnews.go.com/US/utah-doctors-life-lies-unravels-children-convinced-murdered/story?id=63609211
16. Duda, Jeremy, Pleasant Grove Doctor Charged with Sex Abuse, Daily Herald, October 5, 2007, pg. 11.
17. Hogan, Susan, Doctor who drugged his wife, left her to die in bathtub, found dead at Utah prison, Washington Post, April 10, 2017, https://www.washingtonpost.com/news/morning-mix/wp/2017/04/10/doctor-who-drugged-his-wife-left-her-to-die-in-bathtub-found-dead-at-utah-prison/
18. Meyersohn, Jon, After Wife's Death, Dark Secrets of Utah Doctor Revealed, ABC News, February 23, 2011, https://abcnews.go.com/2020/wifes-death-dark-secrets-utah-doctor-revealed/story?id=12982985
19. Hanson, Kurt, MacNeill sentenced to 1-15 years in sex abuse case, Daily Herald, September 15, 2014. https://www.heraldextra.com/news/local/crime-and-courts/macneill-sentenced-to---years-in-sex-abuse-case/article_d8b15225-f78b-58ea-985b-967f3d05bd71.html
20. Foy, Paul, Conviction Follows Family's Pursuit, The Daily Pursuit, November 10, 2013, pg. 4.
21. Associated Press, 2nd mistress testifies in doctor's murder trial, The Daily Spectrum, October 31, 2013, pg. 5.

22. Stewart, James B., Blind Eye: How the Medical Establishment Let a Doctor Get Away with Murder, New York City: Simon & Schuster, 1999.
23. VA Medical Malpractice Information, Michael Swango, http://vamalpractice.info/wp-content/uploads/2013/02/Swango-Ohio-license-status_Page_1.jpg
24. Walker, Carson, Swango's 'talent will be wasted', Argus-Leader, Sioux Falls, December 13, 1992, pg. 4.
25. Staff, Swango held in Zimbabwe death probe, Argus-Leader, Sioux Falls, SD, July 25, 1997, pg. 1.
26. Lynch, Loretta, United States District Court, Eastern District of New York, UNITED STATES OF AMERICA against MICHAEL J. SWANGO, also known as "Jackson Kirk" and "David Jackson Adams," Defendant. July 11, 2000. http://vamalpractice.info/wp-content/uploads/2018/01/Swango-Indictment.pdf
27. Jackson T and Smith R., Harold Shipman. BMJ. 2004;328(7433):231.
28. Whalen, William, and Martin, Bruce, **Defending Donald Harvey: The Case of America's Most Notorious Angel-of-Death Serial Killer**, Clerisy Press, 2005.
29. Scott, Jason, 'He just liked killing': Donald Harvey convicted of 37 murders, FOX 19 Now, September 23, 2019. https://www.fox19.com/2019/09/24/he-just-liked-killing-donald-harvey-convicted-murders/
30. Jacon-Duffy, Marais, From The Vault: 'Angel of Death' case was unlike anything seen in Tri-State, WCPO News 9, Cincinnati, March 30, 2017. https://www.wcpo.com/news/crime/angel-of-death-donald-harvey-was-unlike-anything-ever-seen-in-tri-state
31. Cardona, Claire, What you need to know about 'Dr. Death,' Dallas neurosurgeon Christopher Duntsch, The Dallas Morning News, September 25, 2018.

https://www.dallasnews.com/news/crime/2018/09/26/what-you-need-to-know-about-dr-death-dallas-neurosurgeon-christopher-duntsch/

32. Redmond, Caroline, The True Story Of Christopher Duntsch, The Killer Surgeon Known As 'Dr. Death', All That's Interesting, September 12, 2018. https://allthatsinteresting.com/dr-death-christopher-duntsch
33. Shipp, Brett, Dallas doctor arrested after botched surgeries, USA Today, July 22, 2015. https://www.usatoday.com/story/news/crime/2015/07/22/dallas-spine-surgeon-arrested-after-botched-surgeries/30540103/
34. Eiserer, Tanya, Doctor convicted of botched surgery gets life sentence, USA Today, February 21, 2017. https://www.usatoday.com/story/news/nation-now/2017/02/21/doctor-convicted-life-sentence/98185712/
35. Davis, Carol Anne, Women Who Kill: Profiles of Female Serial Killers, Allison & Busby, 2002.
36. Welsh, Stacey, TIMELINE: The case against suspected serial killer Genene Jones, KENS 5, February 22, 2018. https://www.kens5.com/article/syndication/vile-podcast/timeline-the-case-against-suspected-serial-killer-genene-jones/273-484336630
37. Elkind, Peter, The Death Shift, Texas Monthly, August, 1983. https://www.texasmonthly.com/articles/the-death-shift-2/
38. Morris, Jim, Nurse says she's still out of work, indigent, Fort Worth Star-Telegram, Fort Worth, Texas, September 15, 1983, pg. 26.
39. Associated Press, Defense witness: vocational nurse researched drug, Longview News-Journal, Longview, Texas, Feb. 10, 1984, pg. 3.
40. Staff, Nurse's verdict: tears, cheers, Fort Worth Star-Telegram, Fort Worth, Texas, February 16, 1984, pg. 1-2.

41. Staff, Nurse gets 99 years:, The Evening Sun, Baltimore, MD, May 15, 1984, pg. 2.
42. King, Wayne, Hospital deaths of up to 47 children investigated by Texas grand jury, Democrat and Chronicle, February 24, 1983.
43. Associated Press, Ex-Texas nurse pleads guilty in 1981 death of eleven-month-old, Brownsville Herald, Brownsville, Texas, January 17, 2020, pg. A4.
44. Salinas, Rebecca and Venema, Paul, 'Killer nurse' Genene Jones pleads guilty, gets life in prison, KSAT 12 News, January 16, 2020. https://www.ksat.com/news/local/2020/01/16/watch-killer-nurse-genene-jones-makes-plea-deal-in-court-hearing/
45. Bucknall V, Burwaiss S, MacDonald D, Charles K, Clement R. Mirror mirror on the ward, who's the most narcissistic of them all? Pathologic personality traits in health care. CMAJ. 2015;187(18):1359-1363. https://www.ncbi.nlm.nih.gov/pmc/articles/PMC4674404/
46. Camille Lloyd, Nanette K. Gartrell, Psychiatric symptoms in medical students, Comprehensive Psychiatry, Volume 25, Issue 6, 1984, Pages 552-565.
47. Assadi, S.M., Nakhaei, M.R., Najafi, F. et al. Mental health in three generations of Iranian medical students and doctors. Soc Psychiat Epidemiol 42, 57–60 (2007).

Chapter 12

1. Hippocratic Oath, https://www.congress.gov/117/meeting/house/114995/documents/HHRG-117-IF02-20220719-SD007.pdf
2. Hornblum, A., Newman, J., and Dober, G., **Against Their Will: The Secret History of Medical Experimentation on Children in Cold War America**, St. Martin's Press, 2013.

3. Sims, J.M., **The Story of My Life**, D. Appleton and Company, 1884.
4. Wall LL. The controversial Dr. J. Marion Sims (1813-1883). Int Urogynecol J. 31(2020):1299-1303.
5. Drutz, J.E. and Ligon, B.L., Polio: Its History and Its Eradication, Seminars in Pediatric Infectious Diseases, 11 (2000) 280-286.
6. Emrich, J.S. and Richter, C., Polio: Part 1, The American Association of Immunologists, December, 2020. https://www.aai.org/About/History/History-Articles-Keep-for-Hierarchy/Polio-Part-I%E2%80%94Understanding-and-Treating-a-Perplex
7. Center for Disease Control, What is Polio?, https://www.cdc.gov/polio/what-is-polio/index.htm#Transmission
8. Barr, M., The iron lung – a polio patient's story, J R Soc Med 103 (2010) 256-259. https://www.ncbi.nlm.nih.gov/pmc/articles/PMC2878826/
9. Dawson, L., The Salk Polio Vaccine Trial of 1954: risks, randomization and public involvement in research, Clinical Trials, 1 (2004) 122–130.
10. Guglielmi, G., Africa declared free from wild polio – but vaccine-derived strains remain, Nature, August 28, 2020. https://www.nature.com/articles/d41586-020-02501-3
11. Brodie, M. and Park, W.H., Active immunization against poliomyelitis, Journal of the American Medical Association, 105 (1935) 1089-1093.
12. Brodie, M., and Park, W.H., Active immunization against poliomyelitis, American Journal of Public Health, 26 (1936) 119-125.
13. Blume, S. and Geesink, I., A Brief History of Polio Vaccines, Science 288 (2000) 1593-1594. https://science.sciencemag.org/content/288/5471/1593.full

14. Singh, A.E., and Romanowski, B., Syphilis: Review with Emphasis on Clinical, Epidemiologic, and Some Biologic Features, Clinical Microbiology Reviews, 12 (2020) 187-209. https://journals.asm.org/doi/full/10.1128/CMR.12.2.187
15. Tampa, M., Sarbu, I., Matei, C., Benea, V., and Georgescu, S.R., Brief History of Syphilis, Journal of Medicine and Life, 7 (2014) 4-10. https://www.ncbi.nlm.nih.gov/pmc/articles/PMC3956094/
16. Centers for Disease Control and Prevention, Sexually Transmitted Diseases: Syphilis, https://www.cdc.gov/std/syphilis/
17. Roy, B., The Julius Rosenwald Fund Syphilis Seroprevalence Studies, Journal of the National Medical Association, 88 (1996) 315-322.
18. Pence, G., **Medical Ethics: Accounts of the Cases That Shaped and Defined Medical Ethics**, McGraw Hill, 2007.
19. Barrett, L.A., Tuskegee Syphilis Study of 1932–1973 and the Rise of Bioethics as Shown Through Government Documents and Actions, Documents to the People, 47 (2019). https://journals.ala.org/index.php/dttp/article/view/7213/9852
20. Centers for Disease Control and Prevention, The U.S. Public Health Service Syphilis Study at Tuskegee, https://www.cdc.gov/tuskegee/index.html
21. Health and Human Services Office for Human Research Protections, The Belmont Report, https://www.hhs.gov/ohrp/regulations-and-policy/belmont-report/index.html
22. Reynolds, Gretchen, The Stuttering Doctor's 'Monster Study', NY Times, March 16, 2003. https://www.nytimes.com/2003/03/16/magazine/the-stuttering-doctor-s-monster-study.html

23. Ambrose, N.G., and Yairi, E., The Tudor Study: Data and Ethics, Am. J. Speech Language Pathology, 11(2002) 190-203. https://www.gvsu.edu/cms4/asset/F51281F0-00AF-E25A-5BF632E8D4A243C7/monster_study.pdf
24. Tudor, M. An experimental study of the effect of evaluative labeling on speech fluency. Master's thesis, University of Iowa, 1939.
25. Silverman, F.H., The "Monster" Study, J. Fluency Disorders, 13 (1988) 225-231.
26. Krugman, S., The Willowbrook Hepatitis Studies Revisited: Ethical Aspects, Reviews Infectious Diseases, 8 (1986) 157-162.
27. Rivera, Geraldo, Willowbrook: The Last Great Disgrace, News Expose, 1972, https://www.youtube.com/watch?v=bpVEjzO6Dd0
28. Robinson, W.M. and Unruh, B.T., The Hepatitis Experiments at the Willowbrook State School, in The Oxford Textbook of Clinical Research Ethics, Chapter 7, pp. 80-85, Editors: Emanuel, E.J., Grady, C.C., Crouch, R.A., Lie, R.K., Miller, F.G. and Wendler, D.D., Oxford University Press, 2011.
29. Lerner, B.H., Sins of Omission — Cancer Research without Informed Consent, New England J Medicine, 351 (2004) 628-630.
30. Langer, E., Human Experimentation: Cancer Studies at Sloan-Kettering Stir Public Debate on Medical Ethics Science, 143 (1964) 551-553.
31. Beecher, H.K., Ethics and Clinical Research. N Engl J Med 274 (1966) 1354–1360.
32. Caplan, A., **When Medicine Went Mad: Bioethics and the Holocaust**, Humana Press, 1992.

Chapter 13

1. VA, https://www.va.gov/opa/publications/factsheets/fs_americas_wars.pdf
2. Mehra, A., Politics of Participation: Walter Reed's Yellow-Fever Experiments, AMA Journal of Ethics, Virtual Mentor. 2009;11(4):326-330.
3. Chaves-Carballo, Enrique, Clara Maass, Yellow Fever and Human Experimentation, Military Medicine, Volume 178, Issue 5, May 2013, Pages 557–562.
4. BUREAU OF INTERNATIONAL SECURITY AND NONPROLIFERATION, Protocol for the Prohibition of the Use in War of Asphyxiating, Poisonous or Other Gases, and of Bacteriological Methods of Warfare (Geneva Protocol), Archived 2009-2007, https://2009-2017.state.gov/t/isn/4784.htm
5. United Nations Office for Disarmament Affairs, Protocol for the Prohibition of the Use in War of Asphyxiating, Poisonous or Other Gases, and of Bacteriological Methods of Warfare, https://www.un.org/disarmament/wmd/bio/1925-geneva-protocol/
6. Centers for Disease Control, Facts About Sulfur Mustard, https://emergency.cdc.gov/agent/sulfurmustard/basics/facts.asp
7. Centers for Disease Control, Facts About Lewisite, https://emergency.cdc.gov/agent/lewisite/basics/facts.asp
8. Inskeep, S, and Greene, D., Secret World War II Chemical Experiments Tested Troops By Race, June 22, 2015. https://www.npr.org/series/417162462/world-war-ii-secret-mustard-gas-testing
9. Committee on Veterans' Affairs House of Representatives One Hundred Second Congress, Department of Defense

Mustard Gas Testing, Hearing Before the Subcommittee on Compensation, Pension, and Insurance of the Committee on Veterans' Affairs, March 10, 1993. https://archive.org/details/departmentofdefe1993unit/mode/2up?view=theater

10. Institute of Medicine (US) Committee on the Survey of the Health Effects of Mustard Gas and Lewisite; Pechura CM, Rall DP, editors. Veterans at Risk: The Health Effects of Mustard Gas and Lewisite. Washington (DC): National Academies Press (US); 1993. 3, History and Analysis of Mustard Agent and Lewisite Research Programs in the United States. https://www.ncbi.nlm.nih.gov/books/NBK236059/
11. Mustard Gas Experiments, https://nationaltriallaw.com/mustard-gas-experiments/
12. Fessenden, The Trafic Aftrmath of Mustrd Gas Experiments in World Wa II, https://www.smithsonianmag.com/smart-news/americans-who-were-exposed-mustard-gas-wwii-experiments-180957152/
13. Dickerson, C., World War II Secret Mustrd Gas Testing, https://www.npr.org/series/417162462/world-war-ii-secret-mustard-gas-testing
14. Clinton, W.J., Executive Order 12891 of January 15, 1994, Advisory Committee on Human Radiation Experiments. https://www.archives.gov/files/federal-register/executive-orders/pdf/12891.pdf
15. Advisory Committee on Human Radiation Experiments, Final Report, 1994. https://bioethicsarchive.georgetown.edu/achre/final/chap5_2.html
16. Boissoneault, L., A Spoonful of Sugar Helps the Radioactive Oatmeal Go Down, Smithsonian Magazine, March 8, 2017.

https://www.smithsonianmag.com/history/spoonful-sugar-helps-radioactive-oatmeal-go-down-180962424/

17. Hornblum, A., Newman, J., and Dober, G., Against Their Will: The Secret History of Medical Experimentation on Children in Cold War America, St. Martin's Press, 2013.
18. Rodriguez MA, García R. First, do no harm: the US sexually transmitted disease experiments in Guatemala. Am J Public Health. 2013 Dec;103(12):2122-6.
19. Presidential Commission for the Study of Bioethical Issues, "Ethically Impossible" STD Research in Guatemala from 1946 to 1948, 2011, https://law.stanford.edu/wp-content/uploads/2011/09/EthicallyImpossible_PCSBI_110913.pdf
20. Hornblum, A.M., Acres of Skin: Human Experiments at Holmesburg Prison, U.S. Dept of Justice, https://www.ojp.gov/ncjrs/virtual-library/abstracts/acres-skin-human-experiments-holmesburg-prison
21. Sarai, Tamar, Holmesburg Prison's medical experiments are Philadelphia's 'lasting shame', May 15, 2023, https://prismreports.org/2023/05/15/philadelphia-holmesburg-prison-experiments/
22. Sagner, D. and Cohen, M., Lasting Scars: Albert Kligman and the Holmesburg Prison Experiments, Daily Penn, 2021, https://dailypenn.github.io/old.projects.34st.com/2021/street-features/
23. Smith, J., TESTIMONY ON MEDICAL EXPERIMENTATION IN HOLMESBURG PRISON PRESENTED BY JOSEPH SMITH BEFORE THE PENNSYLVANIA HOUSE JUDICIARY COMMITTEE'S SUBCOMMITTEE ON CRIMES AND CORRECTIONS, 1999, https://www.legis.state.pa.us/WU01/LI/TR/Transcripts/1999_0118_0014_TSTMNY.pdf

24. The College of Physicians of Philadelphia, College Statement Regarding the Holmesburg Prison Experiments.

Chapter 14

1. Grossmann, Atina, German Women Doctors from Berlin to New York: Maternity and Modernity in Weimar and in Exile, Feminist Studies, 19 (1993) 65-88.
2. Marwell, David G., **Mengele: Unmasking the "Angel of Death**, W.W. Norton and Company, 2025.
3. Ohler, Norman, **Blitzed: Drugs in the Third Reich**, Mariner Books, 2018.
4. United States Holocaust Memorial Museum, How Many People did the Nazis Murder? https://encyclopedia.ushmm.org/content/en/article/documenting-numbers-of-victims-of-the-holocaust-and-nazi-persecution
5. United States Holocaust Memorial Museum, Dachau, https://encyclopedia.ushmm.org/content/en/article/dachau
6. Bradsher, Greg, The Nuremberg Laws, Archives Receives Original Nazi Documents That "Legalized" Persecution of Jews, https://www.archives.gov/publications/prologue/2010/winter/nuremberg.html
7. Jewish Virtual Library, The Holocaust: The Nuremberg Laws, https://www.jewishvirtuallibrary.org/the-nuremberg-laws
8. MEMORIAL AND MUSEUM AUSCHWITZ-BIRKENAU, http://auschwitz.org/en/
9. https://auschwitz.ru/en/auschwitz/gas/
10. Boozer, Jack S. "Children of Hippocrates: Doctors in Nazi Germany." *The Annals of the American Academy of Political and Social Science*, vol. 450, 1980, pp. 83–97. *JS*

11. Nuremburg Tribunals Prosecutors, Brief: prosecution closing brief against Fritz Fischer and Herta Oberheuser, https://nuremberg.law.harvard.edu/documents/16-brief-prosecution-closing-brief?q=issue:%22Bone%2C+muscle%2C+nerve+surgery+experiments%22#p.6
12. Rosenthal, Rolf, Deposition or affidavit concerning the sulfanilamide and bone/muscle/nerve experiments at Ravensbrueck [sulfanilamide and others], https://nuremberg.law.harvard.edu/documents/2029-deposition-or-affidavit-concerning?q=sulfanilamide#p.1
13. Bagatur, E., Nazi Medicine—Part 1: Musculoskeletal Experimentation on Concentration Camp Prisoners During World War II, Clin Orthop Relat Res (2018) 476:1899-1905
14. Rascher, Sigmund, Letter to Heinrich Himmler requesting the use of SS prisoners for high altitude experiments http://nuremberg.law.harvard.edu/documents/30-letter-to-heinrich-himmler?q=author:%22Sigmund+Rascher%22#p.1
15. Transcript for the NMT 2: Milch Case, https://nuremberg.law.harvard.edu/transcripts/2-transcripts-for-nmt-2-milch-case?seq=478&q
16. Sievers, Wolfram, Affidavit concerning the high altitude and freezing experiments, https://nuremberg.law.harvard.edu/documents/3089-affidavit-concerning-the-high?mode=image&q=wolfram+sievers+affidavit+freezing+experiments+type:%22document+image%7Cdocument+full+text%7Ctranscript+full+text%7Cphotograph%22
17. Dawson, McKenzie, After Hitler's pal died, Nazis recreated his injuries in a sick experiment, Ny Times, May 8, 2016 https://nypost.com/2016/05/08/the-women-tortured-by-nazi-doctors-and-the-american-heiress-who-saved-them/
18. Oberheuser, Herta, Affidavit concerning medical experiments conducted at Ravensbrueck and Hohenlychen (sulfanilamide, bone transplants), https://nuremberg.law.harvard.edu/documents/1799-

affidavit-concerning-medical-experiments?mode=image&q=oberheuser+affidavit+sulfanilamide

19. Bagatur, E., Nazi Medicine-Part 1: Musculoskeletal Experimentation on Concentration Camp Prisoners During World War II, Clin. Orthop. Relat. Res. 476 (2018) 1899-1905.
20. Broel-Plater, Ludwika, Transcript for NMT 1: Medical Case, https://nuremberg.law.harvard.edu/transcripts/1-transcript-for-nmt-1-medical-case?seq=819
21. Karolewska, Wladyslawa, Transcript for NMT 1: Medical Case, https://nuremberg.law.harvard.edu/transcripts/1-transcript-for-nmt-1-medical-case?seq=832
22. Mrugowsky, J.E., Report to Dr. Widmann concerning poison bullet experiments, https://nuremberg.law.harvard.edu/documents/1240-report-to-dr-widmann?q=poisoned+bullets#p.1
23. Transcript for NMT 1: Medical Case, https://nuremberg.law.harvard.edu/transcripts/1-transcript-for-nmt-1-medical-case?seq=8829&q=
24. Bauer, Ignaz, Affidavit concerning the seawater experiments at Dachau in 1944 https://nuremberg.law.harvard.edu/documents/89-affidavit-concerning-the-seawater?q=issue:%22Seawater+experiments%22#p.2
25. Buhlick, Herbert, Affidavit concerning the typhus experiments at Buchenwald, https://nuremberg.law.harvard.edu/documents/673-affidavit-concerning-the-typhus?q=typhus+experiments#p.1
26. Transcript for NMT 1: Medical Case, https://nuremberg.law.harvard.edu/transcripts/1-transcript-for-nmt-1-medical-case?seq=10560
27. Mrugowsky, Joachim Ernst A., Report to German medical authorities on the results of typhus vaccine tests,

https://nuremberg.law.harvard.edu/documents/718-report-to-german-medical?q=typhus+experiments#p.1

28. Allen, Arthur, How A Jewish Doctor Duped the Nazis, Politico, July 23, 2014. https://www.politico.com/magazine/story/2014/07/lice-doctor-lviv-nazi-germany-109255
29. HLS Nuremberg Trials Project, Transcript for NMT 1: Medical Case, https://nuremberg.law.harvard.edu/transcripts/1-transcript-for-nmt-1-medical-case?seq=53&q=yellow+fever
30. Ding, Erwin, Work journal of the typhus (and other) vaccine project at Buchenwald, 1941-1945, https://nuremberg.law.harvard.edu/documents/1548-work-journal-of-the-typhus?q=ding+diary#p.3
31. Sievers, Wolfram, Memorandum to Rudolf Brandt concerning Lost gas research at Natzweiler https://nuremberg.law.harvard.edu/documents/1077-memorandum-to-rudolf-brandt?q=lost#p.1
32. Grawitz, Ernst, Letter to Heinrich Himmler with a report on experiments in the poison gas (Lost) program at Sachsenhausen, https://nuremberg.law.harvard.edu/documents/1230-letter-to-heinrich-himmler?q=lost#p.1
33. Transcript for IMT: Trial for Major War Criminals, https://nuremberg.law.harvard.edu/transcripts/7-transcript-for-imt-trial-of-major-war-criminals?seq=15304
34. Transcript for NMT: Medical, https://nuremberg.law.harvard.edu/transcripts/1-transcript-for-nmt-1-medical%20case?seq=51
35. Seidelman, W.E., The Holocaust: Medicine and Murder in the Third Reich, Jewish Virtual Library, https://www.jewishvirtuallibrary.org/medicine-and-murder-in-the-third-reich
36. Annas, G. and Utley, E.R., **The Nazi Doctors and the Nuremberg Code : Human Rights in Human Experimentation: Human Rights in Human Experimentation**, Oxford University Press, 1992.

37. Winkelmann A, Freïberger I. Anatomist Hermann Stieve used Nazi victims for his research on the menstrual cycle, but did he also fabricate facts? Arch Gynecol Obstet. 306 (2022) 1287-1298.
38. Weindling P. "Cleansing" anatomical collections: The politics of removing specimens from German anatomical and medical collections 1988-92. Ann Anat. 194 (2012) 237-42.
39. Taylor, Telford, The Doctors' Trial, https://www.courts.wa.gov/library/exhibit2008/The%20Doctors%20Trial%20display%20board.pdf
40. Ghooi RB. The Nuremberg Code-A critique. Perspect Clin Res. 2011;2(2):72-76. doi:10.4103/2229-3485.80371
41. German Guidelines for Human Experimentation 1931.
42. https://artandersonmd.com/1931_german_hu_guidelines.html

Chapter 15

1. Posner, Gerald L and Ware, John., **Mengele: the Complete Story**, Cooper Square Press, 2000.
2. Underwood, Louis A., **The Angel of Death: Inside the Mind of Mengele: "The Chilling Truth Behind Auschwitz's Most Feared Doctor and the Global Hunt that Failed to Catch Him."**, Louis Underwood, 2025.
3. Marwell, David G., **Mengele: Unmasking the "Angel of Death"**, W.W. Norton and Company, 2020.
4. Nyiszli, M., **Auschwitz: A Doctor's Eyewitness Account**, N. Margareta Nyiszli, New York, 1960.
5. Adelsberger, Lucie, and Ray, Susan, **Auschwitz: a Doctor's Story**, Northeastern University Press, 2006.

Chapter 16

1. Seaman, Louis Livingston, **The Real Triumph of Japan: the Conquest of the Silent Foe**. New York, D. Appleton and Co., 1907.
2. https://www.historians.org/about-aha-and-membership/aha-history-and-archives/gi-roundtable-series/pamphlets/em-15-what-shall-be-done-about-japan-after-victory-(1945)/why-did-japan-choose-war
3. Gold, Hal, Japan's Infamous Unit 731, **Firsthand Accounts of Japan's Wartime Human Experimentation Program**. Tokyo, Tuttle Publishing, 2019.
4. Harris, Sheldon, **Chapter 16: Japanese Biomedical Experimentation During the World-War-II Era, Military Medical Ethics, Volume 2**., Ed: L.R. Beam and T.E. Sparacino, 2003.
5. Zhang, Kaiyuan, and MacKinnis, Donald, 2001. **Eyewitness to Massacre: American Missionaries Bear Witness to Japanese Atrocities in Nanjing**, Routledge, 2015.
6. Wakabayashi, Bob Tadashi, The Nanking 100-Man Killing Contest Debate: War Guilt amid Fabricated Illusions, 1971-75., Journal of Japanese Studies, 26 (2000) 307–340.
7. Shudo, H., **Nanking Massacre: Fact Versus Fiction, A Historian's Quest for Truth**, Sekai Shuppan, Inc., 2005.
8. Geneva Convention, https://www.history.com/topics/world-war-ii/geneva-convention
9. Documentary Unit 731—Nightmare in Manchuria, /www.imdb.com/title/tt0224980/
10. Yamada,Oto, **Materials on the Trial of Former Servicemen of the Japanese Army Charged with Manufacturing and Employing Bacteriological Weapons**, Foreign Languages Publishing House, Moscow, 1950.

11. Sky City Gallery, Unit 731, http://www.skycitygallery.com/japan/japan.html#unit731
12. Kristof ND., A Japanese generation haunted by its past. New York Times, Jan. 22, 1997.
13. Riedel S. Biological warfare and bioterrorism: a historical review. Proc (Bayl Univ Med Cent). 17 (2004) 400-406.
14. Mikesh, Robert C., **Japan's World War II Balloon Attacks on North America**, Smithsonian Institution Press, Washington DC, 1973.
15. China Mustard Gas Crisis, NY Times, August 19, 2003.
16. Chan, Jenny, **Marutas of Unit 731: Human Experimentation of the Forgotten Asian Auschwitz (Uncovering Unit 731)**, 2020.
17. National Archives Catalog, Ishii, Shiro, https://catalog.archives.gov/id/139353926
18. French, Howard, Japanese Veteran Testifies in War Atrocity Lawsuit, New York Times, Dec. 21, 2000, https://www.nytimes.com/2000/12/21/world/japanese-veteran-testifies-in-war-atrocity-lawsuit.html

Chapter 17

1. ACS, How the American Cancer Society Funds Research, https://www.cancer.org/research/how-american-cancer-society-research-funding-works.html
2. NIH, Budget, https://www.nih.gov/about-nih/what-we-do/budget
3. International Committee of Medical Journal Editors, Defining the Role of Author and Contributors, http://www.icmje.org/recommendations/browse/roles-and-responsibilities/defining-the-role-of-authors-and-contributors.html
4. Matawlie RH, Arjun Sharma JR, de Rooij JD, Sardjoe Mishre G, Huygen FJ, Gadjradj PS. Honorary authorship in

high-impact journals in anaesthesia and pain medicine. British Journal of Pain. 15 (2021) 246-248.

5. Noruzi, A., Takkenberg, J.J.M., Kayapa, B., Verhemel, A., Gadjradj, P.S., Honorary authorship in cardiothoracic surgery, The Journal of Thoracic and Cardiovascular Surgery, 161 (2021) 156-162.
6. Office of Research Integrity, RCR Casebook: Research Misconduct, https://ori.hhs.gov/rcr-casebook-research-misconduct
7. Fanelli D., How Many Scientists Fabricate and Falsify Research? A Systematic Review and Meta-Analysis of Survey Data. PLoS ONE 4 (2009) e5738. https://journals.plos.org/plosone/article?id=10.1371/journal.pone.0005738
8. Wakefield, Andrew J., et al. Crohn's disease: pathogenesis and persistent measles virus infection. Gastroenterology 108 (1995) 911-916.
9. Wakefield AJ, Murch SH, Anthony A, Linnell J, Casson DM, Malik M, et al. Ileal-lymphoid-nodular hyperplasia, non-specific colitis, and pervasive developmental disorder in children. Lancet. 351 (1998) 637–641.
10. Riedel, S. (2005). **Edward Jenner and the History of Smallpox and Vaccination.** Baylor University Medical Center Proceedings, 18 (2005) 21–25.
11. Anderson, A., What is Variolation?, WebMD, April, 2022, https://www.webmd.com/vaccines/what-is-variolation
12. Smithsonian American History Museum, Measles, Mumps, and Rubella Vaccines, https://www.si.edu/spotlight/antibody-initiative/mmr
13. CDC Measles, Measles Cases and Outbreaks, Oct. 15, 2025, https://www.cdc.gov/measles/data-research/index.html
14. WHO, Measles, Nov. 14, 2024, https://www.who.int.news-room/fact-sheets/detail/measles

15. CDC Mumps, Mumps Cases and Outbreaks, Sept. 12, 2025, https://www.cdc.gov/mumps/outbreaks/index.html
16. Blakemore, E., It's Official: Rubella Has Been Eradicated From the Americas, April 30, 2015. https://www.smithsonianmag.com/smart-news/its-official-rubella-has-been-eradicated-americas-180955137/
17. CDC 2, Measles, Mumps, and Rubella (MMR), Vaccination: What Everyone Should Know, https://www.cdc.gov/vaccines/vpd/mmr/public/
18. Hyman, S.L., *et. al.*, Identification, Evaluation, and Management of Children With Autism Spectrum Disorder, Pediatrics, 145 (2020), electronic journal, https://publications.aap.org/pediatrics/article/145/1/e20193447/36917/Identification-Evaluation-and-Management-of#sec-8
19. National Institute of Mental Health, Autism Spectrum Disorder, https://www.nimh.nih.gov/health/topics/autism-spectrum-disorders-asd
20. Deer, Brian, **The Doctor Who Fooled the World: Science, Deception, and the War on Vaccines**, John Hopkins University Press,
21. Taylor B, Miller E, Farrington CP, Petropoulos MC, Favot-Mayaud I, Li J, Waight PA. Autism and measles, mumps, and rubella vaccine: no epidemiological evidence for a causal association. Lancet. Jun 12 (1999) 2026-9.
22. Peltola, H., *et. al.*, No evidence for measles, mumps, and rubella vaccine-associated inflammatory bowel disease or autism in a 14 year prospective study. The Lancet Research Letters, 351 (1998) 1327-1328.
23. Chen, R.T., and DeStefano, F., Vaccine adverse events: causal or coincidental? The Lancet, 351 (1998) 611-612.
24. DeStefano, Frank, and Robert T. Chen. Negative association between MMR and autism. The Lancet 353. (1999): 1987-1988.

25. Kami Kandola, Yves Bergevin, Immunization and child health in developing countries: Canada's response, Paediatrics & Child Health, 5 (2000) 378–380.
26. Murch SH, Anthony A, Casson DH, Malik M, Berelowitz M, Dhillon AP, et al. Retraction of an interpretation. Lancet. 363 (2004) 750.
27. Sathyanarayana Rao, T.S. and Andrade, C., The MMR vaccine and autism: Sensation, refutation, retraction, and fraud, Indian J Psychiatry. 53 (2011) 95–96.
28. Harrison, J.A., Wrong About Vaccine Safety: A Review of Andrew Wakefield's "Callous Disregard", The Open Vaccine Journal, 6 (2013) 9-25.
29. Wakefield, A. and McCarthy, J., Callous Disregard: Autism and Vaccines--The Truth Behind a Tragedy, Skyhorse, 2017.
30. Arlett, P., Bryan, P., Evans, S., A response to 'measles, mumps, rubella vaccine: through a glass, darkly' by Drs. AJ Wakefield and SM Montgomery and published reviewers' comments. Adverse Drug Reactions Toxicology Reviews, 20 (2001) 37-45.
31. Deer, Brian, Revealed: MMR research scandal, The Sunday Times,London, Feb. 22, 2004.
32. Deer, Brian, The Lancet's two days to bury bad news, Brit Med J, 342 (2011)
33. Horton, Richard, A statement by the editors of the Lancet, Lancet 363 (2004) 820-821.
34. Sathyanarayana Rao T.S., and Chittaranjan Andradel, C., The MMR vaccine and autism: Sensation, refutation, retraction, and fraud. Indian J. Psychiatry, 53 (2011) 95-96.
35. Wakefield, A. Enterocolitis, autism and measles virus. Mol Psychiatry 7 (Suppl 2), S44–S46 (2002).
36. Children's Hospital of Philadelphia, https://www.chop.edu/centers-programs/vaccine-education-center/vaccines-and-other-conditions/vaccines-autism
37. Millward G. Vaccinating Britain: Mass vaccination and the public since the Second World War [Internet]. Manchester

(UK): Manchester University Press; 2019. Chapter 5, MMR. https://www.ncbi.nlm.nih.gov/books/NBK545996/

38. Vaccine Coverage Levels – United States, 1962-2009, https://www.cdc.gov/vaccines/pubs/pinkbook/downloads/appendices/g/coverage.pdf
39. Gambrell A, Sundaram M, Bednarczyk RA. Estimating the number of US children susceptible to measles resulting from COVID-19-related vaccination coverage declines. Vaccine. 40 (2022) 4574-4579.
40. WHO, History of the Measles Vaccine, https://www.who.int/news-room/spotlight/history-of-vaccination/history-of-measles-vaccination
41. U.S. Congress, S.Hrg. 119-21 — NOMINATION OF ROBERT F. KENNEDY, JR. TO SERVE AS SECRETARY OF HEALTH AND HUMAN SERVICES, Jan. 30, 2025, https://www.congress.gov/event/119th-congress/senate-event/LC74210/text
42. Gardner, Sophie and Gardner, Lauren, RFK Jr. to fire all members of the CDC's vaccine advisory committee. Politico, June 9, 2025.

    https://www.politico.com/news/2025/06/09/rfk-jr-to-fire-all-members-of-the-cdcs-vaccine-advisory-committee-00395235
43. Firth, Cannon, AMA Calls for Senate Investigation of RFK Jr., Medpage Today, June 10, 2025.
44. Zadrozny, Brandy, 'CDC is over': RFK Jr. lays off over 1,000 employees in Friday night massacre, MSNBC, Oct. 11, 2025, https://www.msnbc.com/msnbc/news/cdc-rfk-jr-shutdown-layoffs-goal-rcna237035
45. FOR IMMEDIATE RELEASE: CDC Immunization Schedule Adopts Individual-Based Decision-Making for COVID-19 and Standalone Vaccination for Chickenpox in Toddlers, October 6, 2025. https://www.hhs.gov/press-room/cdc-immunization-schedule-individual-decision-covid19-standalone-chickenpox-toddlers.html

46. FOR IMMEDIATE RELEASE: CDC Immunization Schedule Adopts Individual-Based Decision-Making for COVID-19 and Standalone Vaccination for Chickenpox in Toddlers, Oct, 6, 2025, https://www.hhs.gov/press-room/cdc-immunization-schedule-individual-decision-covid19-standalone-chickenpox-toddlers.html
47. Lelland, Kate, Medical journals retract "unethical" research, Reuters, 2011, https://www.reuters.com/article/us-journals-retractions-idUSTRE7235J820110304/
48. Retraction Watch, The new retraction record holder is a German anesthesiologist, with 184, 2023, https://retractionwatch.com/category/by-author/joachim-boldt-retractions/
49. Blake, Heidi, Millions of surgery patients at risk in drug research fraud scandal, The Telegraph, March 3, 2011, https://www.telegraph.co.uk/news/health/8360667/Millions-of-surgery-patients-at-risk-in-drug-research-fraud-scandal.html
50. Boldt, J. Knothe, C., Schindler, E., Hammermann, H., Dapper, F., and Hempelmann, G., Volume replacement with hydroxyethyl starch solution in children, British Journal of Anaesthesia 1993; 70: 661-665. RETRACTED 2010
51. Boldt J, Suttner S, Brosch C, Lehmann A, Röhm K, Mengistu A. Cardiopulmonary bypass priming using a high dose of a balanced hydroxyethyl starch versus an albumin-based priming strategy. Anesth Analg 2009;109:1752–62. RETRACTED 2014
52. Shafer, S., Editor's note: notice of retraction, Anesth Analg. 2014 Nov;119(5):1225

Chapter 18

1. Milgram, S., Behavioral Study of Obedience, Journal of Abnormal and Social Psychology, 67 (1963) 371-378.

2. Encina, G.B., Milgram's Experiments on Obedience to Authority, UC Berkeley Rausser College of Natural Resources, https://nature.berkeley.edu/ucce50/ag-labor/7article/article35.htm
3. Zimbardo, P., Haney, C., Banks, W.C., and Jaffe, D., The Stanford Prison Experiment, https://web.stanford.edu/dept/spec_coll/uarch/exhibits/Narration.pdf
4. Zimbardo, Phillip, **The Lucifer Effect: How Good People Turn Evil**, Random House, 2008.
5. Le Texier, T., Debunking the Stanford Prison Experiment, American Psychologist, 74 (2019) 823-839.
6. Le Texier, T., **Investigating the Stanford Prison Experiment: History of a Lie**, Springer, 2024.
7. Haslam, S. A., Reicher, S., & Van Bavel, J. J. (2018, June 27). Rethinking the nature of cruelty: The role of identity leadership in the Stanford Prison Experiment. https://doi.org/10.31234/osf.io/b7crx
8. Zimbardo, P., Tape E: 8612, Stanford Digital Repository, https://purl.stanford.edu/wn708sg0050
9. Zimbardo, P., Quiet Rage: The Stanford Prison Experiment, video, 1992. https://www.schooltube.com/media/Zimbardo-Stanford-Prison-Experiment/1_p0sp2eur
10. Resnick, Brian, The Stanford Prison Experiment was massively influential. We just learned it was a fraud., Vox, https://www.vox.com/2018/6/13/17449118/stanford-prison-experiment-fraud-psychology-replication

About the Author

I was born and raised in Hayward, California, the youngest of eight children. My father started his own company washing windows in the San Francisco Bay Area. My parents raised me and my siblings on the back of his company, Pyramid Building Maintenance, the name derived from the package of Camel cigarettes he smoked regularly.

I went to the University of California at Davis and earned my BS in physiology in 1984 and my PhD in physiology in 1988. I then went to the University of Iowa for a postdoctoral fellow, which I completed at the University of Alabama at Birmingham. In 1992, I joined the faculty at the University of South Alabama Department of Biomedical Sciences in Mobile, Alabama.

At the University of South Alabama, I taught courses in human physiology, neuroscience, and issues in biomedical sciences. I also conducted research on neural processing of pain. My greatest achievement was authoring a textbook, **Principles in Human Physiology**. My greatest joy, however, was serving as the prehealth advisor to students. In this latter capacity, I also served as national President of Alpha Epsilon Delta, the Prehealth Honor Society for eight years. I retired from the University in 2018.

I started writing this book shortly before retiring. Teaching a bioethics-focused course, I found the topic both fascinating and scary. We like to think that our doctors and nurses care about us. But I learned that is not always the case. I still served on the national board of Alpha Epsilon Delta until 2024, and I presented bioethics workshops at the biennial conventions. The continued contact with students made me want to finish this book, believing I still have something worthwhile to share with students. These students were choosing careers that would put them on a pedestal with high expectations. Most were up for the task; a couple, I worry about. But overall, I am proud of my students, some of whom are now my health care provider.

I am now retired, living in Semmes, Alabama, with my husband and cats on the same property where we had picnics and bonfires with the students. I spend time traveling to exotic places for wildlife photography. I have also taken up woodworking and painting. But my students taught me about passion for serving others. And I follow in their footsteps providing service to the community by participating in sea turtle nest protection, animal rescue, and curating the local art show The Semmesonian.

Made in the USA
Monee, IL
04 November 2026

67381209R00282